INTRAPLATE EARTHQUAKES

Intraplate earthquakes occur away from the well-known tectonic plate boundaries at locations around the world. These locations are particularly difficult to predict, and they can cause huge damage and loss of life: devastating earthquakes have levelled Bhuj in India (2001), Tangshan in China (1976), Charleston in the USA (1886), and Basel in Switzerland (1356). The Bhuj earthquake (featured in this book) was the largest intraplate earthquake for three decades and the rich dataset collected has provided unique insight into these events.

This cutting-edge book brings together research from international leading experts in the field, compiling multidisciplinary data on intraplate earthquakes. Each chapter provides a comprehensive review of the spatial and temporal patterns of these earthquakes in a different global location, ranging from Australia, China, India and the Sea of Japan, to Western Europe, Brazil, New Madrid in Central USA, and Eastern Canada. The book explores the similarities and differences between regional features, and the mechanical models required to explain them. A broad range of techniques are discussed, including geological investigations of neotectonic features; combined analyses of seismicity, geological, GPS, and geophysical data; seismic reflection tomography, and more. Case studies in the book demonstrate that techniques and strategies used for seismic hazard assessment of plate boundary earthquakes are not valid for intraplate settings, and that new approaches are required for these regions.

Providing the first global overview of intraplate earthquakes, this is an essential book for academic researchers and professionals in seismology, tectonics, tectonophysics, geodesy, structural geology, earthquake dynamics, and geophysics, as well as structural engineers working in earthquake-prone areas.

PRADEEP TALWANI is a Distinguished Emeritus Professor of Geophysics in the Department of Earth and Ocean Sciences at the University of South Carolina. He has over 30 years' experience researching intraplate earthquakes, particularly in the Eastern United States and Canada, and is a leading authority on the largest intraplate earthquake to strike the Eastern United States in recorded history, the 1886 event at Charleston, South Carolina. Professor Talwani served as Director of the South Carolina Seismic Network from 1990 until 2009, when he retired, and is a Fellow of the Geological Society of America.

INTRAPLATE EARTHQUAKES

Edited by

PRADEEP TALWANI

University of South Carolina

CAMBRIDGE
UNIVERSITY PRESS

University Printing House, Cambridge CB2 8BS, United Kingdom

One Liberty Plaza, 20th Floor, New York, NY 10006, USA

477 Williamstown Road, Port Melbourne, VIC 3207, Australia

4843/24, 2nd Floor, Ansari Road, Daryaganj, Delhi - 110002, India

79 Anson Road, #06-04/06, Singapore 079906

Cambridge University Press is part of the University of Cambridge.

It furthers the University's mission by disseminating knowledge in the pursuit of education, learning and research at the highest international levels of excellence.

www.cambridge.org
Information on this title: www.cambridge.org/9781108447898

© Cambridge University Press 2014

First published 2014
First paperback edition 2017

A catalogue record for this publication is available from the British Library

Library of Congress Cataloging in Publication data
Intraplate earthquakes / edited by Pradeep Talwani.
pages   cm
Includes bibliographical references and index.
ISBN 978-1-107-04038-0 (hardback)
1. Plate tectonics.   2. Seismology.   3. Earthquake prediction.   4. Geology, Structural.
I. Talwani, Pradeep.
QE511.4.I57   2014
551.22 – dc23   2013049913

ISBN 978-1-107-04038-0 Hardback
ISBN 978-1-108-44789-8 Paperback

Contents

Color plates section is found between pages 180 and 181.

Contributors

Pierre Alexandre
Royal Observatory of Belgium, 3 Avenue Circulaire, B-1180 Brussels, Belgium

Trevor Allen
Pacific Geoscience Centre, Geological Survey of Canada, Sidney, BC, Canada

Marcelo Assumpção
University of São Paulo, São Paulo, Brazil

José Roberto Barbosa
University of São Paulo, São Paulo, Brazil

Lucas Barros
University of Brasília, Brasília, Brazil

Hilario Bezzera
Federal University of Rio Grande do Norte, Natal, Brazil

Sanjib Kumar Biswas
201/C, ISM House, Thakur Village, Kandivali (E), Mumbai-400101, India

Thierry Camelbeeck
Royal Observatory of Belgium, 3 Avenue Circulaire, B-1180 Brussels, Belgium

Dan Clark
Earthquake Hazard Section, Geoscience Australia, GPO Box 378, Canberra ACT 2601, Australia

João Carlos Dourado
UNESP (São Paulo State University), Rio Claro, Brazil

Joaquim Ferreira
Federal University of Rio Grande do Norte, Natal, Brazil

David Garcia Moreno
Royal Observatory of Belgium, 3 Avenue Circulaire, B-1180 Brussels, Belgium

Susan E. Hough
U.S. Geological Survey, Pasadena, California 91106, USA

Cheng Jia
Department of Geological Sciences, University of Missouri, Columbia, Missouri 65211, USA and China Earthquake Network Center, China Earthquake Administration, Beijing, 100045, China

Aitaro Kato
Earthquake Research Institute, The University of Tokyo, Japan

Elisabeth Knuts
Royal Observatory of Belgium, 3 Avenue Circulaire, B-1180 Brussels, Belgium

Maurice Lamontagne
Geological Survey of Canada, Ottawa, Ontario, K1A 0E9, Canada

Thomas Lecocq
Royal Observatory of Belgium, 3 Avenue Circulaire, B-1180 Brussels, Belgium

Mian Liu
Department of Geological Sciences, University of Missouri, Columbia, Missouri 65211, USA

Prantik Mandal
National Geophysical Research Institute (Council of Scientific and Industrial Research), Hyderabad, A.P., India

Andrew McPherson
Earthquake Hazard Section, Geoscience Australia, GPO Box 378, Canberra ACT 2601, Australia

Eduardo Menezes
Federal University of Rio Grande do Norte, Natal, Brazil

Aderson do Nascimento
Federal University of Rio Grande do Norte, Natal, Brazil

Søren B. Nielsen
University of Aarhus, Denmark

Marlon Pirchiner
University of São Paulo, São Paulo, Brazil

Giorgio Ranalli
Department of Earth Sciences, Carleton University, Ottawa, Ontario, K1S 5B6, Canada

Bal Krishna Rastogi
Institute of Seismological Research, Raisan, Gandhinagar-382 009, Gujarat, India

Luís Carlos Ribotta
IPT (Technological Research Institute), São Paulo, Brazil

Alain Sabbe
Mons University, Faculty of Engineering, 53 rue du Joncquois, B-7000 Mons, Belgium

George Sand França
University of Brasília, Brasília, Brazil

Christian Schiffer
University of Aarhus, Denmark

Randell Stephenson
University of Aberdeen, Scotland, UK

Pradeep Talwani (Editor)
University of South Carolina, USA

Roy Van Arsdale
Department of Earth Sciences, The University of Memphis, Memphis, Tennessee 38152, USA

Hui Wang
Department of Geological Sciences, University of Missouri, Columbia, Missouri 65211, USA and Institute of Earthquake Science, China Earthquake Administration, Beijing, 100036, China

Jiyang Ye
Department of Geological Sciences, University of Missouri, Columbia, Missouri 65211, USA

Preface

I live in South Carolina, the state that bore the brunt of the destructive 1886 Charleston earthquake. For those who live in South Carolina, the earthquake is a distant memory, but it also is a constant reminder of the devastating potential of intraplate earthquakes (IPE). Being a resident, this earthquake and IPE in general became a lifelong interest of mine. The destructive 1975 Haicheng and 1976 Tangshan earthquakes in China further fueled this interest.

In the past four decades, after the development of the plate tectonic theory, there has been considerable progress in understanding the nature of these rare earthquakes. In the United States, the first systematic attempt to place IPE in a global setting began in the late 1970s, together with a proliferation of seismic networks. In the 1980s, to assess the seismic hazard posed by future IPE, the United States nuclear industry sponsored a systematic study of the phenomenon in the Central/Eastern United States (CEUS). Simultaneously, another major forward step was the compilation of the global in situ stress data. The results of these studies led to the identification of various global, geologic and seismological aspects of IPE. After the discovery in the early 1980s of buried sand blows associated with the 1886 and prehistoric Charleston earthquakes, systematic paleoseismological investigations led to the discovery and dating of several prehistoric earthquakes in the CEUS.

In the 1990s, with the collection of increasingly better geological, seismological, paleoseismological and GPS data, several theoretical models were proposed to explain the IPE in the New Madrid seismic zone. Further breakthroughs occurred when aftershocks of moderate earthquakes in the eastern margin of the Sea of Japan and the destructive 2001 Bhuj earthquake in Western India were recorded on dense networks of seismic instruments. Tomographic inversions of these data led to an identification of the causative structures and to a better understanding of the seismogenesis of these earthquakes. Simultaneous to the developments in the United States and Eastern Canada, IPE were being studied in Australia, Brazil, eastern China, India and Western Europe. Following a special session on IPE at the American Geophysical Union meeting in Iguassu Falls, Brazil, in August 2010, it was thought that a global review of IPE would be of wide interest. This book was borne out of that proposition.

I sincerely thank the authors of various chapters for their contributions and their patience in responding to various queries and deadlines along the way.

Finally, I want to thank my wife, Anita, and children, Rohit and Radhika, for their love and continual support.

Pradeep Talwani

1

Introduction

PRADEEP TALWANI

The study of intraplate earthquakes began after the development of plate tectonics theory, which explains the genesis of more than 95% of the global seismic energy release. I use a more liberal definition of intraplate earthquakes as those that occur within a tectonic plate away from a plate boundary as contrasted with the more restrictive stable continental interiors (Johnston, 1989), thereby including earthquakes in locations such as the Sea of Japan. Because of the rarity of large destructive intraplate earthquakes, a great disparity in the local geological conditions where they occur, and the wide variation in the efforts made to study them in different parts of the world, our understanding of their nature has been slow to evolve. However, in the past four decades great progress has been made in monitoring of seismicity globally, as well as in diverse approaches to their study, suitable for local geological and logistical conditions. This book is an attempt to improve our understanding about the nature of intraplate earthquakes by bringing together the results of recent advances. The different chapters in the book, written by experts in their fields, are designed in such a way that each describes a different element of the phenomenon. Together they contribute to a more integrated understanding of intraplate earthquakes.

This book describes intraplate earthquakes in eight regions of the world – from Australia, with a relatively low level of seismicity and a short historical record, to China, with a three millennia long history of destructive earthquakes. It covers regions with moderate seismicity (Western Europe and Brazil), New Madrid in the Central United States with a record of large historical earthquakes, and currently active areas of Eastern Canada, the eastern margin of the Japan Sea and Kutch, in Western India. The results of numerical modeling described in one chapter show how these mid-plate earthquakes occur in response to stresses emanating from plate boundaries. Another chapter presents a unified model to explain how these stresses build up on discrete structures and lead to such mid-plate earthquakes. Our changing views in approaches to seismic hazard analysis and conclusions and insights gained from the various chapters are discussed in the final two chapters.

Compared with the other intraplate regions, the historical record of seismicity in Australia spans little more than two centuries, and the instrumental record of seismicity extends

Intraplate Earthquakes, ed. Pradeep Talwani. Published by Cambridge University Press.

over a quarter of that time. Therefore, for planners responsible for estimating seismic hazards, the concern is whether this short record of historical and instrumental seismicity including twenty earthquakes of M > 6.0 and seven with surface rupture is truly representative of the long-term behavior of the seismogenic structures in the continent. Because of a paucity of dense seismic networks, the earthquake locations are not well constrained, and a very small fraction of the estimated depths are accurate enough to make tectonic associations, thus limiting the ability to accurately identify and outline the buried seismogenic features.

Clark, McPherson, and Allen (Chapter 2) solve the problem of a short historical and poorly constrained seismicity record by using a well-preserved geological record to extend the prehistorical record of past earthquake activity to more than 10 Ma. They note, "Australia is one of the lowest, flattest, most arid and slowly eroding continents on Earth. Accordingly, large parts of Australia are favorable for the preservation of tectono-geomorphic features, such as fault scarps, for tens of thousands to millions of years." They use a two-pronged approach in their search for records of paleo and historical (collectively, morphogenic) earthquakes that have ruptured faults during the past millennium. A geomorphological study based on aerial photographs was used to select sites and disruption localities for the second phase of the investigations, which involves trenching for paleoseismological investigations. Evidence of more than 300 fault scarps and other geomorphic features dating back to more than 10 Ma was found in the geological record.

Based on the geology and crustal setting, the authors recognized six neotectonic domains in Australia. For each of those, they analyzed the geomorphic features based on their fault length and vertical displacement. They interpreted the results in terms of seismological parameters such as the maximum magnitude, earthquake frequency, and the temporal pattern of the occurrence of morphogenic earthquakes in different neotectonic domains. They conclude that the catalog of historical seismicity significantly underestimates the large earthquake potential in the continent.

Brazil spans most of the mid-plate area in South America and little has been published on the seismicity characteristics of this large, stable continental region. Two magnitude 6 earthquakes are the largest events in the revised catalog of historical events (from about mid nineteenth century) and instrumental data (mainly from the past 60 years). Assumpção and colleagues (Chapter 3) discuss the distribution of seismicity and possible correlations with the main geological provinces and geophysical features. Despite the short period of the catalog, which may not be representative of the long-term seismicity (paleoseismological studies, for example, indicate magnitudes could reach 7, based on liquefaction features in northeastern Brazil), interesting correlations have been found. A trend of higher seismicity was found in areas presumed to have thinned lithosphere, as also near craton edges (around craton keels), even for magnitudes down to 3.5, similar to global correlations for larger earthquakes. Also, the continental shelf is more seismically active than the average intraplate region, even for low magnitudes. Flexural stresses associated with gravity highs were also found to be an important factor in helping explain Brazilian intraplate seismicity. The new multidisciplinary Brazilian dataset should provide a basis for comparisons with other intraplate regions and to test models developed in other continents.

Accurate determination of the nature and configuration of seismogenic structures responsible for the seismic hazard of a region is particularly important, especially when it includes highly populated locations. Such is the case in Eastern Canada where the metropolitan areas of Ottawa, Montreal, and Quebec all lie in the St. Lawrence Rift System (SLRS), the most active seismic region in eastern North America. Lamontagne and Renalli address this issue in Chapter 4. The NE–SW-trending SLRS, which extends for more than ~1,200 kilometers, is a failed paleo-rift and attached to it are the NW–SE-oriented aulacogens, the Ottawa–Bonnechère and the Saguenay grabens. The structure of the SLRS is further complicated by a Devonian age, 50 to 60 kilometers broad meteor impact structure lying over the SLRS and centered around Charlevoix. The impact structure has further shattered the rocks in the SLRS and produced cross faults.

The authors use an integrated approach in assessing the seismic hazard for this region – critical to the Canadian economy. Starting with historical accounts of European settlers in the seventeenth century, there is evidence of several $M > 6.0$ earthquakes in the SLRS. Instrumentally recorded seismicity identified three main seismic zones: the Charlevoix seismic zone along the SLRS, the northwest-trending Western Quebec seismic zone, and the Lower St. Lawrence seismic zone. Outside these zones there is low-level background seismic activity. Most of the hypocenters are located in the middle to upper lower crust, although a deeper event (~29 km) did occur in the Saguenay graben in 1989. Traditionally, the seismic hazard has been associated with larger events in the rift structure. However, the authors argue that moderate events with $M > 4.5$ can also pose a seismic hazard because of low attenuation, thick unconsolidated post-glacial marine clay soils that amplify ground motions, and the presence of old buildings in these zones. They suggest that most of the moderate earthquakes are associated with local causes rather than with the rift structures and such causes should be considered in evaluating the hazard. They identify these as locally weakened faults, elevated pore pressures, and local variations in the stress level.

In the past decade, as newer and better seismicity data began to accumulate, questions emerged regarding the stationarity of seismicity in intraplate regions, a common assumption in the practice of seismic hazard analysis carried over from studies of interplate regions. In most intraplate regions, the historical data and sporadic paleoseismological investigations are inadequate to accurately describe the spatiotemporal pattern of past seismicity needed to assess the assumption of stationarity of seismic sources. In Chapter 5, Liu, Wang, Ye, and Jia address this question in their study of the intraplate earthquakes of North China.

On a continental scale, North China is one of the most seismically active regions in the world, with a historical record of more than 3,000 years. These records document more than 100 large ($M > 6$) earthquakes in the past 2,000 years, including the world's most catastrophic $M \sim 8.3$ 1556 Xuaxian earthquake with a death toll of 830,000 victims. Liu *et al.* describe the spatiotemporal pattern of the historical seismicity and, using the results of intensive geological, geophysical, seismological, and GPS investigations in the past decades, relate the significant earthquakes to specific structures. They show that no $M \geq 7.0$ event ever ruptured the same fault segment twice in the past 2,000 years, and there was "roaming" of large earthquakes between widespread fault systems. Their results demonstrate the complex spatiotemporal patterns of intraplate earthquakes, the problems of long recurrence time and short and incomplete records, and the need for a reassessment of seismic hazard in intraplate regions.

The largest and best studied intraplate event in the past two decades is arguably the January 26, 2001 M 7.7 Bhuj earthquake in Western India. Soon after the earthquake, scientists from several institutes carried out detailed aftershock studies. In the past 10 years, additional deployment of large arrays of seismograph and GPS receivers has been complemented with intensive geophysical and geological investigations providing an incredible modern-day dataset. In Chapter 6, Rastogi, Mandal, and Biswas summarize the results of these multidisciplinary investigations, providing for the first time an opportunity to study the seismogenesis of a major intraplate earthquake. Using seismicity and seismic tomography data they show that the earthquake nucleated at midcrustal depths in the vicinity of a large, high-density, mafic body, in parts of which were pockets of fluids. The large stress drop associated with it spawned a stress pulse that travelled at 25 km/yr. This stress pulse, together with coseismic stress field changes, led to a sequential activation of multiple faults, both to the north and to the south of the epicenter, producing 20 M 4 to 5.1 earthquakes up to a distance of ∼250 km. The most distant M 5 earthquake occurred in 2011. GPS and InSAR studies revealed that, after the dissipation of the coseismic strain in the first few years after the mainshock, the horizontal strain returned to a background value of ∼2–5 mm/yr. However, InSAR and GPS surveys, 7 to 10 years after the mainshock, provided evidence for localized pockets of high vertical strain rates located in the vicinity of faults that had been activated by the seismicity that followed the 2001 mainshock. Insights gathered from this contemporary study of a major earthquake can be used to analyze the sparse data surrounding older intraplate earthquakes.

Arguably, the St. Lawrence Rift System in Eastern Canada and the New Madrid seismic zone (NMSZ) in the Central United States are the most active seismic zones in eastern North America and also the best studied. After a series of M ∼ 7 earthquakes in 1811–1812, the continuing seismicity in the NMSZ has been at a lower magnitude level, and the contemporary seismicity rarely exceeds M 4. The NMSZ has also been a subject of intense, dedicated studies in the past few decades. Although the NMSZ is one of the best studied cases of intraplate earthquakes, there is an absence of consensus about its cause and the contemporary seismic hazard that it poses.

In Chapter 7, Van Arsdale presents a summary of these investigations, including a detailed geological history of the region, and the results of historical and contemporary seismological and paleoseismological investigations. Additional studies that are reviewed include GPS, geophysical, and geological investigations. These abundant data have spawned a plethora of explanations, models, and speculations about the nature of the seismic hazard in the NMSZ. Van Arsdale carefully presents these often conflicting explanations and how they impact our current assessment of the seismic hazard. One of the suggestions based on geological data is that the contemporary seismicity began in the Holocene, and that the current GPS data indicate very low strain rates and a general absence of a major seismic hazard: a conclusion potentially at odds with the paleoseismological record and alternative explanations of the GPS data. Currently, these data have been explained from a local perspective. I anticipate that new explanations and models will emerge when these assorted studies are evaluated from the global perspective described in this book.

In Western Europe, large earthquakes have been few and far between, but, when they did occur, even the moderate ~M 5 events left a long-lasting impression. In Chapter 8, Camelbeeck and co-authors present a method of quantitative assessment of the seismic hazard based on a detailed evaluation of the hazard caused by past earthquakes, especially those in the past decade. They use the damage caused by past earthquakes as an important source of information about future seismic vulnerability of a region. They use a computerized, rich archival database that includes heritage records about the architectural style and kind of building – church, castle, or ordinary home – the construction style and material, and information about the underlying soils, to develop a seven-point "damage characteristics" scale. They illustrate their methodology with an evaluation of the damage following the Central Belgium earthquake of 1828. Their results show that, in other intraplate regions where detailed macroscopic data are available for the rare, moderate earthquake, it is possible to assess the vulnerability and potential seismic hazard to a region posed by future earthquakes.

Although a spatial association of intraplate earthquakes with rifts has long been suggested, the identification of a causal seismogenic structure within the rift has so far proved elusive because of a lack of detailed seismicity, geological, and geophysical data. In Chapter 9, Kato describes the results of using a dense seismic network deployed immediately after three large intraplate earthquakes along the eastern margin of the Japan Sea. Analysis of the seismicity data together with detailed seismic tomographic studies led to the discovery of a buried rift with stepwise and tilted block structures, together with a buried rift pillow in the lower crust with associated weakening fluids. The seismicity was caused by the reactivation of the pre-existing faults within the ancient rift system. The rift pillow appears to have acted as a stress concentrator, and the fluids as weakening agents. The identification of the seismogenic features responsible for the M 6+ earthquakes provided validation of models suggested for the genesis of these intraplate earthquakes.

A significant percentage of intraplate earthquakes are associated with ancient rift zones and occur hundreds to thousands of kilometers from plate boundaries. One of the important questions is explaining the source(s) of stresses that are responsible for their occurrence. Nielsen, Stephenson and Schiffer address this question in Chapter 10. They note that the locations of the intraplate earthquakes in Western Europe are also the locations of stress inversion of rift basins. They use numerical thermo-mechanical models of stress inversion of basins to show that these intraplate earthquakes in Western Europe result from the interaction of the stresses transmitted from plate boundaries through the lithosphere and the perturbing "potential energy" stresses. These potential energy stresses are derived from models of lateral variation in the present-day density structure of the lithosphere, and lateral pressure variations in the mantle below the lithosphere due to density contrasts and related convection. They show that the main source of intraplate stress is that derived from plate boundaries, and "primary inversion" occurs only when it combines favorably with potential energy stresses in combination with the existence of favorably oriented pre-existing lithosphere-scale weaknesses.

Talwani (Chapter 11) presents a unified model, based on an integration of the results of earlier studies aimed at explaining how stresses transmitted from plate boundaries cause intraplate earthquakes. These stresses were found to cause local stress accumulation on discrete structures, which were identified as local stress concentrators. These local stress concentrators are located in both the upper and lower crust within the rift, and their reactivation occurs in the present-day compressional stress field in the form of earthquakes. Commonly observed local stress concentrators are favorably oriented (relative to the regional stress field) fault bends and fault intersections, flanks of shallow plutons, buried rift pillows, and restraining stepovers. Stress build-up associated with one or more local stress concentrators interacts with and produces a potentially detectable local rotation of the regional stress field with wavelengths of tens to hundreds of kilometers. A local rotation of the regional stress field provides evidence of local stress increase and thus potentially suggests the location of future intraplate earthquakes. This model provides a framework for potentially testing for and assigning the cause of intraplate earthquakes at other locations in the world.

One of the important elements in the study of earthquakes in any region is to assess the potential seismic hazard posed by future earthquakes. The traditional approach used in intraplate regions, especially in the construction of critical facilities, is probabilistic seismic hazard analysis; a methodology imported from plate boundary regions. As better seismicity and other data become available for intraplate regions, it is becoming apparent that some of the assumptions on which the standard seismic hazard analysis is based may not be valid for intraplate regions, especially the assumption of stationarity of the seismic source. Hough (Chapter 12) discusses the problems of applying the standard seismic hazard methodology in the case of the Central and Eastern United States, especially in the case of the New Madrid seismic zone. Among some of the problems noted by Hough are that GPS results show little or no resolvable deformation in the source zones, and paleoseismological evidence suggests large Holocene and late Pleistocene earthquakes in other regions. In addition, there is evidence for short-term temporal clustering of seismicity with longer-term migration, and estimating earthquake rates from short historical records is also problematic. The results of this study serve as a warning to scientists and engineers tasked with assessing seismic hazards in intraplate regions about the wisdom of using standard probabilistic seismic hazard analysis in intraplate regions.

In the final chapter, Talwani presents a synthesis of conclusions from a study of the chapters in this book. These new data and case histories from different parts of the world support earlier ideas of a global pattern in the location of the larger intraplate earthquakes in old rift structures. Various case histories show evidence that large earthquakes do not recur at the same locations but in fact jump from one fault to another. The different chapters illustrate both the similarities and the differences in the nature of intraplate earthquakes in different parts of the world and located in different geological terranes. The chapter concludes with lessons learnt and suggestions for future studies.

Reference

Johnston, A. C. (1989). The seismicity of 'Stable Continental Interiors'. In *Earthquakes at North-Atlantic Passive Margins: Neotectonic and Post-Glacial Rebound*, ed. S. Gregersen and P. W. Basham. Boston, MA: Kluwer Academic, pp. 299–327.

2

Intraplate earthquakes in Australia

DAN CLARK, ANDREW MCPHERSON, AND TREVOR ALLEN

Abstract

Relative to other intraplate areas of the world, Australia has a short recorded history of seismicity, spanning only a couple of centuries. As a consequence, there is significant uncertainty as to whether patterns evident in the contemporary seismic record are representative of the longer term, or constitute a bias resulting from the short sampling period. This problem can, in part, be overcome by validation against Australia's rich record of morphogenic earthquakes – Australia boasts arguably the richest late Neogene to Quaternary faulting record of any stable continental region. Long-term patterns in large earthquake occurrence, both temporal and spatial, can be deduced from the landscape record and used to inform contemporary earthquake hazard science. Seismicity source parameters such as large earthquake recurrence and magnitude vary across the Australian continent, and can be interpreted in a framework of large-scale neotectonic domains defined on the basis of geology and crustal setting. Temporal and spatial clustering of earthquakes is apparent at the scale of a single fault, and at the 1,000 km scale of a domain. The utility of the domains approach, which ties seismicity characteristics to crustal architecture and geology, is that behaviours can be extrapolated from well-characterised regions to poorly known analogous regions, both within Australia and worldwide.

2.1 Introduction

The Australian continent resides entirely within the Indo-Australian Plate, and is classified as a Stable Continental Region (SCR) in terms of its plate tectonic setting and seismicity (Johnston *et al.*, 1994; Schulte and Mooney, 2005). Such settings host approximately 0.2% of the world's seismic moment release (Johnston, 1994), and moderate to large earthquakes are rare. Analysis of focal mechanisms from SCR crust (Zoback, 1992;

Intraplate Earthquakes, ed. Pradeep Talwani. Published by Cambridge University Press. © Cambridge University Press 2014.

Reinecker *et al.*, 2003; Schulte and Mooney, 2005) indicates that compressive stress regimes dominate within continental interiors, with maximum compressive stresses oriented predominantly in accordance with absolute plate motion (see also Richardson, 1992). Australia is anomalous in this respect. While earthquake focal mechanisms suggest that the crustal stress field at seismogenic depths in Australia is everywhere compressive, with significant strike-slip components along the northwest margin (NWSSZ, Figure 2.1) and in the Flinders/Mount Lofty Ranges (FRSZ, Figure 2.1) (Leonard *et al.*, 2002; Clark *et al.*, 2003), stress orientations in southern Australia are typically not parallel to the north–northeast-directed plate motion vector (Coblentz *et al.*, 1995; Hillis and Reynolds, 2000, 2003; Sandiford *et al.*, 2004, 2005; Hillis *et al.*, 2008).

In the southern half of the continent, the maximum horizontal stress orientation (S_{Hmax}) is essentially east–west in western and central Australia and rotates to northwest–southeast in eastern Australia (Figure 2.1). In the northern half of the continent, the stress field transitions from the generally east–west trend in the south, to a broadly northeast–southwest trend. To a first order these regional stress orientations are not influenced by tectonic terrane, crustal thickness, heat flow, regional structural trends, geological age, or by the depth at which orientations are sampled (e.g., Hillis *et al.*, 2008; Sandiford and Quigley, 2009; Holford *et al.*, 2011). The trends have been satisfactorily modelled in terms of a balance between plate driving and resisting torques generated at the margins of the Indo-Australian Plate (Cloetingh and Wortel, 1986; Coblentz *et al.*, 1995, 1998; Reynolds *et al.*, 2002; Burbidge, 2004; Sandiford *et al.*, 2004; Dyksterhuis and Müller, 2008) (Figure 2.2).

Stratigraphic relationships establish that fault-related and presumably seismogenic deformation consistent with the present stress field commenced in the late Miocene, in the interval 10–6 Ma (Dickinson *et al.*, 2002; Sandiford *et al.*, 2004; Keep and Haig, 2010), as the result of complex evolving plate boundary conditions (Sandiford and Quigley, 2009). It has been proposed that the onset of deformation at specific locations may reflect rising stress levels related to the combination of all plate boundary forcings (Sandiford and Quigley, 2009), with variations in the thermal structure of the Australian lithosphere influencing the localisation of deformation at the regional or terrane scale (e.g., Celerier *et al.*, 2005; Sandiford and Egholm, 2008; Holford *et al.*, 2011). Analysis of the spatial and temporal patterns of contemporary seismicity might therefore be improved by studying the extraordinarily rich Neogene to Quaternary record of seismicity in the Australian landscape (e.g., Clark *et al.*, 2012). In the sections that follow, earthquake occurrence in Australia is therefore examined from both the historic and prehistoric (Neogene and Quaternary) records.

2.2 Two centuries of earthquake observations in Australia

Between 1788, when European colonists reported the first earthquake felt in Australia (Historical Records of Australia, 1914), and the early 1900s, newspaper articles were the main source of information about earthquakes (e.g., Malpas, 1991), and continued to be important until the 1960s (McCue, 2004). The first seismographs in Australia for

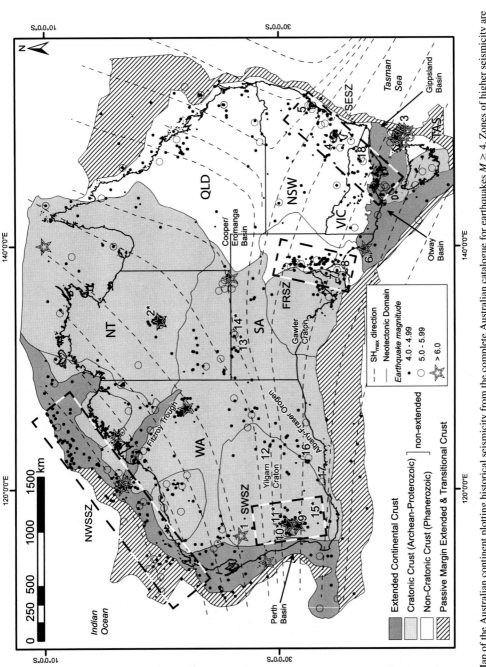

Figure 2.1 Map of the Australian continent plotting historical seismicity from the complete Australian catalogue for earthquakes $M \geq 4$. Zones of higher seismicity are shown: SWSZ, Southwest Seismic Zone; NWSSZ, Northwest Shelf Seismic Zone; FRSZ, Flinders Ranges Seismic Zone; SESZ, Southeast Seismic Zone. Significant earthquakes or earthquake sequences mentioned in text are indicated by numbers: 1, Meeberrie; 2, Tennant Creek; 3, offshore northeast Tasmania; 4, Dalton-Gunning; 5, Newcastle; 6, Beachport; 7, Warooka; 8, Adelaide; 9, Meckering; 10, Calingiri; 11, Cadoux; 12, Kalgoorlie-Boulder; 13, Ernabella; 14, Marryat Creek; 15, Katanning; 16, Norseman; 17, Ravensthorpe; 18, Mount Hotham. Numbers with asterisks indicate confirmed surface-rupturing earthquakes. The maximum horizontal stress (S_{Hmax}) direction is indicated by fine dashed lines. The continent is broadly divided on the basis of geology and tectonic history, showing the dominance of non-extended cratonic crust in the central and western parts, non-extended non-cratonic crust in the east, and extended crust around much of the continental margin.

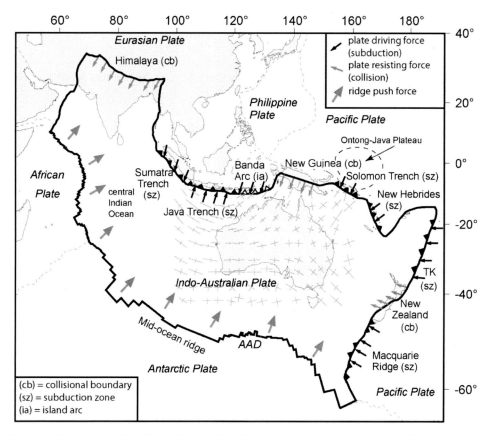

Figure 2.2 Indo-Australian Plate with plate boundary forces and orientation of modelled maximum and minimum horizontal stresses used in the finite element stress modelling of Reynolds *et al.* (2002). Much of the southern part of the continent has an east–west-oriented maximum horizontal compressive stress oriented at a high angle to the NNE-oriented plate velocity vector. Solid triangles indicate the direction of subduction and open triangles delineate the Banda Arc. TK, Tonga–Kermadec Trench; AAD, Australian–Antarctic discordance. (Modified after Reynolds *et al.*, 2002 and Hillis *et al.*, 2008.)

which records remain were four Milne seismographs established between 1901 and 1909 (Doyle and Underwood, 1965; McCue, 2004). The seismograph network remained sparse until 1962–3, after which time the Australian catalogue could be considered complete for magnitude (*M*) 5.0 and above earthquakes, and instrumental records largely replaced felt reports as the primary information source (McCue, 2004; Leonard, 2008). Doyle *et al.* (1968) summarised the history of earthquakes and seismological research in Australia up to 1966, and prepared a comprehensive reference list of historical earthquake sources. The authors tabulated over 160 Australian earthquakes, which formed the core of the preliminary digital Australian earthquake catalogue (Denham *et al.*, 1975). Despite enhancements resulting in the lowering of the completeness magnitude for the national catalogue to

M 3.5+ by the 1980s, the Australian National Seismic Network (ANSN – http://www. ga.gov.au/earthquakes/seismicSearch.do) remains relatively sparse, with no more than 70 seismic stations suitable for general earthquake monitoring in a continent of a similar size to the conterminous United States or Western Europe (Leonard, 2008). This results in generally poor constraints being placed on the location of the more than 36,000 onshore earthquakes recorded in the current Australian earthquake catalogue (Allen *et al.*, 2012a).

Approximately 20 earthquakes of $M > 6.0$ have been recorded in Australian continental crust, at the rate of one every ~6–8 years (McCue, 1990; Leonard, 2008). On average, less than one $M > 5.0$ event has occurred per year (Leonard, 2008). The largest instrumentally recorded earthquake occurred in 1941 near Meeberrie, Western Australia (Everingham, 1982) (Figure 2.1 – 1). The event is associated with a surface wave magnitude of M_S 6.8 (Allen *et al.*, 2012a). In January 1988 three large "mainshock" earthquakes of magnitude M_S 6.3–6.7 (Choy and Bowman, 1990) occurred in a 12-hour period near Tennant Creek in the Northern Territory (Jones *et al.*, 1991) (Figure 2.1 – 2). The events were preceded by a year-long foreshock sequence, and followed by a continuing aftershock sequence (e.g., Bowman *et al.*, 1990). The most notable pre-instrumental earthquake sequence occurred in Bass Strait, off the northeast coast of Tasmania (Figure 2.1 – 3), in the 1880s–1890s. Over 2,400 events were felt during that period (Ripper, 1963), four of which caused damage in Tasmania, and two of which are estimated from felt effects to rival the Meeberrie earthquake in magnitude (Michael-Leiba, 1989). The Dalton–Gunning region of southeast Australia (Figure 2.1 – 4) continues to experience a less intense earthquake sequence (comprising more than 600 recorded events) that began with an M_L 5.6 event in the 1930s (Cleary, 1967; Denham *et al.*, 1981; Michael-Leiba *et al.*, 1994).

Earthquake clusters (e.g., Dalton–Gunning) are considered to be distinct from swarms, which are defined as occurring within a limited volume, lasting over a period from hours to months, with the largest event occurring well after the swarm commences, and not having a magnitude significantly greater than the second largest event (~0.5 magnitude unit; Gibson *et al.*, 1994; Dent, 2008). Geomechanical testing suggests that earthquake swarms occur preferentially in regions of extremely heterogeneous geological structure (Mogi, 1963). Increasing regional stress manifests as high stress concentration around numerous cracks and faults within the structured volume, resulting in failure on many local fractures at low stress. This has the effect of reducing the probability of failure on a single large fracture. Earthquake swarms are an important component of Australian seismicity, and can represent a large percentage of events in earthquake catalogues, particularly in regions of granitic geology (Dent, 2008). A non-exhaustive list of 42 earthquake swarms was compiled by Dent (2008, 2009), which includes perhaps the most significant of recent swarms, the 2001–2 Burakin Swarm (Leonard, 2002, 2003) in the Southwest Seismic Zone (SWSZ; Doyle, 1971) of Western Australia (Figure 2.1). This swarm involved six events in the magnitude range M_W 4.0–4.6 (Allen *et al.*, 2006) with over 18,000 smaller events, all purportedly occurring within a volume of ~5 km diameter (Leonard, 2003). The centre of activity for the swarm occurs approximately 20 km north of the surface rupture relating to the 1979 Cadoux M_S 6.0 earthquake (Figure 2.1 – 11). Several other swarms in the area (Kalannie, Beacon)

Table 2.1 *Documented surface-rupturing earthquakes in Australia. Refer to Figure 2.1 for locations.*

Year	Location	Magnitude (M_W)	Reference(s)
1968	Meckering, WA	6.6	Gordon and Lewis (1980), Vogfjörd and Langston (1987), Johnston (1994)
1970	Calingiri, WA	5.5	Gordon and Lewis (1980), Johnston (1994)
1979	Cadoux, WA	6.1	Lewis *et al.* (1981), Fredrich *et al.* (1988)
1986	Marryat Creek, SA	5.8	Fredrich *et al.* (1988), Machette *et al.* (1993)
1988	Tennant Creek, NT*	6.3–6.6	Bowman *et al.* (1990), Chung *et al.* (1992), Crone *et al.* (1992)
2008	Katanning, WA	4.7	Dawson *et al.* (2008)
2012	Ernabella, SA	5.4	Clark *et al.* (2013)

* Tennant Creek involved a series of three consecutive events within one day.

suggest that the events might reflect continuing crustal adjustment relating to the Cadoux rupture (Dent, 2008, 2009). However, many other swarm sequences are apparently unrelated to larger events (Gibson *et al.*, 1994; Dent, 2008).

On 27 December 1989 an M_L 5.6 earthquake occurred near Newcastle, NSW (Figure 2.1 – 5), resulting in 13 fatalities and significant damage. The event was one of the most destructive and costly natural disasters to have occurred in Australia (McCue *et al.*, 1990; Sinadinovski *et al.*, 2002; Insurance Council of Australia, 2013), but not the first event to affect a populated centre. The 1897 Beachport, 1902 Warooka, and 1954 Adelaide events (McCue, 1975; Greenhalgh and Singh, 1988; Love, 1996) (Figure 2.1 – 6, 7, 8) caused damage in South Australia; the Warooka event also claimed two lives (McCue, 1975). The 1968 Meckering, 1970 Calingiri (Gordon and Lewis, 1980), 1979 Cadoux (Lewis *et al.*, 1981), and 2010 Kalgoorlie–Boulder (Bathgate *et al.*, 2010) (Figure 2.1 – 9, 10, 11, 12) events all caused damage in Western Australia, as did the 1988 Tennant Creek sequence in the Northern Territory (Bowman, 1992) (Figure 2.1 – 2).

Several of the damaging earthquakes noted previously also ruptured the ground surface. If the three 1988 Tennant Creek mainshocks (Bowman, 1988; Choy and Bowman, 1990; Bowman, 1992; Crone *et al.*, 1992) are treated as a single scarp-forming event, then the 23 March 2012 M_W 5.4 Ernabella earthquake (Clark *et al.*, 2013) (Figure 2.1 – 13) represents the seventh historic Australian event that can be unequivocally associated with surface rupture (Figure 2.1; Table 2.1). All seven of these events ruptured non-extended SCR cratonic crust (Figure 2.1) and involved a dominant reverse component to motion (Gordon and Lewis, 1980; Lewis *et al.*, 1981; Crone *et al.*, 1992; Machette *et al.*, 1993; Dawson *et al.*, 2008). As such, they are similar to the 1989 Ungava (Canada) (Adams *et al.*, 1991, 1992; Bent, 1994) and the 1993 Killari Latur (India) (Rajendran *et al.*, 1996, 2001; Seeber *et al.*, 1996; Rajendran, 2000) earthquakes.

In addition to the few documented surface-rupturing events, isoseismal maps have been compiled for almost 400 historic Australian earthquakes (Everingham, 1982; Rynn *et al.*, 1987; McCue, 1996). McCue (1980) derived a relationship between Richter magnitude (*M*) and the radius of a circle of equivalent area to the Modified Mercalli Intensity III isoseismal for Australian earthquakes with recorded magnitudes ranging from 3.6 to 7.0 (reproduced in McCue, 2004). This permitted magnitudes to be assigned for pre-instrumental earthquakes or recent earthquakes that were widely felt but either too close to or too far from the nearest seismograph to measure the magnitude (McCue, 2004).

2.2.1 Mechanism, geographic distribution, and strain rate

The Australian crustal stress regime is generally considered to be compressive (Denham *et al.*, 1981; Hillis and Reynolds, 2000, 2003). This assessment is supported by the majority of earthquake focal mechanisms, which range from thrust to oblique strike-slip (e.g., Leonard *et al.*, 2002; Keep *et al.*, 2012). Notable exceptions, with a dominant normal component, are the 1985 Norseman and 2001 Ravensthorpe earthquakes (Figure 2.1 – 16, 17), both located along the Albany–Fraser Orogen margin of the Yilgarn Craton, with the latter being anomalously deep at 18 km (McCue, 1989; Clark, 2004). The 1966 Mount Hotham earthquake in the SESZ (Denham *et al.*, 1982) (Figure 2.1 – 18), several of the Burakin Swarm events in the SWSZ (Leonard *et al.*, 2002), and two Tennant Creek aftershocks (Clark and Leonard, 2003) also have a dominant normal component to their focal mechanisms. Note that Leonard *et al.* (2002) incorrectly transposes the P and T axes, and dilatant and compressional quadrants, for the Norseman 1985 event (cf. McCue, 1989).

Assuming a maximum possible magnitude (M_{max}) of ~7.0 (refer also to Section 2.5), the maximum seismogenic strain rate estimates for the continent (averaged over the last ~50 years of complete data) are $\sim 10^{-17}$ to 10^{-16} s^{-1} (Leonard, 2008; Braun *et al.*, 2009; Sandiford and Quigley, 2009). Comparable results (i.e., 10^{-17} s^{-1}) are obtained from thin-plate finite element modelling using plate boundary conditions, heat flow, stress, and geodetic data as inputs (Burbidge, 2004). This has been equated to an east–west shortening rate across southern Australia of approximately 0.3–0.4 mm/yr (Leonard, 2008), which compares to estimates based upon laser ranging and geodetic GPS of 0.65–3.0 ± 2.0 mm/yr (Smith and Kolenkiewicz, 1990; Tregonning, 2003; Leonard, 2008).

However, earthquake epicentres are not randomly distributed across the Australian continent in time or space (Denham, 1988; Leonard, 2008; Sinadinovski and McCue, 2010). At the continental scale, concentrations of epicentres occur in four major "seismogenic zones" (Figure 2.1) (Hillis *et al.*, 2008; Leonard, 2008; Sandiford and Egholm, 2008), with the continental margins also demonstrating a comparatively high number of epicentres relative to the interior (Sandiford and Egholm, 2008). Strain rates calculated for the higher seismicity zones are no more than 10^{-16} to 10^{-15} s^{-1} (Leonard, 2008; Braun *et al.*, 2009). Sandiford and Quigley (2009) provide an example from the Flinders Ranges Seismic Zone (FRSZ, Figure 2.1), where the bulk strain rate of 10^{-16} s^{-1} implies a total shortening of

~250 m/Ma across the ~100 km wide zone, which could be accommodated by 10 faults with slip rates consistent with the current topography and neotectonic faulting record (e.g., Sandiford, 2003b; Quigley *et al.*, 2006; Clark *et al.*, 2011a). Braun *et al.* (2009) estimate that half of the current relief of the Flinders Ranges, and a non-negligible proportion of the relief in the Eastern Highlands (SESZ, Figure 2.1), might plausibly have been built since the inception of the current stress regime. In contrast, similar seismogenic strain rates calculated for the SWSZ (Leonard, 2008; Braun *et al.*, 2009) (Figure 2.1) are not consistent with the present low-relief landscape (Clark, 2010), the distribution of paleo-earthquake fault scarps (Leonard, 2008; Leonard and Clark, 2011), or GPS strain rates (Leonard *et al.*, 2007). This implies migration of the locus of seismicity with time (Leonard, 2008; Leonard and Clark, 2011), an assertion supported by evidence suggesting that the rate of seismic activity in the SWSZ has increased dramatically in only the last 50 years (Michael-Leiba, 1987; Leonard and Clark, 2011).

2.2.2 Seismogenic depth

Depth constraint for Australian earthquakes is generally poor. Leonard (2008) selected a subset of the Australian earthquake catalogue for detailed hypocentral depth analysis using only depth values whose uncertainty was either less than the depth itself, or less than 5 km. More than 75% of epicentres from the current earthquake catalogue are rejected using these criteria. The selected data correlate with the regions of highest station density in private networks, and in the ANSN (i.e., southern Australia).

The concentration of epicentres in the SWSZ (Figure 2.1) occurs in non-extended cratonic crust. The Leonard (2008) dataset shows that earthquakes in this region are typically very shallow (see also Figure 2.3), with 95% of events found to have occurred within the upper 5 km of crust, consistent with previous analyses (Doyle, 1971; Everingham and Smith, 1979; Denham, 1988; McCue, 1990). This region has produced four surface-rupturing earthquakes in the last five decades (Figure 2.1; Table 2.1). The largest of these earthquakes (1968 Meckering; Figure 2.1 – 9) is calculated to have initiated at 1.5 km depth and to have ruptured both upwards to the surface and down to ~6 km depth (Langston, 1987; Fredrich *et al.*, 1988), while the smallest (2008 Katanning; Figure 2.1 – 15) ruptured from ~0.6 km depth to the surface (Dawson *et al.*, 2008). Earthquake swarms are also particularly prevalent in the SWSZ, with well-located depths being in the upper 2–3 km (Leonard, 2002, 2003). Analysis of the characteristics of modern and prehistoric fault scarps (Clark, 2010; Leonard and Clark, 2011; Clark *et al.*, 2012) suggests that large earthquake characteristics in the SWSZ (e.g., shallow depth) are typical of what might be expected throughout non-extended cratonic crust in Australia. However, Australia's deepest known earthquakes are also from this crustal setting: the 1989 Uluru earthquake (Figure 2.3 – 1), with a calculated depth of 31 km (Michael-Leiba *et al.*, 1994), and a 1992 earthquake beneath the Arafura Sea, north of the Northern Territory coastline (Figure 2.3 – 2), which had a depth of 39 km (McCue and Michael-Leiba, 1993).

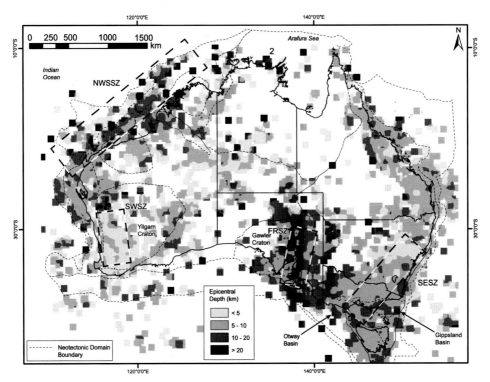

Figure 2.3 Interpolated surface of variation in earthquake hypocentral depths across the Australian continent constructed using the ARCGIS Point Statistics tool (plotted value is the mean of values within a moving 3 degree search box, output cell size 0.25 degrees). Events with a catalogued depth of greater than 50 km are not considered. The earthquake catalogue used is from Allen *et al.* (2012a), and is not filtered according to hypocentral uncertainty (cf. Leonard, 2008). Higher seismicity zones are shown (cf. Figure 2.1). Locations referred to in text are marked. Significant earthquakes mentioned in text are indicated by numbers: 1, Uluru; 2, Banda Sea. Note the greater depth of events in the FRSZ and NWSSZ compared to the SWSZ and SESZ.

Despite being one of the most seismogenic regions of Australia (Leonard, 2008; Sandiford and Egholm, 2008), the Northwest Shelf Seismic Zone (NWSSZ – Figure 2.1) is characterized by extremely poorly constrained hypocentres, with depth uncertainties typically significantly larger than the depth values (e.g., Revets *et al.*, 2009; Allen *et al.*, 2012a). Consequently, this region was not analysed in detail by Leonard (2008). However, earthquake hypocentres tend to be deeper in this region of extended continental crust than in the adjoining non-extended cratonic crust (Figure 2.2). Approximately 40% of events recorded in the NWSSZ occur at depths greater than 10 km.

The FRSZ in South Australia consistently produces deeper earthquakes than other parts of the continent (Figure 2.3), with over 75% of well-constrained events occurring at depths greater than 10 km (Leonard, 2008). A temporary seismograph deployment (September to December 2003) suggested spatial variation in the depth distribution, with a predominance of hypocentres in the 10–20 km depth range in the central Flinders Ranges, and less than 10 km depth in the southern Flinders Ranges (Cummins *et al.*, 2009), consistent with previous work (Greenhalgh *et al.*, 1994). The high proportion of hypocentres with depths greater than 10 km suggests that the depth to brittle–ductile transition in the Flinders Ranges is deeper than might be inferred from heat flow data (cf. Celerier *et al.*, 2005; Holford *et al.*, 2011).

Within the Southeast Seismic Zone (SESZ – Figure 2.3), and non-cratonic eastern Australia in general, hypocentral depth estimates are bimodal (Close and Seeber, 2007), ranging between very shallow (<5 km; 30% of data) and mid-crustal depths (10–20 km; 40% of data) (Allen *et al.*, 2012). Aftershocks tend to be very shallow and numerous, perhaps accounting for the shallower mode (Gibson *et al.*, 1981; Leonard, 2008). Within a sub-zone that encompasses the Pliocene to Holocene Newer Volcanic Province (e.g., Sutton *et al.*, 1977; Sheard, 1995), mid-crustal earthquakes are generally deeper, with 95% of recorded events occurring in the range 9–17 km (Leonard, 2008).

The Otway and Gippsland basins of southeast Australia (Figure 2.3) form part of an aulacogen developed in non-cratonic Paleozoic crust that was extended by rifting in the late Mesozoic–early Cenozoic. Earthquake hypocentres are deep in this region compared to those in the non-extended parts of the same Paleozoic basement province that is more typical of the SESZ. More than 70% of hypocentres in the extended basins are in the 10–25 km depth range (Allen *et al.*, 2012a), exemplified by the well-located 19 June 2012 M_L 5.4 Moe earthquake at a depth of 17 km (Sandiford and Gibson, 2012). It is probable that the contribution from these basins accounts for the greater depth element present in the southern southeast Australia (SEA-S) zone of Leonard (2008), as he does not distinguish between these two geological settings.

2.2.3 Attenuation and scaling relations

Several studies have shown that earthquakes in SCRs, such as Australia, are generally felt at larger distances than earthquakes in active tectonic regions (McCue, 1990; Frankel, 1994; Bakun and McGarr, 2002; Atkinson and Wald, 2007; Wald *et al.*, 2011). This is because the seismic energy propagates more efficiently through cold, relatively homogeneous continental crust, which is less susceptible to anelastic and scattering effects. It is recognised from analysis of isoseismal radii (Gaull *et al.*, 1990) that attenuation of seismic wave energy varies transversely across the Australian continent, with relatively low attenuation in the Archaean and Proterozoic terranes of western and central Australia and higher attenuation in the younger Phanerozoic terranes of eastern Australia (cf. Figure 2.1). Moreover, it is also observed that ground-motion amplitudes at large distances (>100 km) for an earthquake of given magnitude are lower in southeast Australia (SESZ) than in eastern North America (ENA) based on macroseismic intensities (Bakun and McGarr, 2002) and instrumental data (Allen and Atkinson, 2007). The higher attenuation observed in SESZ at distances greater than ~100 km is likely to be due to the broad crustal velocity gradient (Collins *et al.*, 2003), which allows dispersion of Lg-wave energy into the upper mantle (e.g., Bowman and Kennett, 1991; Atkinson and Mereu, 1992). However, there appears to be no discernible difference between the attenuation of Australian and ENA ground motions in the distance range of engineering significance (i.e., <100 km; Allen and Atkinson, 2007; Allen, 2012). Furthermore, attenuation rates at these near-source distances are also comparable to those in active tectonic regions (e.g., Hanks and Johnston, 1992; Atkinson and Morrison, 2009; Campbell, 2011).

Rupture dimensions have been modelled for only two of Australia's large earthquake sequences – the 1968 Meckering earthquake and the 1988 Tennant Creek sequence (Somerville *et al.*, 2010) – both of which occurred in the cratonic western and central regions of Australia respectively (cf. Figure 2.1; Table 2.1). Significant simplifications were required to model the subsurface rupture geometries of both events, which involved complex surface rupture on intersecting structural elements of varying orientation (e.g., Crone *et al.*, 1992; Dentith *et al.*, 2009). While this might limit the confidence that can be placed in the rupture models, earthquake scaling relations developed from the models compare favourably with empirically derived relations for intraplate dip-slip earthquakes (Johnston, 1994), and self-similar scaling relations (Leonard, 2010). Following the 2012 Ernabella earthquake, updated relations between seismic moment and surface rupture length were developed, and suggest that earthquakes in nonextended cratonic Australia produce longer surface ruptures than is estimated from published scaling relations for earthquakes larger than $M_w \sim 6.5$ (Clark *et al.*, 2013). At present there are no estimates of the rupture dimensions of any earthquakes in the non-cratonic or extended crust regions of eastern Australia.

2.3 A long-term landscape record of large (morphogenic) earthquakes

Australia is one of the lowest, flattest, most arid and slowly eroding continents on Earth. Accordingly, large parts of Australia are favourable for the preservation of tectono-geomorphic features, such as fault scarps, for tens of thousands to millions of years (e.g., Quigley *et al.*, 2010). In regions with extremely low erosion rates, such as the Nullarbor Plain (Figure 2.4), it has been claimed that a morphogenic earthquake record spanning the last 15 Ma has been preserved essentially intact (Hillis *et al.*, 2008). On this basis, Australia boasts arguably the richest Late Neogene to Quaternary faulting record in all of the world's SCR crust (Sandiford, 2003b; Quigley *et al.*, 2010; Clark *et al.*, 2011a, 2012). Over 300 features (mainly fault scarps and folds) suspected or known to have been displaced under the current crustal stress regime have been identified and recorded (Clark *et al.*, 2011a, 2012) (Figure 2.4). The majority of these features, by virtue of their length and/or vertical displacement (see Section 3.1), are likely to reflect multiple surface-rupturing events (cf. Leonard, 2010). This remarkable archive, while undoubtedly incomplete, has the potential to extend the historic record of seismicity to a timescale commensurate with the recurrence time of large earthquakes in this intraplate setting.

Variation in fault scarp length, vertical displacement, proximity to other faults, and relationship to topography permits further division of this neotectonic record according

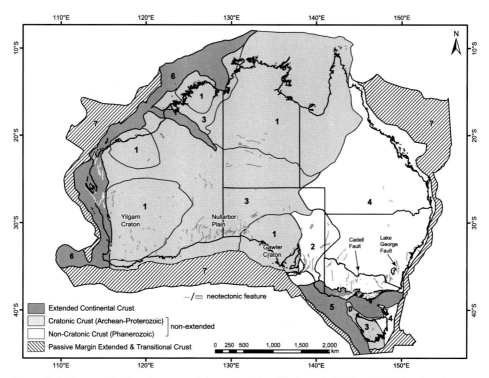

Figure 2.4 Neotectonic domains map of Australia (after Clark *et al.*, 2011a, 2012) showing features known or suspected of hosting displacement under the current stress field (grey lines). Neotectonic domains are indicated by numbers: 1, Precambrian Craton and Non-reactivated Proterozoic Crust; 2, Sprigg Orogen; 3, Reactivated Proterozoic Crust; 4, Eastern Australian Phanerozoic Accretionary Terranes; 5 Eastern Extended Continental Crust; 6, Western Extended Continental Crust; 7, Passive Margin Extended Continental and Transitional Crust. Shading and hatching as per Figure 2.1 shows the domains in their broader context of non-extended cratonic crust in the central and western parts, non-extended non-cratonic crust in the east, and extended margins around much of the continent.

to fault character. Six onshore "neotectonic domains" are recognised, with an additional offshore domain proposed by analogy with the eastern United States (Clark *et al.*, 2011a, 2012) (Figure 2.4). Each domain relates to a distinct underlying crustal type and architecture, broadly considered to represent non-extended cratonic and non-cratonic crust, as well as extended crustal settings (Figure 2.4) (cf. Johnston *et al.*, 1994). Herein domains will be referred to using the abbreviation Dx, where x is the number of the domain (e.g., D1 is Domain 1).

The Australian landscape reveals a marked disparity in its crustal response to imposed tectonic stresses. For example, neotectonic faults in western and central Australia, which are founded in cratonic crust (Figure 2.4 – D1, D3), are typically spaced more than a scarp length apart, not associated with historic seismicity, and displace a low, undulating landscape by less than 10 m (e.g., Clark, 2010). In contrast, faults in non-cratonic crust of the Mount Lofty and Flinders Ranges (Figure 2.4 – D2) are relatively closely spaced, commonly associated with historic seismicity, and displace Neogene strata by up to a few hundred metres (e.g., Sandiford, 2003b; Quigley *et al.*, 2006). Further east, within non-cratonic Australia (Figure 2.4 – D4), faults are commonly found in *en echelon* arrangements, with fault complexes extending for several hundred kilometres along strike (Beavis, 1960, 1962; Moye *et al.*, 1963). As little as 15 m (e.g., Canavan, 1988; Robson and Webb, 2011; McPherson *et al.*, 2012b) and as much as several hundred metres of neotectonic displacement has been documented on several of these features (Beavis, 1960, 1962; Moye *et al.*, 1963; Abell, 1985), and there is no clear relationship to historic seismicity. Faults within the extended continental crust fringing the Australian continental margin (Figure 2.4 – D5, D6), a remnant of the Cretaceous breakup of the supercontinent Gondwana (e.g., Veevers, 2000), have some of the larger throws of any Australian neotectonic faults (greater than 100 m; Holdgate *et al.*, 2003). Extensional structural architecture is largely preserved (e.g., Hill *et al.*, 1994), and neotectonic faults are often spatially associated with earthquake epicentres.

2.3.1 *Variation in fault scarp length and vertical displacement*

The neotectonic data compiled by Clark *et al.* (2012) contain two semi-quantitative variables useful for characterising fault behaviour – length and vertical displacement. The population distributions for Australian fault scarp length and vertical displacement data are presented in Figure 2.5a, b. Fault length is defined as the along-strike distance (tail to tail) of discrete geomorphic features (most often fault scarps) that are considered to represent one or more surface-rupturing earthquake events. Vertical displacement is the vertical separation across a topographic scarp or fold. Many of the fault-length values reported by Clark *et al.* (2012) might be expected to be underestimates, as vertical displacement tapers towards the tails of ruptures, resulting in these scarp sections being less discoverable in digital elevation data: the primary tool for their identification and characterisation (Clark, 2010; Clark *et al.*, 2011a, 2012). It is also reasonable to assume that relatively short scarps are under-represented as they are less discoverable. This factor may explain the positive skew in both length and height data (Figure 2.5a, b). Furthermore, the resolution and noise content of digital elevation data from various sources might be expected to affect the precision of both length and vertical displacement measurements.

Interpolated surfaces for Australian neotectonic data demonstrate the spatial variation in these fault parameters (Figure 2.5c, d), and these spatial patterns are borne out in the statistics for each of the neotectonic domains (Figure 2.5e, f). The cratonic domains (D1, D3) are characterised by the lowest vertical displacement values (Figure 2.5d, f). In view of the extremely low erosion rates in these parts of Australia (Bierman and Caffee, 2002; Belton *et al.*, 2004; Chappell, 2006), this is indicative of very low time-averaged rates of morphogenic seismicity. Scarp lengths in D3 are up to 100 km greater than in D1 (Figure 2.5e), raising the possibility that relief in the Proterozoic mobile belts is built in fewer, larger earthquakes. Five fault scarps have been subject to detailed paleoseismological investigation in this crustal type: (1) Meckering (Clark *et al.*, 2011a), (2) Hyden (Crone

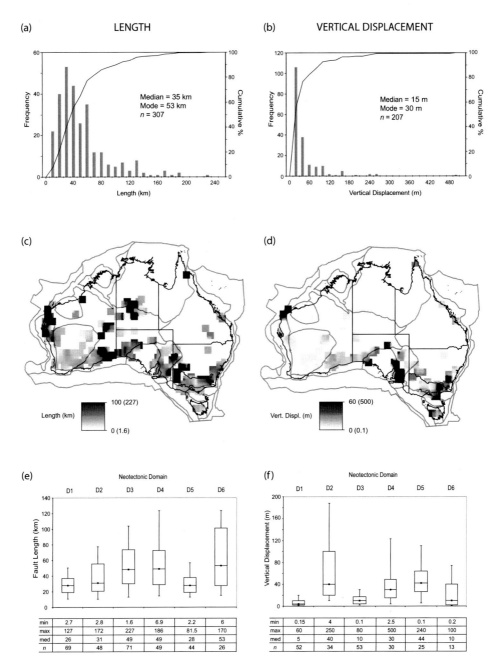

Figure 2.5 Population distributions for Australian neotectonic (a) fault length and (b) vertical displacement data, with accompanying plots showing the spatial distribution of (c) fault length and (d) vertical displacement data across the Australian continent. Interpolated surfaces in parts (a) and (b) are constructed using the same method as Figure 2.3. Grey lines indicate neotectonic domain boundaries (cf. Figure 2.4). Data values have been scaled for gridding – values in parentheses in the legend indicate the full range of underlying data. Box and whisker plots of (e) fault length and (f) vertical displacement data, binned by neotectonic domain. Boxes denote 75th and 25th percentiles, central point indicates median value, and whiskers define 90th and 10th percentiles. Data table shows minimum, maximum, median, and number of data points (*n*) for each domain.

et al., 2003; Clark *et al.*, 2008), (3) Dumbleyung and (4) Lort River (Estrada, 2009), and (5) Lake Edgar (Clark *et al.*, 2011b). These features range from 30 to 40 km in length (cf. Clark, 2010) with maximum single-event vertical displacements in the order of 1.2–3.1 m. Recurrence data for morphogenic earthquake events is restricted to the most recent one to three events on each fault, and indicates inter-event intervals of ~10–40 ka (Figure 2.6).

Non-cratonic (D2, D4) and extended (D5, D6) domains are characterised by comparatively large vertical displacements (Figure 2.5d, f), which scaling relationships (e.g., Wells and Coppersmith, 1994; Leonard, 2010; Leonard and Clark, 2011) imply must have built as the result of multiple seismic cycles. There is, therefore, less certainty as to whether scarp lengths are representative of single-event ruptures, or are the product of segmented rupture. A positively skewed fault-length data distribution (Figure 2.5a) might plausibly reflect segmented rupture behaviour. The longest fault scarp in non-cratonic eastern Australia (D4) that has been subject to paleoseismological investigation is the Cadell Fault (Figure 2.4). Evidence from abandoned fluvial and tectonic landforms (e.g., Bowler and Harford, 1966; Rutherfurd and Kenyon, 2005) is consistent with seismic rupture of the entire 80 km scarp length, potentially involving 2–4 m of uplift (Clark *et al.*, 2011a; McPherson *et al.*, 2012a). While the timing of individual seismic events is poorly constrained, the average recurrence interval within the 70–20 ka active period (assuming full-length rupture) may have been as little as ~8 ka. Within D2, single-event displacement values of 1.8 m (1.5 m vertical) have been recorded on the Alma and Williamstown-Meadows faults (Clark and McPherson, 2011). The latter fault, which has a mapped length of over 100 km, is associated with a 25 km long single-event scarp.

2.3.2 The influence of crustal type and character on seismic activity rates

Given that the compressive nature of the Australian stress field (Hillis and Reynolds, 2003) results in a predominance of dip-slip faulting (Leonard *et al.*, 2002), long-term qualitative seismic activity rates across the continent might be assessed in terms of neotectonic uplift of the landscape. With respect to the morphogenic earthquake record, this is a function of neotectonic fault slip rate and density.

Long-term slip rates estimated for faults in the cratonic western part of the continent (D1, D3) are typically in the order of ~1 m/Ma (Clark *et al.*, 2008, 2012; Hillis *et al.*, 2008), which is equal to or less than the extant erosion rates (Bierman and Caffee, 2002; Belton *et al.*, 2004; Chappell, 2006). These long-term slip rates are consistent with the low-relief landscape, and contrast with uplift rates estimated from the last several decades of seismicity in the SWSZ, which have been suggested to be in the order of 10 m/Ma (Braun *et al.*, 2009).

Long-term slip rates in non-extended, non-cratonic eastern Australia (D4) are less well constrained. The slip rate on the Cadell Fault, averaged over the life of the current stress field (i.e., ~10 Ma), has been estimated at ~10 m/Ma (Clark *et al.*, 2007). New data from the Lake George Fault (Pillans, 2012) within the Eastern Highlands (Figure 2.4) indicate a

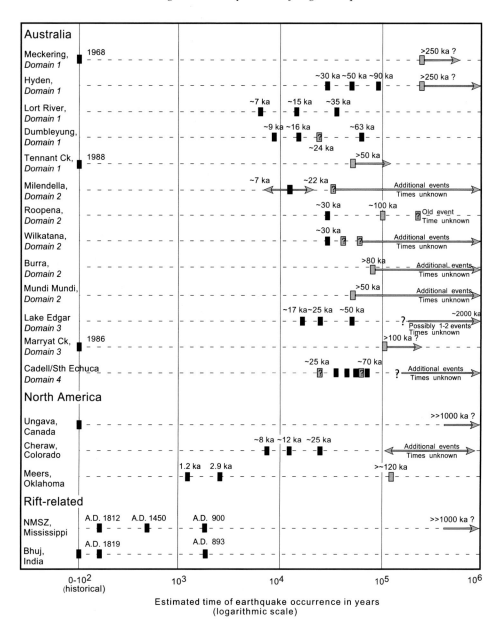

Figure 2.6 Compilation of surface-breaking earthquake recurrence data for SCR settings (Clark *et al.*, 2012, modified after Crone *et al.*, 2003). Lort River and Dumbleyung data from Estrada (2009), Wilkatana, Burra and Mundi Mundi data from Quigley *et al.* (2006), Lake Edgar data from Clark *et al.* (2011b) and Cadell data from Clark *et al.* (2007). Australian examples are labelled with their relevant neotectonic domain (cf. Figure 2.4).

slip rate of ∼50 m/Ma. Higher relief areas of eastern Australia are associated with erosion rates of up to 30–50 m/Ma (Weissel and Seidl, 1998; Heimsath *et al.*, 2000, 2001; Wilkinson *et al.*, 2005; Tomkins *et al.*, 2007). Several authors have suggested that as much as 200 m of relief has been added to the Eastern Highlands over the last ∼10 Ma (Sandiford, 2003b; Holdgate *et al.*, 2008; Braun *et al.*, 2009). Such estimates are consistent with uplift rates estimated from strain rates derived from contemporary seismicity (Braun *et al.*, 2009); however, little unequivocal evidence exists to assign this uplift to active faults (cf. Holdgate *et al.*, 2006). The approximate equivalence of erosion rates and fault slip rates suggests that much of the extant relief may be inherited (e.g., Bishop *et al.*, 1982; Pickett and Bishop, 1992; van der Beek *et al.*, 2001).

Perhaps the most spectacular examples of neotectonism in southeast Australia are the inverted Mesozoic Otway and Gippsland Basins (D5) (see Figure 2.1) (Holdgate *et al.*, 2003, 2007; Sandiford, 2003a), which formed by extension of non-cratonic (D4) crust. Within the Gippsland Basin in particular, the Cretaceous basin deeps now form the topographic highs at elevations of 200–300 m above sea level. Preliminary cosmogenic radionuclide ages obtained on overlying folded alluvial sediments (Holdgate *et al.*, 2007) suggest uplift rates of 60 m/Ma and greater on the major relief-forming faults (McPherson *et al.*, 2009; Clark *et al.*, 2011a, 2012). The relatively high fault density, combined with high slip rates and frequent contemporary seismicity (e.g., Leonard, 2008), identify these basins as among the most actively deforming parts of the continent (compare with the Flinders Ranges – D2).

Slip rates on faults underlying folds in the passive margin basins that dominate the Northwest Shelf region (D6) (NWSSZ, Figure 2.1), are poorly constrained. While locally having resulted in the uplift of Miocene marine deposits to elevations of over 100 m above sea level (e.g., van de Graaff *et al.*, 1976), neotectonic uplift is typically more modest than in the eastern aulacogen and passive margin basins (e.g., D5).

Long-term (i.e., averaged over several million years) vertical slip rates are known from some of the range-bounding faults of the Flinders and Mount Lofty Ranges (D2). These typically vary between ∼20 and 50 m/Ma (Bourman *et al.*, 1999; Belperio *et al.*, 2002; Sandiford, 2003b; Celerier *et al.*, 2005; Quigley *et al.*, 2006). Bedrock erosion rates have been recorded at up to 122 m/Ma, but average around 40 m/Ma (Bierman and Caffee, 2002; Chappell, 2006; Quigley *et al.*, 2007a, b), allowing that relief is being produced within the ranges. Several authors suggest that up to half of the ∼800 m relief has been built in the current stress regime (Sandiford, 2003b; Quigley *et al.*, 2007c; Braun *et al.*, 2009). Seismic reflection data indicate that the structural architecture mapped at the surface, corresponding to the Paleozoic Adelaide Fold Belt which developed over the inversion axis of a Neoproterozoic rift basin (Jenkins and Sandiford, 1992; Paul *et al.*, 1999), extends to depths of 10–12 km beneath the ranges (Flöttmann and Cockshell, 1996). However, much of the seismicity recorded in the region occurs below this depth (Leonard, 2008), potentially in cratonic crust relating to the eastern margin of the Gawler Craton (cf. Figure 2.1). Therefore, it is unclear how the surface-faulting record might be related, if at all, to most of the instrumental seismicity (cf. Braun *et al.*, 2009).

In general, greater topographic expression associated with faults and fault systems occurring in extended crust relative to non-extended crust suggests a higher rate of seismic activity in the extended setting, consistent with observations worldwide (e.g., Johnston, 1994; Cloetingh *et al.*, 2008; Mooney *et al.*, 2012). Using the same reasoning, non-cratonic crust might be expected to have a higher rate of seismic activity than cratonic crust (cf. Figures 2.4 and 2.5), by virtue of there being no relief generation in cratonic crust. This distinction, together with the variation in fault character between domains, should be recognised in attempts to identify analogous systems worldwide.

2.4 Patterns in earthquake occurrence

The record of contemporary seismicity in Australia suggests that earthquake epicentres are spatially and temporally clustered (Denham, 1988; Leonard, 2008; Sinadinovski and McCue, 2010). As discussed in Section 2.1, concentrations of epicentres in the historic catalogue are dominated by four "seismogenic zones" (Figure 2.1) (Hillis *et al.*, 2008; Leonard, 2008; Sandiford and Egholm, 2008). However, whether the short record of contemporary seismicity is representative of time periods significantly longer than the observation window has not been statistically or empirically tested in the Australian context (cf. Kafka, 2002). The persistence of patterns in the short historic catalogue can be assessed at much longer timescales by comparison with the record of morphogenic seismicity. Evidence from the paleo-record (Crone *et al.*, 2003; Clark *et al.*, 2011a, 2012), which essentially captures events of $\sim M > 5.5$, suggests that large earthquake occurrence within Australia exhibits both spatial and temporal clustering (Section 2.3). Temporal patterns in large SCR earthquake occurrence may be inferred at the scale of a single fault (Crone *et al.*, 1997, 2003; Clark *et al.*, 2008), of groups of faults (Leonard and Clark, 2011), and at the domain scale (Holdgate *et al.*, 2003; Sandiford, 2003b; Paine *et al.*, 2004; Braun *et al.*, 2009; Clark *et al.*, 2011a, 2012).

For example, the distribution of fault scarps in cratonic southwest Western Australia (D1) (Figure 2.4), together with the seismogenic strain arguments referred to previously (cf. Leonard *et al.*, 2007; Leonard, 2008; Braun *et al.*, 2009), infer that seismicity in the SWSZ represents only the current locus of activity, rather than a zone of long-lived activity (cf. Sandiford and Egholm, 2008). As most fault scarps in other parts of Australia are not associated with historic seismicity (e.g., Crone *et al.*, 2003; Clark, 2010), a similar rule may apply.

There is no precedent in cratonic Australia (neotectonic domains 1 and 3 of Clark *et al.,* 2012) to indicate how long seismicity will persist. However, a large proportion of contemporary seismicity in the SWSZ is thought to relate to the Calingiri, Cadoux, and Meckering earthquakes (Leonard, 2008). If aftershock activity relating to surface-rupturing earthquakes is used as a measure of the longevity of activity in a region, then the work of Stein and Liu (2009) suggests that a millennial timescale might be applicable. Liu *et al.* (2011) and Liu and Wang (2012) present evidence for migration of the locus of seismicity at a centennial timescale in active intraplate northern China (Kusky *et al.*, 2007; Wheeler, 2011) with no recurrence over the 2,000–3,000-year window captured by the Chinese record of historical seismicity. As crustal deformation rates in this 1,000 km × 1,000 km region of China are in the order of 1–2 mm/yr (Liu *et al.*, 2011), such data might be considered as a minimum seismicity migration rate for Australia at the neotectonic domain scale.

There is some indication that the temporal clustering behaviour emerging from single-fault studies in non-cratonic Australia may be symptomatic of a larger picture of the more-or-less continuous tectonic activity from the late Miocene to Recent being punctuated by "pulses" of activity in specific deforming regions (e.g., Quigley *et al.*, 2010; Clark *et al.*, 2012). For example, major deformation episodes are constrained to the interval 6–4 Ma in southwest Victoria (D4) (Paine *et al.*, 2004) and 2–1 Ma in the Otway Ranges (D5) (Sandiford, 2003a). An episode of deformation ceased at 1.0 Ma in the offshore Gippsland Basin (D5) although it continued onshore until ~250 ka (Holdgate *et al.*, 2003). Holdgate *et al.* (2008) present evidence from the southeast highlands that resurrects the idea of a punctuated post-Eocene Kosciuszko Uplift event (Browne, 1967) that continued into the late Pliocene and possibly the Pleistocene.

The Mount Lofty and Flinders Ranges (D2) are perhaps an exception to this rule. The FRSZ has been associated with diffuse earthquake activity throughout the historic era (Greenhalgh and Singh, 1988; Greenhalgh *et al.*, 1994). While epicentres cannot in most cases be confidently associated with neotectonic faults, strain rates and uplift rates estimated from the seismic catalogue are approximately consistent with the number of neotectonic faults and their paleoseismologically derived slip rates (e.g., Sandiford, 2003b; Braun *et al.*, 2009). Hence, it is possible that seismicity is a long-lived (millions of years) process in this crustal setting.

While the suite of neotectonic fault behaviours may vary across Australia, as implied by the neotectonic domains model, one individual fault characteristic appears to be common to most Australian intraplate faults studied – active periods comprising a finite number of events are separated by typically much longer periods of quiescence (Crone *et al.*, 1997, 2003; Clark *et al.*, 2007, 2008, 2011a, 2012; Estrada, 2009) (Figure 2.6). Data and modelling from elsewhere in the world identify similar episodic behaviour on faults with low slip rates (e.g., Wallace, 1987; Ritz *et al.*, 1995; Marco *et al.*, 1996; Friedrich *et al.*, 2003) and suggest that the time between successive clusters of events (deformation phases) is highly variable but significantly longer than the inter-event times between successive earthquakes within an active phase (Marco *et al.*, 1996; Stein *et al.*, 1997; Chéry *et al.*, 2001; Chéry and Vernant, 2006; Li *et al.*, 2009). This characteristic has been referred to as Wallace-type behaviour (see Wallace, 1987), and may be conceptualised using a model similar to that proposed by Friedrich *et al.* (2003) (Figure 2.7).

The sparse data available in Australia suggests that an active period (e.g., t_1, t_2, t_3 in Figure 2.7) in cratonic central or western Australia (D1) might comprise as few as two or three events (e.g., Hyden – Crone *et al.*, 2003; Clark *et al.*, 2008), with interseismic intervals between large events in an active period in the order of 20–40 ka (Crone *et al.*,

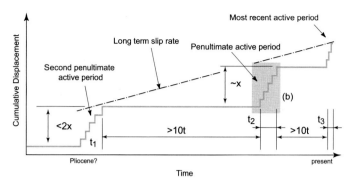

Figure 2.7 Generalised fault-slip diagram for Australian SCR faults based upon data from the Cadell Fault (Clark *et al.*, 2007; McPherson *et al.*, 2012a). Three active periods of fault growth (earthquake occurrence) are denoted by t_1, t_2, and t_3. These active periods are relatively short-lived, and each is composed of only a few ruptures (*c.* <6 events per active period). Inter-event times between successive ruptures within an active period may range up to several thousands to several tens of thousands of years. We adopt a characteristic earthquake rupture model based upon paleoseismic data from the Cadell (Clark *et al.*, 2007; McPherson *et al.*, 2012a) and Lake Edgar (Clark *et al.*, 2011b) faults. Long quiescent periods separate the active periods, and the length of the quiescent periods can range from many tens of thousands of years to greater than a million years.

1997, 2003; Clark *et al.*, 2008; Estrada, 2009). In non-cratonic eastern Australia (D4), recently acquired data on the Cadell Fault (Clark *et al.*, 2007; McPherson *et al.*, 2012a) indicate more frequent rupture, with up to six morphogenic events in the interval *c.* 70–25 ka. It is inferred that three uplift events of similar magnitude had occurred on the Cadell Fault prior to the diversion of the Murray and Goulburn Rivers at *c.* 45 ka (Bowler and Harford, 1966; Bowler, 1978), suggesting that these events are likely to have been spaced thousands of years apart, similar to the Meers and Cheraw faults in the western Central United States (Crone and Luza, 1990; Crone and Machette, 1995). This contrasts with the New Madrid seismic zone in the intraplate Central United States, where sequences of large earthquakes have occurred on average every 500 years for at least the last two seismic cycles in the current active period (Tuttle *et al.*, 2002). While Talwani and Shaeffer (2001) propose a similar recurrence for earthquakes large enough to produce liquefaction in the South Carolina Coastal Plain, it has not been determined whether their data reflect rupture on a single fault or multiple faults. Quiescent intervals can be sufficiently prolonged (hundreds of thousands to millions of years), in the western and central parts of Australia in particular, that most or all relief relating to an active period might be removed by erosion prior to the next active period (Crone *et al.*, 2003; Clark *et al.*, 2007, 2008, 2011a).

2.5 Maximum magnitude earthquake

Large earthquakes are so infrequent in SCRs such as Australia that the data distributions upon which recurrence and M_{max} estimates are based are heavily skewed towards magnitudes below M_W 5.0 (e.g., Leonard, 2008), and thus require significant extrapolation up to magnitudes at which the most damaging ground-shaking might be expected (Wheeler, 2009a, b). Australia is uniquely suited to assessing the validity of M_{max} estimates derived from the instrumental record of seismicity by virtue of its extraordinary neotectonic and paleo-earthquake record spanning many tens of thousands of years (e.g., Clark *et al.*, 2011a).

Where a fault has been studied paleoseismologically, single-event scarp lengths and single-event displacements can provide two independent estimates of the characteristic magnitude (cf. Schwartz and Coppersmith, 1984) for that fault, with scarp length being considered the more reliable measure (e.g., Wells and Coppersmith, 1994; Hemphill-Haley and Weldon, 1999; Leonard, 2010). Where a fault has not been studied in detail the "characteristic" rupture of the entire scarp length during each morphogenic event, subject to appropriate caveats regarding segmentation and dip, might be assumed in order to estimate future large earthquake potential (e.g., Hemphill-Haley and Weldon, 1999; Stirling *et al.*, 2002; Wheeler, 2009a).

In most regions of Australia the neotectonic record is either incomplete, or under-explored (Sandiford, 2003b; Clark *et al.*, 2011a). As such, it is not possible to assert with confidence that future large earthquakes will be restricted to known scarps/faults with documented source characteristics. It is therefore desirable to aggregate characteristic event magnitudes from a group of faults with similar source characteristics from within a region of interest (e.g., neotectonic domain or aggregation of domains [Leonard *et al.*, 2012; Clark *et al.*, 2012]) to provide an estimate of M_{max}. This can be achieved with varying levels of confidence, depending upon the estimated completeness of the paleo-record.

2.5.1 Scarp length as a proxy for paleo-earthquake magnitude

A relatively large area of 10 m resolution DEM data in the southwest corner of the Yilgarn Craton (D1) (Figure 2.4) allowed for mapping of fault scarps in unprecedented detail (Clark, 2010). Based upon an assessment of erosion and landscape modification rates, it was estimated that most scarps representing events of $M_W \geq 6.5$ that had occurred in the last ~100 ka were captured in the catalogue (Leonard and Clark, 2011). The scaling relations of Leonard (2010) were used to develop a paleo-seismicity catalogue comprising 65 events (Leonard and Clark, 2006, 2011), which was subsequently combined with the catalogue derived from instrumental seismicity. The data were found to exhibit typical truncated Gutenberg–Richter recurrence characteristics with a slope (*b*) of 0.9–1.0 between M_W 6.5 and 6.9. A rapid roll off in recurrence occurred above M_W 6.9, with an asymptotic value of M_W 7.25 ± 0.1 considered to be the M_{max} for non-extended cratonic crust typified by the Yilgarn Craton (Leonard and Clark, 2006, 2011). A less well constrained M_{max} of M_W 7.65 ± 0.1 was determined for scarps representing extended crust in the rift basins flanking the western margin of the continent (D6).

In most regions of Australia the neotectonic catalogue is far from complete, the preservation time of seismogenic features in the landscape is uncertain, and single-event scarp

Table 2.2 *Ninetieth (90th) percentile fault scarp length values for each neotectonic domain of Clark et al. (2012) and corresponding characteristic earthquake magnitude (M_W) values determined using the scaling relation of Leonard (2010) assuming a 45° dipping fault and a seismogenic depth of 15 km*

Domain	Length [90%] (km)	Magnitude (M_W)
1	51	7.3[*]
2	78	7.4
3	104	7.6
4	124	7.6
5	57	7.3
6	124	7.6
All data	101	7.6

[*]The cratonic non-extended relation of Clark *et al.* (2013) obtains a value for D1 of M_w 6.9.

length and slip data are largely absent. Without these data, key assumptions of the curve-fitting approach to M_{max} estimation described above are not satisfied. However, scarp length variation across the continent (Figure 2.5) does imply variation in characteristic earthquake magnitude (and M_{max}).

Across Australia there is a strong positive skew in the length data distribution, with 90% of scarps less than 101 km in length (Figure 2.5a; Table 2.2). Assuming a generic fault dip of 45° and a seismogenic depth of 15 km (Collins *et al.*, 2003), the 90th percentile value for length corresponds to a value of M_W 7.6 (cf. Leonard, 2010) (Table 2.2). Without paleoseismic data on the longest scarps it is difficult to assess the validity of this estimate as a value for M_{max}. The longest scarp with paleoseismological evidence consistent with entire length rupture is the 80 km long (~M_W 7.4) Cadell Fault scarp in eastern Australia (D4) (Clark *et al.*, 2007, 2011a; McPherson *et al.*, 2012a); however, rupture segmentation is plausible for intraplate faults of several tens of kilometres in length or greater (e.g., Machette *et al.*, 1991). The modal scarp length value of 53 km for all neotectonic features (Figure 2.5a) corresponds to an earthquake of magnitude ~M_W 7.3.

Fault scarp length data for all individual domains (Figure 2.5e) show a similar positive skew to the aggregated population data distribution (Figure 2.5a), which highlights the same uncertainties with respect to possible fault segmentation. The 90% values for fault length provide magnitude estimates in the range of M_W 7.3–7.6 (Table 2.2). The values for D1 and D3 are within error of the M_{max} estimated by Leonard and Clark (2011). Within D2 the only known single-event scarp is ~28 km long (M_W 6.8–6.9) and relates to the greater than 100 km long Williamstown-Meadows Fault (Clark and McPherson, 2011).

However, paleoseismological evidence exists elsewhere in D2 for very large single-event displacements (>7 m – Quigley *et al.*, 2006; Reid, 2007; Clark *et al.*, 2011a, 2012), consistent with modelled event magnitudes of M_W 7.3–7.5 (Somerville *et al.*, 2008). Very long 90th percentile values for D3 and D4 require validation in terms of equivalence to M_{max} as these scarps also tend to have accumulated significant neotectonic throw (tens of metres), implying multiple morphogenic events. The dominance of folding as opposed to discrete faulting in D5 has thus far prevented estimation of single-event rupture lengths and displacements (Clark *et al.*, 2011a).

As expected, overall indications are that the historic catalogue of seismicity significantly underestimates the large earthquake potential (and, by proxy, M_{max}) in most regions of Australia. To a first order at least, the sub-division of the continent into domains on the basis of geology and tectonic history (Johnston, 1994; Clark *et al.*, 2011a, 2012) provides useful insights into variations in faulting character that can facilitate the interpretation of the neotectonic and historic catalogues. The use of fault-length data from the neotectonic catalogue provides reasonable preliminary estimates of M_{max} when applied in conjunction with appropriate scaling relations (e.g., Leonard, 2010; Clark *et al.*, 2013). Analysis of the current neotectonic catalogue for Australia suggests that a range of M_{max} values of M_W ~ 7.0–7.6 could reasonably encompass all geological and tectonic settings continent-wide (Clark *et al.*, 2011a, 2012; Leonard, 2012; Burbidge, 2012).

2.6 Implications for SCR analogue studies: factors important in earthquake localisation

Analogues between the Australian neotectonic domains (Clark *et al.*, 2011a, 2012) and SCR crust elsewhere in the world (cf. Johnston, 1994) are readily apparent. For example, poly-phase deformation of a compressional nature is a common feature in the post-rift evolution of many extended passive margins and rifts (D5/D6 analogues) (van Arsdale, 2000; Balasubrahmanyan, 2006; Cloetingh *et al.*, 2008). Archean cratonic nuclei fringed by Paleoproterozoic mobile belts (D1 analogues) make up a large portion of the geology of Peninsular India (Kroner and Cordani, 2003) and North America (Hoffman, 1989). Meso- and Neoproterozoic mobile belts involved in the accretion of the supercontinent Rodinia (D3 analogues) are found worldwide (e.g., Collins and Pisarevsky, 2005; Cawood and Buchan, 2007), as are Phanerozoic accretionary terranes associated with the amalgamation of the supercontinent Gondwana (D4 analogues) (e.g., Hoffman, 1989). The temporal clustering of morphogenic earthquake events seen in studies on Australian faults (Quigley *et al.*, 2010; Clark *et al.*, 2011a, 2012) also appears to be mirrored in the Central and Eastern United States (Crone and Luza, 1990; Crone *et al.*, 1997; Cox *et al.*, 2006). Inferences made regarding mechanisms responsible for localising intraplate seismicity in Australia might then be assessed in terms of their crustal and lithospheric setting, and tested on analogous crust elsewhere in the world.

2.6.1 Mechanical and thermal influences

It has been proposed that intraplate regions with higher seismic potential have pre-existing zones of weakness (Sykes, 1978; Talwani and Rajendran, 1991; Stuart *et al.*, 1997; Kenner and Segall, 2000; Dentith and Featherstone, 2003), intersecting faults (Talwani, 1988, 1999), elevated heat flow (Liu and Zoback, 1997; Celerier *et al.*, 2005; Hillis *et al.*, 2008; Holford *et al.*, 2011), crustal anomalies (Campbell, 1978; Kenner and Segall, 2000; Gangopadhyay and Talwani, 2003; Pandey *et al.*, 2008; Assumpção and Sacek, 2013) or can be identified by crustal boundaries inferred from potential field data (Langenheim and Hildenbrand, 1997; Lamontagne *et al.*, 2003; van Lanen and Mooney, 2007; Dentith *et al.*, 2009). It is likely that the variety of models reflects the range of mechanisms that are operating, and although all of the above mechanisms have demonstrated local applicability in the Australian context, counter-examples are abundant. This is particularly the case where models relying on thermal mechanisms for strain localisation have been proposed.

Sandiford and Egholm (2008) argue that enhanced seismicity along some parts of the Australian continental margin is a consequence of thermal weakening due to steady-state heat flow across the lithospheric thickness steps between oceanic and continental crust (e.g., SESZ, Figure 2.1) and extended continental and non-extended cratonic crust (e.g., SWSZ, Figure 2.1) (Fishwick *et al.*, 2008; Kennett *et al.*, 2013). While the hypothesis is intuitively appealing, in Australia and along the eastern seaboard of North America (Wheeler and Frankel, 2000), several lines of evidence suggest that the contribution of this mechanism to the continental seismicity budget over geological timescales is minor.

For example, east of the large lithospheric thickness step on the western boundary of the SWSZ (i.e., across the Darling Fault), fault scarps are randomly distributed (Clark, 2010) and the topography predicted if instrumental seismic moment release rates are extrapolated to million-year timescales (Braun *et al.*, 2009) is absent. Given extremely low rates of bedrock erosion (Bierman and Caffee, 2002; Belton *et al.*, 2004; Chappell, 2006), this finding implies that the locus of seismicity in the SWSZ is transitory, rather than responding to steady-state heat flow at the margin. Furthermore, very little seismicity, paleo- or instrumental, can be correlated with the dramatic transition from cratonic to non-cratonic lithospheric thickness in eastern Australia (i.e., the Tasman Line, Figures 2.1 and 2.3), and significant instrumental seismicity proximal to the eastern seaboard (the SESZ) is not associated with a large lithospheric thickness step (Fishwick *et al.*, 2008).

More generally, Holford *et al.* (2011) propose that the thermal properties of the crust and upper mantle exert a regional-scale (100–1000 km) modulating control on which parts of the Australian lithosphere undergo (seismogenic) failure and which parts experience relatively less deformation (cf. Celerier *et al.*, 2005; Sandiford and Quigley, 2009). Specifically, these authors invoke relatively high heat flow to explain localisation of seismic moment release and deformation in the Flinders Ranges Seismic Zone (FRSZ, Figure 2.1) compared to the flanking Murray Basin and Nullarbor Plain. The correlation is imperfect; the heat flow anomaly does not extend to the Gippsland and Otway Basins (Densley *et al.*, 2000;

Holdgate *et al.*, 2003; Sandiford, 2003a) (Figure 2.1), which are manifestly amongst the fastest deforming regions on the Australian continent (Sandiford, 2003a, b; Clark *et al.*, 2011a, 2012). Actively inverting basins of the Northwest Shelf (NWSSZ, Figure 2.1) are also not associated with significant heat flow anomalies (cf. He and Middleton, 2002). Furthermore, the heat flow anomaly is most pronounced in the Cooper/Eromanga Basin, a region sparse in seismicity (Figure 2.1) and devoid of any known neotectonic features or tectonic uplift.

At the sub-regional scale, a range of factors have been proposed to explain localisation of historic seismicity. However, it seems clear that, on timescales of thousands to tens of thousands of years, seismic potential is determined by factors at a scale much larger than a single fault or region, and might instead relate more strongly to continental-scale lithospheric and crustal architecture, and the age of that architecture, as first proposed by Johnston *et al.* (1994).

2.6.2 Structural architectural influences

In the context of structural architecture, intraplate seismicity worldwide is considered to be concentrated at rifted margins (Stein *et al.*, 1989; Wheeler, 1995, 1996; Sandiford and Egholm, 2008; Cloetingh *et al.*, 2008; Etheridge *et al.*, 1991; Talwani and Schaeffer, 2001), interior rifts [aulacogens] (Johnston, 1994; Wheeler, 1995; Gangopadhyay and Talwani, 2003; Schulte and Mooney, 2005; Sinha and Mohanty, 2012) and at the margins of cratons (e.g., Lenardic *et al.*, 2000; Mazzotti, 2007; Sloan *et al.*, 2011; Craig *et al.*, 2011; Mooney *et al.*, 2012). A consequence of the relatively cold geotherm characterising cratonic areas is that there is no decoupling between crustal and mantle deformation (Braun *et al.*, 2009). This implies that, over geological timescales, deformation of the upper crust must be spatially uniform (e.g., Clark, 2010), resulting in little strain localisation, and hence minimal topography. Predictably, the historic record of seismicity in Australia is a poor guide as to where localisation of seismic activity occurs over geological timescales, with significant concentrations of seismicity occurring within cratons (e.g., SWSZ, Figure 2.1), and far from extended crust or craton boundaries (e.g., parts of the SESZ; see also Adams *et al.* [1992] and Rajendran *et al.* [1996]).

The major tenet of the Johnston *et al.* (1994) SCR model – that extended crust is more seismically active than non-extended crust, and that within the non-extended class non-cratonic crust is more active than cratonic crust – appears to hold true for the Australian neotectonic record if the uplift rate implied by vertical neotectonic fault displacement is taken as a proxy for seismic activity (Figure 2.4d, f). Australian neotectonic data (Clark *et al.*, 2011a, 2012) permit further sub-division of the extended and non-extended crustal classes proposed by Johnston *et al.* (1994). In extended crust domains (e.g., D5, D6), the age of major rifting appears to be important in terms of the record of neotectonic activity. Paleozoic intracratonic rifts (e.g., the Fitzroy Trough [Drummond *et al.*, 1991]; cf. Figure 2.1) and passive margin components (e.g., Perth Basin [Crostella and Backhouse,

2000; Norvick, 2004]) preserve less evidence for neotectonic strain localisation than those rifted in the Mesozoic (e.g., Gippsland and Otway basins). As such, the long-term seismic potential of aulacogens impinging along the eastern margin of the North American continent (e.g., Ottawa Rift, Saguenay graben, Reelfoot Rift, Southern Oklahoma aulacogen; Wheeler and Frankel [2000]) might be assessed based upon the proportion of Mesozoic extensional reactivation of these largely Paleozoic structures. The distinction becomes even clearer when one considers most Precambrian rifts (included in D1 and D3). The major exception to this rule is the FRSZ (D2), which formed as a Precambrian rift, but was subsequently extensively reactivated under compression in the Paleozoic (Flöttmann and James, 1997). The Appalachian Orogen (Wheeler, 1996), including the St. Lawrence seismic zone (Vlahovic *et al.*, 2003) at its northern margin, occupies a similar crustal setting between Precambrian mobile belts (D3 equivalent) and Phanerozoic accretionary terranes (D4 equivalent).

Considerable uncertainty remains as to the long-term seismic potential of non-extended, non-cratonic crust (D4), which typically comprises Phanerozoic accretionary terranes (e.g., Hoffman, 1989; Glen, 2005). Within the Australian continent, much of this crustal type is associated with little positive relief, and so might be thought of as having low seismic potential (e.g., Sandiford, 2003b). However, the eastern part of D4 is punctuated by the Eastern Highlands. While it is widely considered that most of the relief of the highlands relates to the opening of the Tasman Sea in the Cretaceous (e.g., Norvick and Smith, 2001), debate remains as to how much relief has been added from the Neogene to Recent (Holdgate *et al.*, 2008, 2011; VandenBerg, 2010). The few known faults are poorly documented, but appear to have similar characteristics to those documented from the relief-poor western parts of the domain (compare the Lake George Fault [Abell, 1981, 1985] to the Cadell Fault [Clark *et al.*, 2007; McPherson *et al.*, 2012a]). Braun *et al.* (2009) hypothesise that deformation in eastern Australia is determined by the lithospheric strength (or rigidity) and hence controlled by lithospheric structures, in contrast to the western part of the continent where deformation is localised by crustal structures, because the underlying lithospheric mantle is almost ubiquitously strong.

A fundamental implication of the neotectonic domains model presented by Clark *et al.* (2011a, 2012), building upon the pioneering work of Johnston *et al.* (1994), is that intraplate fault characteristics (and by inference, seismicity) are not universal in their applicability in analogue studies. Careful choice of subject regions, structures, or faults within analogous crust of similar stress field character is required to extrapolate meaningfully to imperfectly characterised areas.

2.7 Conclusions

Australia has a short recorded history of seismicity, spanning only a couple of centuries (Leonard, 2008). As a consequence, there is significant uncertainty as to whether patterns evident in the contemporary seismic record are representative of the longer term. This uncertainty can, in part, be overcome by validation against Australia's rich record of morphogenic earthquakes (Quigley *et al.*, 2010; Clark *et al.*, 2012). Long-term patterns in large earthquake occurrence, both temporal and spatial, can be deduced from the landscape record and used to inform contemporary earthquake hazard science. Seismicity source parameters such as large earthquake recurrence and magnitude vary across the Australian continent, and can be interpreted in a framework of large-scale neotectonic domains defined on the basis of geology and crustal setting (Clark *et al.*, 2012). Temporal and spatial clustering of earthquakes is apparent at the scale of a single fault, and at the 1,000 km scale of a domain. The utility of the domains approach, which ties seismicity characteristics to crustal architecture and geology, is that behaviours might be extrapolated from well-characterised regions to poorly known analogous regions, both within Australia and worldwide.

Acknowledgements

This paper is published with the permission of the CEO of Geoscience Australia. The authors would like to thank Marcelo Assumpção and Pradeep Talwani for providing review comments that improved the manuscript. Mark Leonard and David Burbidge are also thanked for their constructive reviews.

References

Abell, R. S. (1981). Notes to accompany the 1:25 000 scale geological field compilation sheets of the CANBERRA 1:100 000 sheet area. *Bureau of Mineral Resources, Geology and Geophysics Record 1981/6.*

Abell, R. S. (1985). Geology of the Lake George Basin, NSW *Bureau of Mineral Resources, Geology and Geophysics Record 1985/4.*

Adams, J., R. Wetmiller, H. Hasegawa, and J. Drysdale (1991). The first surface faulting from a historical earthquake in North America. *Nature*, 352, 617–619.

Adams, J., J. Percival, R. Wetmiller, J. Drysdale, and P. Robertson (1992). Geological controls on the 1989 Ungava surface rupture: a preliminary interpretation. *Geological Survey of Canada Paper*, 92C, 147–155.

Allen, T. I. (2012). Ground motion prediction equations. In *The 2012 Australian Earthquake Hazard Map*, ed. D. R. Burbidge. *Geoscience Australia Record*, 2012/71, pp. 44–53.

Allen, T. I., and G. M. Atkinson (2007). Comparison of earthquake source spectra and attenuation in eastern North America and southeastern Australia. *Bulletin of the Seismological Society of America*, 97, 1350–1354.

Allen, T. I., M. Leonard, and C. D. N. Collins (2012). Revised Australian Earthquake Catalogue. In *The 2012 Australian Earthquake Hazard Map*, ed. D. R. Burbidge. *Geoscience Australia Record*, 2012/71, pp. 5–16.

Assumpção, M., and V. Sacek (2013). Intra-plate seismicity and flexural stresses in central Brazil. *Geophysical Research Letters*, 40, 1–5.

Atkinson, G. M., and R. F. Mereu (1992). The shape of ground motion attenuation curves in southeastern Canada. *Bulletin of the Seismological Society of America*, 82, 2014–2031.

Atkinson, G. M., and M. Morrison (2009). Observations on regional variability in ground-motion amplitudes for small-to-moderate earthquakes in North America. *Bulletin of the Seismological Society of America*, 99, 2393–2409.

Atkinson, G. M., and D. J. Wald (2007). "Did You Feel It?" intensity data: a surprisingly good measure of earthquake ground motion. *Seismological Research Letters*, 78, 362–368.

Bakun, W. H., and A. McGarr (2002). Differences in attenuation among the stable continental regions. *Geophysical Research Letters*, 29, 2121, doi:10.1029/2002GL015457.

Balasubrahmanyan, M. N. (2006). *Geology and Tectonics of India: An Overview*. International Association for Gondwana Research, IAGR Memoir 9.

Bathgate, J., H. Glanville, and C. Collins (2010). The Kalgoorlie earthquake of 20 April 2010 and its aftershock sequence. *Australian Earthquake Engineering Society 2010 Conference*, Perth, Paper 54.

Beavis, F. C. (1960). The Tawonga Fault, northeast Victoria. *Proceedings of the Royal Society of Victoria*, 72, 95–100.

Beavis, F. C. (1962). The geology of the Kiewa area. *Proceedings of the Royal Society of Victoria*, 75, 349–410.

Belperio, A. P., N. Harvey, and R. P. Bourman (2002). Spatial and temporal variability in the Holocene sea-level record of the South Australian coastline. *Sedimentary Geology*, 150, 153–169.

Belton, D. X., R. W. Brown, B. P. Kohn, D. Fink, and K. A. Farley (2004). Quantitative resolution of the debate over antiquity of the central Australian landscape: implications for the tectonic and geomorphic stability of cratonic interiors. *Earth and Planetary Science Letters*, 219, 21–34.

Bent, A. L. (1994). The 1989 (Ms 6.3) Ungava, Quebec, earthquake: a complex intraplate event. *Bulletin of the Seismological Society of America*, 84, 1075–1088.

Bierman, P. R., and M. Caffee (2002). Cosmogenic exposure and erosion history of Australian bedrock landforms. *Geological Society of America Bulletin*, 114, 787–803.

Bishop, P., P. Hunt, and P. W. Schmidt (1982). Limits to the age of the Lapstone Monocline, N.S.W.; a palaeomagnetic study. *Journal of the Geological Society of Australia*, 29, 319–326.

Bourman, R. P., A. P. Belperio, C. V. Murray-Wallace, and J. H. Cann (1999). A last interglacial embayment fill at Normanville, South Australia, and its neotectonic implications. *Transactions of the Royal Society of South Australia*, 123, 1–15.

Bowler, J. M. (1978). Quaternary climate and tectonics in the evolution of the Riverine Plain, southeastern Australia. In *Landform Evolution in Australasia*, ed. J. L. Davies, and M. A. J. Williams. Canberra: ANU Press, pp. 70–112.

Bowler, J. M., and L. B. Harford (1966). Quaternary tectonics and the evolution of the Riverine Plain near Echuca, Victoria. *Journal of the Geological Society of Australia*, 13, 339–354.

Bowman, J. R. (1988). Constraints on locations of large intraplate earthquakes in the Northern Territory, Australia, from observations at the Warramunga seismic array. *Geophysical Research Letters*, 15, 1475–1478.

Bowman, J. R. (1992). The 1988 Tennant Creek, Northern Territory, earthquakes: a synthesis. *Australian Journal of Earth Sciences*, 39, 651–669.

Bowman, J. R., and B. L. N. Kennett (1991). Propagation of *Lg* waves in the north Australian craton: influence of crustal velocity gradients. *Bulletin of the Seismological Society of America*, 81, 592–610.

Bowman, J. R., G. Gibson, and T. Jones (1990). Aftershocks of the 1988 January 22 Tennant Creek, Australia intraplate earthquakes: evidence for a complex thrust-fault geometry. *International Geophysical Journal*, 100, 87–97.

Braun, J., D. Burbidge, F. Gesto, *et al.* (2009). Constraints on the current rate of deformation and surface uplift of the Australian continent from a new seismic database and low-T thermochronological data. *Australian Journal of Earth Sciences*, 56, 99–110.

Browne, W. R. (1967). Geomorphology of the Kosciusko block and its north and south extensions. *Proceedings of the Linnean Society of New South Wales*, 92, 117–144.

Burbidge, D. R. (2004). Thin plate neotectonic models of the Australian plate. *Journal of Geophysical Research*, 109, B10405, doi:10.1029/2004JB003156.

Burbidge, D. R. (editor) (2012). *The 2012 Australian Earthquake Hazard Map, Geoscience Australia, Record 2012/71.*

Campbell, D. L. (1978). Investigation of the stress-concentration mechanism for intraplate earthquakes. *Geophysical Research Letters*, 5, 477–479.

Campbell, K. W. (2011). Ground motion simulation using the hybrid empirical method: issues and insights. *Geotechnical Geological and Earthquake Engineering*, 14, 81–95.

Canavan, F. (1988). Deep lead gold deposits of Victoria. *Geological Society of Victoria Bulletin*, 62.

Cawood, P. A., and C. Buchan (2007). Linking accretionary orogenesis with supercontinent assembly. *Earth Science Reviews*, 82, 217–256.

Celerier, J., M. Sandiford, D. L. Hansen, and M. Quigley (2005). Modes of active intraplate deformation, Flinders Ranges, Australia. *Tectonics*, 24, doi:10.029/2004&C001679.

Chappell, J. (2006). Australian landscape processes measured with cosmogenic nuclides. In *Regolith Geochronology and Landscape Evolution*, ed. B. Pillans. Perth: CRC LEME, pp. 19–26.

Chéry, J., S. Carretier, and J. F. Ritz (2001). Postseismic stress transfer explains time clustering of large earthquakes in Mongolia. *Earth and Planetary Science Letters*, 194, 277–286.

Chéry, J., and P. Vernant (2006). Lithospheric elasticity promotes episodic fault activity. *Earth and Planetary Science Letters*, 243, 211–217.

Choy, G. L., and J. R. Bowman (1990). Rupture process of a multiple main shock sequence; analysis of teleseismic, local, and field observations of the Tennant Creek, Australia, earthquakes of January 22, 1988. *Journal of Geophysical Research B, Solid Earth and Planets*, 95, 6867–6882.

Chung, W.-Y., A. C. Johnston, and J. Pujol (1992). *A Global Study of Stable Continental Intraplate Earthquakes: Focal Mechanisms, Source-Scaling Relations, and Seismotectonics*. Final Report, EPRI RP-2556–54, Electric Power Research Institute, Palo Alto, California.

Clark, D. (2004). Seismicity along the northern foreland of the Albany-Fraser Orogen, Geoscience Australia (Earthquake Hazard Section) unpublished report.

Clark, D. (2010). Identification of Quaternary scarps in southwest and central west Western Australia using DEM-based hill shading: application to seismic hazard assessment and neotectonics. *International Journal of Remote Sensing*, 31, 6297–6325.

Clark, D. J., and M. Leonard (2003). Principal stress orientations from multiple focal plane solutions: new insight in to the Australian intraplate stress field. In *Evolution and Dynamics of the Australian Plate*, ed. R. R. Hillis, and D. Muller. Geological Society of Australia Special Publication 22 and Geological Society of America Special Paper 372, 91–105.

Clark, D., and A. McPherson (2011). Large earthquake recurrence in the Adelaide region:a palaeoseismological perspective. *Australian Earthquake Engineering Society 2011 Conference*, 18–20 November, Barossa Valley, South Australia, Paper 11.

Clark, D., M. Dentith, and M. Leonard (2003). Linking earthquakes to geology: contemporary deformation controlled by ancient structure. *Geological Society of Australia, Abstracts*, 72, 90.

Clark, D., R. Van Dissen, M. Cupper, C. Collins, and A. Prendergast (2007). Temporal clustering of surface ruptures on stable continental region faults: a case study from the Cadell Fault scarp, south eastern Australia. *Australian Earthquake Engineering Society Conference*, Wollongong, Australian Earthquake Engineering Society.

Clark, D., M. Dentith, K. H. Wyrwoll, *et al.* (2008). The Hyden fault scarp, Western Australia: paleoseismic evidence for repeated Quaternary displacement in an intracratonic setting. *Australian Journal of Earth Sciences*, 55, 379–395.

Clark, D., A. McPherson, and C. D. N. Collins (2011a). *Australia's Seismogenic Neotectonic Record: A Case for Heterogeneous Intraplate Deformation, Geoscience Australia Record*, 2011/11.

Clark, D. J., M. Cupper, M. Sandiford, and K. Kiernan (2011b). Style and timing of late Quaternary faulting on the Lake Edgar Fault, southwest Tasmania, Australia: implications for hazard assessment in intracratonic areas. In *Geological Criteria for Evaluating Seismicity Revisited: Forty Years of Paleoseismic Investigations and the Natural Record of Past Earthquakes,* ed. F. A. Audemard, M. A. Michetti, and J. P. McCalpin. Geological Society of America Special Publication 479, pp. 109–131.

Clark, D., A. McPherson, and R. Van Dissen (2012). Long-term behaviour of Australian Stable Continental Region (SCR) faults. *Tectonophysics*, 566–567, 1–30.

Clark, D., A. McPherson, T. Allen, and M. De Kool (2013). Co-seismic surface deformation relating to the March 23, 2012 MW 5.4 Ernabella (Pukatja) earthquake, central Australia: implications for cratonic fault scaling relations. *Bulletin of the Seismological Society of America*, 104, doi:10.1785/0120120361.

Cleary, J. R. (1967). The seismicity of the Gunning and surrounding areas, 1958–1961. *Journal of the Geological Society of Australia*, 14, 23–29.

Cloetingh, S., F. Beekman, P. A. Ziegler, J.-D. Van Wees, and D. Sokoutis (2008). Post-rift compressional reactivation potential of passive margins and extensional basins. In *The Nature and Origin of Compression in Passive Margins*, ed. H. Johnson, A. G. Doré, R. W. Gatliff *et al.* Geological Society, London, Special Publication 306, pp. 27–70.

Cloetingh, S., and R. Wortel (1986). Stress in the Indo-Australian plate. *Tectonophysics*, 132, 49–67.

Close, D., and L. Seeber (2007). Shallow seismicity in stable continental regions. *Seismological Research Letters*, 78, 554–562.

Coblentz, D. D., M. Sandiford, R. M. Richardson, Z. Shaohua, and R. Hillis (1995). The origins of the intraplate stress field in continental Australia. *Earth and Planetary Science Letters*, 133, 299–309.

Coblentz, D. D., S. Zhou, R. R. Hillis, R. M. Richardson, and M. Sandiford (1998). Topography boundary forces and the Indo-Australian intraplate stress field. *Journal of Geophysical Research*, 103, 919–938.

Collins, A. S., and S. A. Pisarevsky (2005). Amalgamating eastern Gondwana: the evolution of the Circum-Indian Orogens. *Earth Science Reviews*, 71, 229–270.

Collins, C. D. N., B. J. Drummond, and M. G. Nicoll (2003). Crustal thickness patterns in the Australian continent. In *Evolution and Dynamics of the Australian Plate*, ed. R. R. Hillis, and R. D. Müller. Geological Society of Australia Special Publication 22 and Geological Society of America Special Paper 372, pp. 121–128.

Cox, R. T., J. Cherryhomes, J. B. Harris, and D. Larsen (2006). Paleoseismology of the southeastern Reelfoot rift in western Tennessee and implications for intraplate fault zone evolution. *Tectonics*, 25, 1–17.

Craig, T. J., J. A. Jackson, K. Priestly, and D. McKenzie (2011). Earthquake distribution patterns in Africa: their relationship to variations in lithospheric and geologic structure, and their rheological implications. *Geophysical Journal International*, 185, 403–434.

Crone, A. J., and K. V. Luza (1990). Style and timing of Holocene surface faulting on the Meers Fault, southwestern Oklahoma. *Geological Society of America Bulletin*, 102(1), 1–17.

Crone, A. J., and M. N. Machette (1995). Holocene faulting on the Cheraw Fault, southeastern Colorado: another hazardous late Quaternary fault in the stable continental interior. *Eos Supplement*, 76, F362.

Crone, A. J., M. N. Machette, and J. R. Bowman (1992). Geologic investigations of the 1988 Tennant Creek, Australia, earthquakes: implications for paleoseismicity in stable continental regions. *Bulletin U.S. Geological Survey*, 2032-A.

Crone, A. J., M. N. Machette, and J. R. Bowman (1997). Episodic nature of earthquake activity in stable continental regions revealed by palaeoseismicity studies of Australian and North American Quaternary faults. *Australian Journal of Earth Sciences*, 44, 203–214.

Crone, A. J., P. M. de Martini, M. N. Machette, K. Okumura, and J. R. Prescott (2003). Paleoseismicity of two historically quiescent faults in Australia: implications for fault behavior in stable continental regions. *Bulletin of the Seismological Society of America*, 93, 1913–1934.

Crostella, A., and J. Backhouse (2000). Geology and petroleum exploration of the central and southern Perth Basin, Western Australia. *Western Australia Geological Survey, Report*, 57, 85.

Cummins, P., C. Collins, and A. Bullock (2009). Monitoring of earthquakes in the Flinders Ranges, South Australia, using a temporary seismometer deployment. Geoscience Australia Publication (digital), https://www.ga.gov.au/products/servlet/controller?event=GEOCAT_DETAILS&catno=69525.

Dawson, J., P. Cummins, P. Tregoning, and M. Leonard (2008). Shallow intraplate earthquakes in Western Australia observed by Interferometric Synthetic Aperture Radar. *Journal of Geophysical Research*, 113, doi:10.1029/2008JB005807.

Denham, D. (1988). Australian seismicity: the puzzle of the not so stable continent. *Seismological Research Letters*, 59, 235–240.

Denham, D., G. R. Small, J. R. Cleary, *et al.* (1975). Australian earthquakes (1897–1972), *Search*, 6, 34–36.

Denham, D., J. Weekes, and C. Krayshek (1981). Earthquake evidence for compressive stress in the southeast Australian crust. *Journal of the Geological Society of Australia*, 28, 323–332.

Denham, D., G. Gibson, R. S. Smith, and R. Underwood (1982). Source mechanisms and strong ground motion from the 1982 Wonnangatta and the 1966 Mount Hotham earthquakes. *Australian Journal of Earth Sciences*, 32, 37–46.

Densley, M. R., R. R. Hillis, and J. E. P. Redfearn (2000). Quantification of uplift in the Carnarvon Basin based on interval velocities. *Australian Journal of Earth Sciences*, 47, 111–122.

Dent, V. F. (2008). Graphical representation of some recent Australian earthquake swarms, and some generalisations on swarm characteristics. *Australian Earthquake Engineering Society Conference Proceedings*, Ballarat, Paper 23.

Dent, V. F. (2009). The Beacon, WA earthquake swarm commencing January 2009. *Australian Earthquake Engineering Society Conference*, Newcastle, poster presentation.

Dentith, M. C., and W. E. Featherstone (2003). Controls on intra-plate seismicity in southwestern Australia. *Tectonophysics*, 376, 167–184.

Dentith, M. C., D. J. Clark, and W. E. Featherstone (2009). Aeromagnetic mapping of Precambrian geological structures that controlled the 1968 Meckering Earthquake (Ms 6.8): implications for intraplate seismicity in Western Australia. *Tectonophysics*, 475, 544–553.

Dickinson, J. A., M. W. Wallace, G. R. Holdgate, S. J. Gallagher, and L. Thomas (2002). Origin and timing of the Miocene-Pliocene unconformity in southeast Australia. *Journal of Sedimentary Research*, 72, 288–303.

Doyle, H. A. (1971). Seismicity and structure in Australia. *Bulletin of the Royal Society of New Zealand*, 9, 149–152.

Doyle, H. A., and R. Underwood (1965). Seismological stations in Australia. *Australian Journal of Earth Sciences*, 28, 40–43.

Doyle, H. A., I. B. Everingham, and D. J. Sutton (1968). Seismicity of the Australian continent. *Journal of the Geological Society of Australia*, 15, 295–312.

Drummond, B. J., M. J. Sexton, T. J. Barton, and R. D. Shaw (1991). The nature of faulting along the margins of the Fitzroy Trough, Canning Basin, and implications for the tectonic development of the Trough. *Exploration Geophysics*, 22, 111–115.

Dyksterhuis, S., and R. D. Müller (2008). Cause and evolution of intraplate orogeny in Australia. *Geology*, 36, 495–498.

Estrada, B. (2009). Neotectonic and palaeoseismological studies in the southwest of Western Australia. Ph.D. thesis (unpublished), School of Earth and Environment, The University of Western Australia, Perth.

Etheridge, M. A., H. W. S. McQueen, and K. Lambeck (1991). The role of intraplate stress in Tertiary (and Mesozoic) deformation of the Australian continent and its margins: a key factor in petroleum trap formation. *Exploration Geophysics*, 22, 123–128.

Everingham, I. B. (1982). Atlas of isoseismal maps of Australian earthquakes. *Burea of Mineral Resources Bulletin*, 214.

Everingham, I. B., and R. S. Smith (1979). Implications of fault-plane solutions for Australian earthquakes on 4 July 1977, 6 May 1978 and 25 November 1978. *BMR Journal of Australian Geology and Geophysics*, 4, 297–301.

Fishwick, S., M. Heintz, B. L. N. Kennett, A. M. Reading, and K. Yoshizawa (2008). Steps in lithospheric thickness within eastern Australia, evidence from surface wave tomography. *Tectonics*, 27, doi:10.1029/2007TC002116.

Flöttmann, T., and C. D. Cockshell (1996). Palaeozoic basins of southern South Australia: new insights into their structural history from regional seismic data. *Australian Journal of Earth Sciences*, 43, 45–55.

Flöttmann, T., and P. James (1997). Influence of basin architecture on the style of inversion and fold-thrust belt tectonics: the southern Adelaide Fold-Thrust Belt, South Australia. *Journal of Structural Geology*, 19, 1093–1110.

Frankel, A. (1994). Implications of felt area–magnitude relations for earthquake scaling and the average frequency of perceptible ground motion. *Bulletin of the Seismological Society of America*, 84, 462–465.

Fredrich, J., R. McCaffrey, and D. Denham (1988). Source parameters of seven large Australian earthquakes determined by body waveform inversion. *Geophysical Journal*, 95, 1–13.

Friedrich, A. M., B. P. Wernicke, N. A. Niemi, R. A. Bennett, and J. L. Davis (2003). Comparison of geodetic and geologic data from the Wasatch region, Utah, and implications for the spectral character of Earth deformation at periods of 10 to 10 million years. *Journal of Geophysical Research*, 108 (B4), doi:10.1029/2001JB000682.

Gangopadhyay, A., and P. Talwani (2003). Symptomatic features of intraplate earthquakes. *Seismological Research Letters*, 74, 863–883.

Gaull, B. A., H. Kagami, H. Taniguchi, *et al.* (1990). Preliminary results of the microzonation of the Perth Metropolitan area using microtremor spectral ratios. *Bureau of Mineral Resources, Geology and Geophysics Record*, 1990/44.

Gibson, G., V. Wesson, and R. Cuthbertson (1981). Seismicity of Victoria to 1980. *Journal of the Geological Society of Australia*, 28, 341–356.

Gibson, G., V. Wesson, and T. Jones (1994). The Eugowra NSW earthquake swarm of 1994, *Australian Earthquake Engineering Society Conference Proceedings*, Canberra, ACT, pp. 67–76.

Glen, R. A. (2005). The Tasmanides of Eastern Australia. In *Terrane Processes at the Margins of Gondwana*, ed. A. P. M. Vaughan, P. T. Leat, and R. J. Pankhurst. Geological Society, London, Special Publication 246, pp. 23–96.

Gordon, F. R., and J. D. Lewis (1980). The Meckering and Calingiri earthquakes October 1968 and March 1970. *Western Australia Geological Survey Bulletin*, 126.

Greenhalgh, S. A., and R. Singh (1988). The seismicity of the Adelaide Geosyncline, South Australia. *Bulletin of the Seismological Society of America*, 78, 243–263.

Greenhalgh, S. A., D. Love, K. Malpas, and R. McDougall (1994). South Australian seismicity, 1980–1992. *Australian Journal of Earth Sciences*, 41, 483–495.

Hanks, T. C., and A. C. Johnston (1992). Common features of the excitation and propagation of strong ground motion for North American earthquakes. *Bulletin of the Seismological Society of America*, 82, 1–23.

He, S., and M. Middleton (2002). Heat flow and thermal maturity modelling in the Northern Carnarvon Basin, North West Shelf, Australia. *Marine and Petroleum Geology*, 19, 1073–1088.

Heimsath, A. M., J. Chappell, W. E. Dietrich, K. Nishiizumi, and R. C. Finkel (2000). Soil production on a retreating escarpment in southeastern Australia. *Geology*, 28, 787–790.

Heimsath, A. M., J. Chappell, W. E. Dietrich, K. Nishiizumi, and R. C. Finkel (2001). Late Quaternary erosion in southeastern Australia: a field example using cosmogenic isotopes. *Quaternary International*, 83–85, 169–185.

Hemphill-Haley, M. A., and R. J. Weldon (1999). Estimating prehistoric earthquake magnitude from point measurements of surface rupture. *Bulletin of the Seismological Society of America*, 89, 1264–1279.

Hill, K. A., G. T. Cooper, M. J. Richardson, and C. J. Lavin (1994). Structural framework of the eastern Otway Basin: inversion and interaction between two major structural provinces. *Exploration Geophysics*, 25, 79–87.

Hillis, R. R., and S. D. Reynolds (2000). The Australian stress map. *Journal of the Geological Society of London*, 157, 915–921.

Hillis, R. R., and S. D. Reynolds (2003). In situ stress field of Australia. In *Evolution and Dynamics of the Australian Plate*, ed. R. R. Hillis, and D. Muller. Geological Society of Australia Special Publication 22 and Geological Society of America Special Paper 372, pp. 49–58.

Hillis, R. R., M. Sandiford, S. D. Reynolds, and M. C. Quigley (2008). Present-day stresses, seismicity and Neogene-to-Recent tectonics of Australia's 'passive' margins: intraplate deformation controlled by plate boundary forces. In *The Nature and Origin of Compression in Passive Margins*, ed. H. Johnson, A. G. Doré, R. W. Gatliff *et al*. Geological Society, London, Special Publication 306, 71–90.

Historical Records of Australia (1914). The Library Committee of the Commonwealth Parliament, Vol. 1, Series 1, p. 50.

Hoffman, P. F. (1989). Precambrian geology and tectonic history of North America. In *The Geology of North America – Vol. A, The Geology of North America: An Overview*, ed. A. W. Bally, and A. R. Palmer. Boulder, Colorado: Geological Society of America, pp. 447–512.

Holdgate, G. R., M. W. Wallace, S. J. Gallagher, *et al*. (2003). Plio-Pleistocene tectonics and eustasy in the Gippsland Basin, southeast Australia: evidence from magnetic imagery and marine geological data. *Australian Journal of Earth Sciences*, 50, 403–426.

Holdgate, G. R., M. W. Wallace, S. J. Gallagher, *et al*. (2006). Cenozoic fault control on 'deep lead' palaeoriver systems, Central Highlands, Victoria. *Australian Journal of Earth Sciences*, 53, 445–468.

Holdgate, G. R., I. Cartwright, D. T. Blackburn, *et al*. (2007). The Middle Miocene Yallourn coal seam: the last coal in Australia. *International Journal of Coal Geology*, 70, 95–115.

Holdgate, G. R., M. W. Wallace, S. J. Gallagher, B. E. Wagstaff, and D. Moore (2008). No mountains to snow on: major post-Eocene uplift of the East Victoria Highlands; evidence from Cenozoic deposits. *Australian Journal of Earth Sciences*, 55, 211–234.

Holdgate, G. R., M. W. Wallace, S. J. Gallagher, B. E. Wagstaff, and D. Moore (2011). Paleogene basalts prove early uplift of Victoria's Eastern Uplands. *Australian Journal of Earth Sciences*, 58, 95–99.

Holford, S. P., R. R. Hillis, M. Hand, and M. Sandiford (2011). Thermal weakening localizes intraplate deformation along the southern Australian continental margin. *Earth and Planetary Science Letters*, 305, 207–214.

Insurance Council of Australia (2013). *Historical Disaster Statistics*, http://www.insurancecouncil.com.au/industry-statistics-data/disaster-statistics/historical-disaster-statistics, retrieved 16 December 2013.

Jenkins, R. J. F., and M. Sandiford (1992). Observations on the tectonic evolution of the southern Adelaide Fold Belt. *Tectonophysics*, 214, 27–36.

Johnston, A. C. (1994). Seismotectonic interpretations and conclusions from the stable continental region seismicity database. In *The Earthquakes of Stable Continental Regions: Vol. 1, Assessment of Large Earthquake Potential*, ed. A. C. Johnston, K. J. Coppersmith, L. R. Kanter, and C. A. Cornell. Palo Alto, CA: Electric Power Research Institute, 4-1–4-103.

Johnston, A. C., K. J. Coppersmith, L. R. Kanter, and C. A. Cornell (1994). *The Earthquakes of Stable Continental Regions: Vol. 1, Assessment of Large Earthquake Potential.* Electric Power Research Institute Report, TR102261V1.

Jones, T. D., G. Gibson, K. F. McCue, *et al.* (1991). Three large intraplate earthquakes near Tennant Creek, Northern Territory, on 22 January 1988. *BMR Journal of Australian Geology and Geophysics*, 12, 339–343.

Kafka, A. L. (2002). Statistical analysis of the hypothesis that seismicity delineates areas where future large earthquakes are likely to occur in the Central and Eastern United States. *Seismological Research Letters*, 73, 992–1003.

Keep, M., and D. W. Haig (2010). Deformation and exhumation in Timor: distinct stages of a young orogeny. *Tectonophysics*, 483, 93–111.

Keep, M., J. Hengesh, and B. Whitney (2012). Natural seismicity and tectonic geomorphology reveal regional transpressive strain in northwestern Australia. *Australian Journal of Earth Sciences*, 59, 341–354.

Kenner, S. J., and P. Segall (2000). A mechanical model for intraplate earthquakes: application to the New Madrid seismic zone. *Science*, 289, 2329–2332.

Kennett, B. L. N., A. Fitchner, S. Fishwick, and K. Yoshizawa (2013). Australian Seismological Reference Model (AuSREM): mantle component. *Geophysical Journal International*, 192, 871–887.

Kroner, A., and U. Cordani (2003). African, southern Indian and South American cratons were not part of the Rodinia supercontinent: evidence from field relationships and geochronology. *Tectonophysics*, 375, 325–352.

Kusky, T. M., B. F. Windley, and M.-G. Zhai (2007). Tectonic evolution of the North China Block: from orogen to craton to orogen. In *Mesozoic Sub-Continental Lithospheric Thinning Under Eastern Asia,* Geological Society, London, Special Publication 280, 1–34.

Lamontagne, M., P. Keating, and S. Perreault (2003). Seismotectonic characteristics of the Lower St. Lawrence seismic zone, Quebec: insights from geology, magnetics, gravity, and seismics. *Canadian Journal of Earth Sciences*, 40, 317–336.

Langenheim, V. E., and T. G. Hildenbrand (1997). Commerce geophysical lineament: its source, geometry, and relation to the Reelfoot rift and New Madrid seismic zone. *Geological Society of America Bulletin*, 109, 580–595.

Langston, C. A. (1987). Depth of faulting during the 1968 Meckering, Australia, earthquake sequence determined from waveform analysis of local seismograms. *Journal of Geophysical Research*, 92(B11), 11561–11574.

Lenardic, A., L. Moresi, and H. Muhlhaus (2000). The role of mobile belts for the longevity of deep cratonic lithosphere: the crumple zone model. *Geophysical Research Letters*, 27, 1235–1238.

Leonard, M. (2002). The Burrakin WA earthquake sequence Sept 2000 – June 2002. *Australian Earthquake Engineering Society Conference: Total Risk Management in the Privatised Era*, Adelaide, Paper 22.

Leonard, M. (2003). Respite leaves Burakin quaking in anticipation. *Ausgeo News*, 70, 5–7.

Leonard, M. (2008). One hundred years of earthquake recording in Australia. *Bulletin of the Seismological Society of America*, 98, 1458–1470.

Leonard, M. (2010). Earthquake fault scaling: self-consistent relating of rupture length, width, average displacement, and moment release. *Bulletin of the Seismological Society of America*, 100, 1971–1988.

Leonard, M. (2012). Earthquake recurrence parameterisation. In *The 2012 Australian Earthquake Hazard Map*, ed. D. R. Burbidge. *Geoscience Australia Record*, 2012/71, 27–43.

Leonard, M., and D. Clark (2006). Reconciling neotectonic and seismic recurrence rates in SW WA. *Proceedings of the Australian Earthquake Engineering Society Conference*, Paper 19, 19–24.

Leonard, M., and D. Clark (2011). A record of stable continental region earthquakes from Western Australia spanning the late Pleistocene: insights for contemporary seismicity. *Earth and Planetary Science Letters*, 309, 207–212.

Leonard, M., I. D. Ripper, and Y. Li (2002). Australian earthquake fault plane solutions. *Geoscience Australia Record*, 2002/19.

Leonard, M., D. Darby, and G. Hu (2007). GPS-geodetic monitoring of the South West Seismic Zone of Western Australia: progress after two observation epochs in 2002 and 2006. *Proceedings of the Australian Earthquake Engineering Society Conference*, Wollongong, Paper 28.

Leonard, M., A. McPherson, and D. Clark (2012). Source zonation. In *The 2012 Australian Earthquake Hazard Map*, ed. D. R. Burbidge. *Geoscience Australia Record*, 2012/71, 17–26.

Lewis, J. D., N. A. Daetwyler, J. A. Bunting, and J. S. Moncrieff (1981). The Cadoux earthquake. *Western Australia, Geological Survey Report*, 1981/11, 133.

Li, Q., M. Liu, and S. Stein (2009). Spatiotemporal complexity of continental intraplate seismicity: insights from geodynamic modeling and implications for seismic hazard estimation. *Bulletin of the Seismological Society of America*, 99, 52–60.

Liu, L., and M. D. Zoback (1997). Lithospheric strength and intraplate seismicity in the New Madrid seismic zone. *Tectonics*, 16, 585–595.

Liu, M., S. Stein, and H. Wang (2011). 2000 years of migrating earthquakes in North China: how earthquakes in midcontinents differ from those at plate boundaries. *Lithosphere*, doi: 10.1130/L129.1.

Liu, M., and H. Wang (2012). Roaming earthquakes in China highlight midcontinental hazards. *Eos, Transactions, American Geophysical Union*, 93, 453–454.

Love, D. (1996). *Seismic Hazard and Microzonation of the Adelaide Metropolitan Area*, South Australia, Department of Mines and Energy, Report Book 96/27.

Machette, M. N., S. F. Personius, A. R. Nelson, D. P. Schwartz, and W. R. Lund (1991). The Wasatch fault zone, Utah: segmentation and history of Holocene earthquakes. *Journal of Structural Geology*, 13, 137–149.

Machette, M. N., A. J. Crone, and J. R. Bowman (1993). Geologic investigations of the 1986 Marryat Creek, Australia, earthquake: implications for paleoseismicity in stable continental regions. *Bulletin U.S. Geological Survey*, 2032-B.

Malpas, K. L. (1991). Seismic risk in South Australia. BSc (Hons) thesis (unpublished). Department of Earth Sciences, Flinders University of South Australia, Adelaide.

Marco, S., M. Stein, and A. Agnon (1996). Long term earthquake clustering: a 50,000-year paleoseismic record in the Dead Sea Graben. *Journal of Geophysical Research*, 101, 6179–6191.

Mazzotti, S. (2007). Geodynamic models for earthquake studies in intraplate North America. In *Continental Intraplate Earthquakes: Science, Hazard, and Policy Issues*, ed. S. Stein, and S. Mazzotti. Geological Society of America Special Paper 425, pp. 17–33.

McCue, K. (1990). Australia's large earthquakes and recent fault scarps. *Journal of Structural Geology*, 12, 761–766.

McCue, K. (1996). Atlas of isoseismal maps of Australian earthquakes: Part 3. *AGSO Record*, 1995/44.

McCue, K. (2004). Australia: historical earthquake studies. *Annals of Geophysics*, 47, 387–397.

McCue, K., V. Wesson, and G. Gibson (1990). The Newcastle, New South Wales, earthquake of 28 December 1989. *BMR Journal of Australian Geology and Geophysics*, 11, 559–567.

McCue, K. F. (1975). *Seismicity and Seismic Risk in South Australia*. University of Adelaide Report, ADP 137.

McCue, K. F. (1980). Magnitudes of some early earthquakes in Southeastern Australia. *Search*, 11(3), 78–80.

McCue, K. F. (1989). *Australian Seismological Report 1985*, Bureau of Mineral Resources, Geology & Geophysics Report 285.

McCue, K. F., and M. O. Michael-Leiba (1993). Australia's deepest known earthquake. *Seismology Research Letters*, 64(34), 201–206.

McPherson, A., D. Clark, and T. Barrows (2009). Long-term slip rates for onshore faults in the Gippsland Basin, south-eastern Australia. *7th International Conference on Geomorphology (ANZIAG)*, Melbourne, Australia.

McPherson, A., D. Clark, M. Cupper, C. D. N. Collins, and G. Nelson (2012a). The Cadell Fault: a record of long-term fault behaviour in south-eastern Australia. *Proceedings Second Australian Regolith Geoscientists Association Conference/22nd Australian Clay Minerals Society Conference*, Mildura, Victoria, February 7–9, 2012, pp. 7–16.

McPherson, A., D. Clark, M. MacPhail, and M. Cupper (2012b). Neotectonism on the eastern Australian passive margin: evidence from the Lapstone Structural Complex. *Proceedings 15th Australia New Zealand Geomorphology Group Conference*, Bundanoon, NSW, December 2–7, 2012.

Michael-Leiba, M. O. (1987). Temporal variation in seismicity of the southwest seismic zone, Western Australia: implications for earthquake risk assessment. *BMR Journal of Australian Geology and Geophysics*, 10, 133–137.

Michael-Leiba, M. (1989). Macroseismic effects, locations and magnitudes of some early Tasmanian earthquakes. *BMR Journal of Australian Geology and Geophysics*, 11, 89–99.

Michael-Leiba, M., D. Love, K. McCue, and G. Gibson (1994). The Uluru (Ayers Rock), Australia, earthquake of 28 May 1989. *Bulletin of the Seismological Society of America*, 84, 209–214.

Mogi, K. (1963). Some discussions on aftershocks, foreshocks and earthquake swarms: the fracture of a semi-infinite body caused by an inner stress origin and its relation to the earthquake phenomena, 3. *Bulletin of the Earthquake Research Institute, Tokyo University*, 41, 615–658.

Mooney, W. D., J. Ritsema, and Y. K. Hwang (2012). Crustal seismicity and the earthquake catalog maximum moment magnitude (M_{cmax}) in stable continental regions (SCRs): correlation with the seismic velocity of the lithosphere. *Earth and Planetary Science Letters*, 357–358, 78–83.

Moye, D. G., K. R. Sharp, and D. H. Stapledon (1963). *Geology of the Snowy Mountains Region*. Snowy Mountains Hydro-electric Authority, Cooma (unpublished: held in Geological Survey of New South Wales Library).

Norvick, M. S. (2004). Tectonic and stratigraphic history of the Perth Basin. *Geoscience Australia, Record*, 2004/16.

Norvick, M. S., and M. A. Smith (2001). Mapping the plate tectonic reconstruction of southern and southeastern Australia and implications for petroleum systems. *APEA Journal*, 41, 15–35.

Paine, M. D., D. A. Bennetts, J. A. Webb, and V. J. Morand (2004). Nature and extent of Pliocene strandlines in southwestern Victoria and their application to Late Neogene tectonics. *Australian Journal of Earth Sciences*, 51, 407–422.

Pandey, O. P., K. Chandrakala, G. Parthasarathy, P. R. Reddy, and G. K. Reddy (2008). Upwarped high velocity mafic crust, subsurface tectonics and causes of intra plate Latur-Killari (M 6.2) and Koyna (M 6.3) earthquakes, India: a comparative study. *Journal of Asian Earth Sciences*, doi: 10.1016/j.jseaes.2008.11.014.

Paul, E., T. Flottmann, and M. Sandiford (1999). Structural geometry and controls on basement-involved deformation in the northern Flinders Ranges, Adelaide Fold Belt, South Australia. *Australian Journal of Earth Sciences*, 46, 343–354.

Pickett, J. W., and P. Bishop (1992). Aspects of landscape evolution in the Lapstone Monocline area, New South Wales. *Australian Journal of Earth Sciences*, 39, 21–28.

Pillans, B. (2012). How old is Lake George? *Australian and New Zealand Geomorphology Group 15th Conference*, Bundanoon, NSW, December 2–7, 2012, Program and Abstracts 79.

Quigley, M. C., M. L. Cupper, and M. Sandiford (2006). Quaternary faults of south-central Australia: palaeoseismicity, slip rates and origin. *Australian Journal of Earth Sciences*, 53, 285–301.

Quigley, M., M. Sandiford, A. Alimanovic, and L. K. Fifield (2007a). Landscape responses to intraplate tectonism: quantitative constraints from [10]Be abundances. *Earth and Planetary Science Letters*, 261, 120–133.

Quigley, M., M. Sandiford, K. Fifield, and A. Alimanovic (2007b). Bedrock erosion and relief production in the northern Flinders Ranges, Australia. *Earth Surface Processes and Landforms*, 32, 929–944.

Quigley, M. C., M. Sandiford, and M. L. Cupper (2007c). Distinguishing tectonic from climatic controls on range-front sedimentation, Flinders Ranges, South Australia. *Basin Research*, doi: 10.1111/j.1365–2117.2007.00336.x.

Quigley, M., D. Clark, and M. Sandiford (2010). Late Cenozoic tectonic geomorphology of Australia. In *Australian Landscapes*, ed. P. Bishop and B. Pillans. Geological Society, London, Special Publication 346, 243–265.

Rajendran, C. P. (2000). Using geological data for earthquake studies: a perspective from peninsular India. *Current Science*, 79, 1251–1258.

Rajendran, C. P., K. Rajendran, and B. John (1996). The 1993 Killari (Latur), central India, earthquake: an example of fault reactivation in the Precambrian crust. *Geology*, 24, 651–654.

Rajendran, K., C. P. Rajendran, M. Thakkar, and M. P. Tuttle (2001). The 2001 Kutch (Bhuj) earthquake: coseismic surface features and their significance. *Current Science*, 80, 1397–1405.

Reid, B. (2007). A Neotectonic study of the Willunga and Milendella Faults, South Australia. B.Sc. (Hons) thesis, School of Earth Sciences,The University of Melbourne, Melbourne.

Reinecker, J., O. Heidbach, and B. Mueller (2003). The 2003 release of the world stress map (available online at www.world-stress-map.org).

Revets, S. A., M. Keep, and B. L. N. Kennett (2009). NW Australian intraplate seismicity and stress regime. *Journal of Geophysical Research*, 114, B10305.

Reynolds, S. D., D. D. Coblentz, and R. R. Hillis (2002). Tectonic forces controlling the regional intraplate stress field in continental Australia: results from new finite element modeling. *Journal of Geophysical Research, Solid Earth*, 107(B7) 2131, 10.1029/2001JB000408.

Richardson, R. M. (1992). Ridge forces, absolute plate motions, and the intraplate stress field. *Journal of Geophysical Research*, 97, 11739–11749.

Ripper, I. D. (1963). Local and regional earthquakes recorded by the Tasmania Seismic Net. B.Sc thesis, University of Tasmania, Hobart.

Ritz, J. F., E. T. Brown, D. L. Bourles, *et al.* (1995). Slip rates along active faults estimated with cosmic-ray-exposure dates: application to the Bogd Fault, Gobi-Altai, Mongolia. *Geology*, 23, 1019–1022.

Robson, T. C., and J. A. Webb (2011). Late Neogene tectonics in northwestern Victoria: evidence from the Late Miocene–Pliocene Loxton Sand. *Australian Journal of Earth Sciences*, 58, 579–586.

Rutherfurd, I. D., and C. Kenyon (2005). Geomorphology of the Barmah-Millewa forest. *Proceedings of the Royal Society of Victoria*, 117, 23–39.

Rynn, J. M. W., D. Denham, S. A. Greenhalgh, *et al.* (1987). Atlas of isoseismal maps of Australian earthquakes, Part 2. *Bureau of Mineral Resources Bulletin*, 222.

Sandiford, D., and G. Gibson (2012). The Moe/Thorpedale earthquake: preliminary investigation. *Proceedings of the Australian Earthquake Engineering Society Conference*, Tweed Heads, Australia, December 7–9, 2012, Paper 25.

Sandiford, M. (2003a). Geomorphic constraints on the late Neogene tectonics of the Otway Ranges. *Australian Journal of Earth Sciences*, 50, 69–80.

Sandiford, M. (2003b). Neotectonics of southeastern Australia: linking the Quaternary faulting record with seismicity and in situ stress. In *Evolution and Dynamics of the Australian Plate*, ed. R. R. Hillis, and D. Muller. Geological Society of Australia Special Publication 22 and Geological Society of America Special Paper 372, 101–113.

Sandiford, M., and D. L. Egholm (2008). Enhanced intraplate seismicity along continental margins: some causes and consequences. *Tectonophysics*, 457, 197–208.

Sandiford, M., and M. C. Quigley (2009). Topo-Oz: insights into the various modes of intraplate deformation in the Australian continent. *Tectonophysics*, 474, 405–416.

Sandiford, M., M. Wallace, and D. Coblentz (2004). Origin of the in situ stress field in southeastern Australia. *Basin Research*, 16, 325–338.

Sandiford, M., D. D. Coblentz, and W. P. Schellart (2005). Evaluating slab-plate coupling in the Indo-Australian plate. *Geology*, 33, 113–116.

Schulte, S. M., and W. D. Mooney (2005). An updated global earthquake catalogue for stable continental regions: reassessing the correlation with ancient rifts. *Geophysical Journal International*, 161, 707–721.

Schwartz, D. P., and K. J. Coppersmith (1984). Fault behaviour and characteristic earthquakes: examples from the Wasatch and San Andreas Fault Zones. *Journal of Geophysical Research*, 89(B7), 5681–5698.

Seeber, L., G. Ekstrom, S. K. Jain, *et al.* (1996). The 1993 Killari earthquake in central India: a new fault in Mesozoic basalt flows? *Journal of Geophysical Research*, 101, 8543–8560.

Sheard, M. J. (1995). Quaternary volcanic activity and volcanic hazards. In *The Geology of South Australia*, Vol. 2, ed. J. F. Drexel, and W. V. Preiss. *Geological Survey of South Australia, Bulletin*, 54, 264–268.

Sinadinovski, C., and K. McCue (2010). Testing the hypothesis that earthquake hazard is uniform across Australia. *Proceedings Australian Earthquake Engineering Society Conference*, Perth, Western Australia, Paper 42.

Sinadinovski, C., T. Jones, D. Stewart, and N. Corby (2002). Earthquake history, regional seismicity and the 1989 Newcastle earthquake. In *Earthquake Risk in Newcastle and Lake Macquarie*, ed. T. Dhu, and T. Jones. Geoscience Australia Record, 2002/15.

Sinha, S., and S. Mohanty (2012). Spatial variation of crustal strain in the Kachchh region, India: implication on the Bhuj earthquake of 2001. *Journal of Geodynamics*, 61, 1–11.

Sloan, R. A., J. A. Jackson, D. McKenzie, and K. Priestley (2011). Earthquake depth distributions in central Asia, and their relations with lithospheric thickness, shortening and extension. *Geophysical Journal International*, 185, 1–29.

Smith, D. E., and R. Kolenkiewicz (1990). Tectonic motion and deformation from satellite laser ranging to LAGEOS. *Journal of Geophysical Research*, 95, 22,013–22,041.

Somerville, P., P. Quijada, H. K. Thio, M. Sandiford, and M. Quigley (2008). Contribution of identified active faults to near fault seismic hazard in the Flinders Ranges. *Australian Earthquake Engineering Society Meeting*, Paper 45.

Somerville, P., R. W. Graves, N. F. Collins, S. G. Song, and S. Ni (2010). *Ground Motion Models for Australian Earthquakes*. Final report to Geoscience Australia, 30 March 2010, URS Corporation, Pasadena, CA.

Stein, R. S., A. A. Barka, and J. H. Dieterich (1997). Progressive failure on the North Anatolian fault since 1939 by earthquake stress triggering. *Geophysical Journal International*, 128, 594–604.

Stein, S., and M. Liu (2009). Long aftershock sequences within continents and implications for earthquake hazard assessment. *Nature*, 462, 87–89.

Stein, S. S., S. Cloentingh, N. H. Sleep, and R. Wortel (1989). Passive margin earthquakes, stresses and rheology. In *Earthquakes at North American Passive Margins: Neotectonics and Postglacial Rebound,* ed. S. Gregersen and P. Basham. NATO ASI Series C, Boston, MA: Kluwer Academic, pp. 231–259.

Stirling, M., D. Rhoades, and K. Berryman (2002). Comparison of earthquake scaling relations derived from data of the instrumental and preinstrumental era. *Bulletin of the Seismological Society of America* 92, 812–830.

Stuart, W. D., T. G. Hildenbrand, and R. W. Simpson (1997). Stressing the New Madrid Seismic Zone by a lower crustal detachment fault. *Journal of Geophysical Research*, 102, 27623–27633.

Sutton, D. J., K. F. McCue, and A. Bugeja (1977). Seismicity of the southeast of South Australia. *Australian Journal of Earth Sciences*, 24, 357–364.

Sykes, L. R. (1978). Intraplate seismicity, reactivation of preexisting zones of weakness, alkaline magmatism, and other tectonism postdating continental fragmentation. *Reviews of Geophysics and Space Physics*, 16, 621–688.

Talwani, P. (1988). The intersection model for intraplate earthquakes. *Seismological Research Letters*, 59, 305–310.

Talwani, P. (1999). Fault geometry and earthquakes in continental interiors. *Tectonophysics*, 305, 371–379.

Talwani, P., and K. Rajendran (1991). Some seismological and geometric features of intraplate earthquakes. *Tectonophysics*, 186, 19–41.

Talwani, P., and W. T. Schaeffer (2001). Recurrence rates of large earthquakes in the South Carolina coastal plain based on paleoliquefaction data. *Journal of Geophysical Research*, 106, 6621–6642.

Tomkins, K. M., G. S. Humphreys, M. T. Wilkinson, *et al.* (2007). Contemporary versus long-term denudation along a passive plate margin: the role of extreme events. *Earth Surface Processes and Landforms*, 32, 1013–1031.

Tregonning, P. (2003). Is the Australian Plate deforming? A space geodetic perspective. In *Evolution and Dynamics of the Australian Plate* ed. R. R. Hillis, and D. Muller. Geological Society of Australia Special Publication 22 and Geological Society of America Special Publication 372, pp. 41–48.

Tuttle, M. P., E. S. Schweig, J. D. Sims, *et al.* (2002). The earthquake potential of the New Madrid seismic zone. *Bulletin of the Seismological Society of America*, 92, 2080–2089.

van Arsdale, R. (2000). Displacement history and slip rate on the Reelfoot fault of the New Madrid seismic zone. *Engineering Geology*, 55, 219–226.

van de Graaff, W. J. E., P. D. Denman, and R. M. Hocking (1976). Emerged Pleistocene marine terraces on Cape Range, Western Australia. *Western Australia Geological Survey Annual Report 1975*, 62–70.

van der Beek, P., A. Pulford, and J. Braun (2001). Cenozoic landscape development in the Blue Mountains (SE Australia): lithological and tectonic controls on rifted margin morphology. *The Journal of Geology*, 109, 35–56.

van Lanen, X., and W. D. Mooney (2007). Integrated geologic and geophysical studies of North American continental intraplate seismicity. In *Continental Intraplate Earthquakes: Science, Hazard, and Policy Issues*, ed. S. Stein, and S. Mazzotti. Geological Society of America Special Paper 425, 101–112.

VandenBerg, A. H. M. (2010). Paleogene basalts prove early uplift of Victoria's Eastern Uplands. *Australian Journal of Earth Sciences*, 57, 291–315.

Veevers, J. J. (2000). *Billion-Year Earth History of Australia and Neighbours in Gondwanaland*. Sydney, Australia: GEMOC Press.

Vlahovic, G., C. Powell, and M. Lamontagne (2003). A three-dimensional P wave velocity model for the Charlevoix seismic zone, Quebec, Canada. *Journal of Geophysical Research*, 108, 2439, doi:10.1029/2002JB002188.

Vogfjörd, K. S., and C. A. Langston (1987). The Meckering earthquake of 14th October 1968: a possible downward propagating rupture. *Bulletin of the Seismological Society of America*, 77, 1558–1578.

Wald, D. J., V. Quitoriano, B. Worden, M. Hopper, and J. W. Dewey (2011). USGS "Did You Feel It?" Internet-based macroseismic intensity maps. *Annals of Geophysics*, 54, 688–707.

Wallace, R. E. (1987). Grouping and migration of surface faulting and variations in slip rates on faults in the Great Basin province. *Bulletin of the Seismological Society of America*, 77, 868–876.

Weissel, J. K., and M. A. Seidl (1998). Inland propagation of erosional escarpments and river profile evolution across the southeast Australian passive continental margin. In *Rivers Over Rock: Fluvial Processes in Bedrock Channels*, ed. K. J. Tinkler and E. E. Wohl. American Geophysical Union, Geophysical Monograph 107, pp. 189–206.

Wells, D. L., and K. J. Coppersmith (1994). New empirical relationships among magnitude, rupture length, rupture width, rupture area, and surface displacement. *Bulletin of the Seismological Society of America*, 84, 974–1002.

Wheeler, R. L. (1995). Earthquakes and the cratonward limit of Iapetan faulting in eastern North America. *Geology*, 23, 105–108.

Wheeler, R. L. (1996). Earthquakes and the southeastern boundary of the intact Iapetan margin in Eastern North America. *Seismological Research Letters*, 67, 77–83.

Wheeler, R. L. (2009a). *Methods of M_{max} Estimation East of the Rocky Mountains*. U.S. Geological Survey Open-File Report 2009–1018.

Wheeler, R. L. (2009b). *Sizes of the Largest Possible Earthquakes in the Central and Eastern United States: Summary of a Workshop, September 8–9, 2008, Golden, Colorado*. U.S. Geological Survey Open-File Report 2009–1263.

Wheeler, R. L. (2011). Reassessment of stable continental regions of Southeast Asia. *Seismological Research Letters*, 82, 971–983.

Wheeler, R. L., and A. Frankel (2000). Geology in the 1996 USGS Seismic-hazard Maps, Central and Eastern United States. *Seismological Research Letters*, 71, 273–282.

Wilkinson, M. T., J. Chappell, G. S. Humphreys, *et al.* (2005). Soil production in heath and forest, Blue Mountains, Australia: influence of lithology and palaeoclimate. *Earth Surface Processes and Landforms*, 30, 923–934.

Zoback, M. D. (1992). First- and second-order patterns of stress in the lithosphere: the world stress map project. *Journal of Geophysical Research*, 97, 11703–11728.

3

Intraplate seismicity in Brazil

MARCELO ASSUMPÇÃO, JOAQUIM FERREIRA, LUCAS BARROS,
HILARIO BEZERRA, GEORGE SAND FRANÇA, JOSÉ ROBERTO BARBOSA,
EDUARDO MENEZES, LUÍS CARLOS RIBOTTA, MARLON PIRCHINER,
ADERSON DO NASCIMENTO, AND JOÃO CARLOS DOURADO

Abstract

We describe the development of the Brazilian earthquake catalogue and the distribution of seismicity in Brazil and neighbouring areas in mid-plate South America. This large mid-plate region is one of the least seismically active stable continental regions (SCR) in the world: the maximum known earthquake had a magnitude of 6.2 m_b and events with magnitudes 5 and above occur with a return period of 4 years. Several seismic zones can be delineated in Brazil, including some along craton edges and in sedimentary basins. Overall, the exposed cratonic regions tend to have half as many earthquakes compared to the average expected rate for all of mid-plate South America. Earthquakes tend to occur in Neoproterozoic foldbelts especially in areas of thin lithosphere, or near craton edges around cratonic keels. Areas with positive isostatic gravity anomalies tend to have more earthquakes, indicating that flexural stresses from uncompensated lithospheric loads are an important factor in explaining the intraplate seismicity. We also found that earthquakes are two to three times more likely to occur within 20 km of mapped neotectonic faults, compared to events at larger distances. On closer examination, however, we observe that most of these events occur near but not directly on the major neotectonic faults. This discrepancy could be explained by the model of stress concentration near intersecting structures. The Brazilian passive margin is also a region of higher than average seismicity. Although clear differences are found between different areas along the passive margin (extended crust in southeast Brazil having especially high seismicity compared to narrow continental shelves elsewhere), overall the Brazilian passive margin has 70% more earthquakes (magnitudes above 3.5) than the average stable continental region.

Intraplate Earthquakes, ed. Pradeep Talwani. Published by Cambridge University Press. © Cambridge University Press 2014.

3.1 Introduction

Explaining the genesis of intraplate seismicity is a great challenge. Several models have been proposed to explain the cause of earthquakes far away from the active plate boundaries. These can be broadly divided into two kinds of models: those involving weakness zones and those involving stress concentrations. Crustal weak zones are usually due to the last major orogenic process and involve ancient rifts or failed rifts (e.g., Sykes, 1978; Johnston and Kanter, 1990; Assumpção, 1998; Schulte and Mooney, 2005), or suture zones (i.e., around craton edges as shown by Mooney *et al.* [2012]). Stress concentrations can arise from lateral density variations (e.g., Stein *et al.*, 1989; Assumpção and Araujo, 1993; Zoback and Richardson, 1996; Assumpção and Sacek, 2013), contrasts of elastic properties (e.g., Campbell, 1978; Stevenson *et al.*, 2006), or fault intersections (Talwani, 1999; Gangopadhyay and Talwani, 2003, 2007). Quite often, weak zones and stress concentrations are linked in the same process. For example, lithospheric thin spots between ancient cratonic areas can be regarded as a weak zone but the seismicity is due to stress concentration in the elastic upper crust (e.g., Assumpção *et al.*, 2004). Evidence for higher seismicity along Phanerozoic suture zones in North America can be interpreted as concentration of stresses and deformation ("crumpling" zones) near craton edges (Lenardic *et al.*, 2000), which are ultimately caused by lateral variations of lithospheric thickness (Mooney *et al.*, 2012).

If lateral density variations are not isostatically compensated, flexural stresses can arise in the upper crust and reach large magnitudes to significantly contribute to intraplate seismicity (e.g., Cloetingh *et al.*, 1984; Stein *et al.*, 1989; Zoback and Richardson, 1996; Assumpção *et al.*, 2011; Assumpção and Sacek, 2013). Flexural stresses can be caused by ice-sheet retreat (e.g., Mazzotti *et al.*, 2005), sediment load in the continental margin (Stein *et al.*, 1989; Assumpção, 1998; Assumpção *et al.*, 2011), or intracrustal loads from past geological processes (e.g., Zoback and Richardson, 1996; Assumpção and Sacek, 2013).

As reviewed by Mazzotti (2007), it is usually agreed that several different factors can contribute to produce an intraplate seismic zone. However, one of the major difficulties with most models for intraplate seismicity is the fact that the same geological/structural features are also present in areas with no current seismic activity. For example, not all continental shelves are equally active, despite having similar geological structures and potentially the same sources of stress. This has contributed to the debate of whether long-term migration of intraplate seismic zones occurs, with significant implications for seismic hazard assessment (Stein *et al.*, 2009; Li *et al.*, 2009), and highlighted the importance of further studies of intraplate mechanisms and a more detailed comparison of seismicity patterns between different regions and continents.

Here we present the spatial distribution of seismicity in Brazil and discuss possible correlations with geological and geophysical features, trying to assess some of the proposed models mentioned above. Assessing the applicability of general models for intraplate seismogenesis is important to better delineate seismic zones in mid-plate South America.

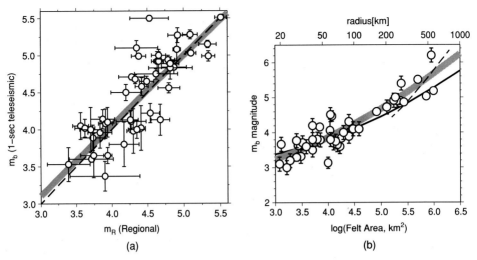

Figure 3.1 (a) Relationship between the regional magnitude scale m_R (Assumpção, 1983) and the teleseismic m_b scale (IASPEI). Error bars are standard deviations of the mean values. Grey line is the m_b:m_R regression, which is not statistically different than $m_b = m_R$ (dashed line). The regional scale is equivalent to the teleseismic magnitude. (b) Empirical relationship between total felt area (A_f) and magnitude. The magnitudes used in this plot are average of m_b and m_R values. The grey line is the regression $m = 2.44 - 0.015\log(A_f) + 0.0922[\log(A_f)]^2$; standard deviation $= 0.35$ magnitude units. The solid thin line is the empirical relation for Central and Eastern United States (Nuttli *et al.*, 1979) using magnitude $m_b(L_g)$. The dashed line is the approximate relationship between m_b and felt area from the worldwide compilation of Johnston *et al.* (1994). Top axis is the equivalent radius for the felt area.

3.2 Earthquake catalogue

The Brazilian earthquake catalogue is based mainly on the compilation of Berrocal *et al.* (1983, 1984) complemented since 1982 by the *Brazilian Seismic Bulletin* prepared jointly by the universities of São Paulo (USP), Brasília (UnB), Rio Grande do Norte (UFRN), and the Technological Research Institute (IPT) of the state of São Paulo. Other institutions (State University of São Paulo, UNESP, and National Observatory, ON) also contribute data to the Brazilian bulletin. Events in a neighbouring area just west of the Brazilian border (also characterized as SCR) are included to help establish correlation with intraplate geological and geophysical features – these events outside Brazil were taken from the literature, the ISC catalogue, as well as located by the Brazilian stations (e.g., Berrocal *et al.*, 1984).

The magnitude scale adopted in the catalogue is the 1-second P-wave teleseismic m_b, especially for events larger than about 5, and the regional magnitude (m_R) developed by Assumpção (1983) for earthquakes recorded between 200 km and 2000 km using the maximum P-wave particle velocity in the period range 0.1 to 1.0 s. Figure 3.1a shows that the regional magnitude m_R is equivalent to the teleseismic m_b scale in the range 3.5 to 5.5.

Magnitudes of historical earthquakes were estimated from felt area, as shown in Figure 3.1b. Both total felt area (inside isoseismal II MM) and A_{IV} (area inside isoseismal IV MM) were used. The relations between felt area and magnitudes are similar to those of the Central and Eastern United States as reported by Nuttli *et al.* (1979), and shown in Figure 3.1b. Our data are also consistent with the global compilation of Johnston *et al.* (1994) and Johnston (1996) for $m_b > 4.5$. An event with magnitude 5 is usually felt up to 200–500 km away. The new regression shown in Figure 3.1b is not significantly different from the one used by Berrocal *et al.* (1984). For historical events, m_b magnitudes can be estimated with a standard error of 0.35 units.

Completeness of the catalogue varies considerably according to the region. For the most populated areas, such as southeastern Brazil, magnitudes above 4.5 (which should be felt to 100 km distance or more; Figure 3.1) are believed to be complete since about 1850. For less populated areas such as the Amazon region, the threshold of 4.5 was only attained in the late 1960s with the installation of WWSSN stations in South America, and the array station in Brasilia, central Brazil.

3.3 Seismicity map

Figure 3.2a shows all epicentres of the Brazilian catalogue with magnitudes above 3.0 (a total of about 800 events) and the main geological provinces in Brazil (Almeida *et al.*, 2000). Our seismic catalogue is relatively recent: the two earliest well-documented events with magnitude above 4 occurred in 1807 ($\sim$5 m_b in northeastern Brazil; Veloso, 2012) and 1861 ($\sim$4.4 m_b in southeastern Brazil). The maximum observed earthquake, with a magnitude of 6.2 m_b, occurred on 31 January 1955, in the northern part of the Parecis basin (Figure 3.2a). There are sparse historical data on a possibly large earthquake in the central Amazon basin in 1690, which could have magnitude $\sim$7 (Veloso, 2014), but no reliable accounts allow a confirmation.

Paleoseismological studies in northeastern Brazil (Bezerra *et al.*, 2005, 2011; Rossetti *et al.*, 2011) have described the collective occurrence of liquefaction structures in Quaternary gravels, which are consistent with magnitudes as high as $\sim$6.0–7.0. These paleoseismological studies are still patchily distributed in Brazil, but show that magnitudes up to 7 m_b should be considered in hazard estimates.

The distribution of epicentres in Figure 3.2a is much affected by the population distribution: the higher population density in southeastern and northeastern Brazil results in a large number of historical events reported in newspapers, old journals, and books. To be able to compare seismicity rates in different parts of the country a more geographically uniform coverage is necessary. For this reason, we filtered the Brazilian catalogue (Figure 3.2a) with time-variable thresholds to produce the "uniform" epicentral map of Figure 3.2b. We selected events with magnitudes higher than 6 since 1940 (probably detectable with the worldwide stations reporting to the ISS bulletin); magnitudes higher than 5.0 since 1962 due to the increased coverage of the WWSS network; magnitudes higher than 4.5 since 1968 due to the installation of the Brasilia array in 1967, the WWSSN

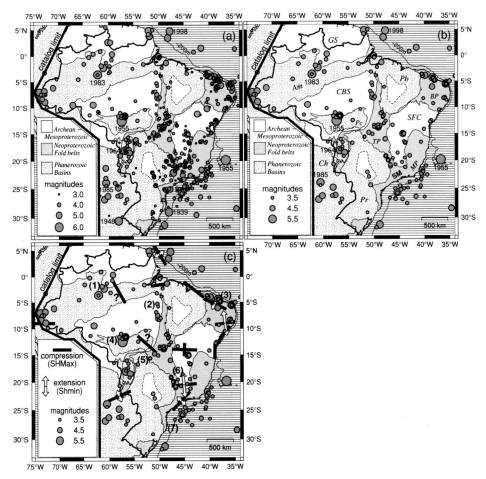

Figure 3.2 Intraplate epicentres in Brazil and neighbouring areas. (a) Raw catalogue, with crustal events only (depth < 50 km), and magnitudes ≥ 3.0. (b) "Uniform" catalogue with time-variable magnitude thresholds. Thick solid line is the approximate limit of the Brazilian catalogue. Shading denotes the main geological provinces in Brazil. Cratonic areas: Guyana shield (GS) and Central Brazil shield (CBS) as part of the Amazon craton; São Francisco craton (SFC); intracratonic basins: Amazon (Am), Parnaiba (Pb), Parecis (Pc), Chaco (Ch), Paraná (Pr) and Pantanal (Pt); Brasiliano fold belt provinces: Tocantins (TP), Borborema (BP), Mantiqueira (MP); SM denotes the "Serra do Mar" coastal ranges in southeast Brazil. Dashed white areas are proposed cratonic blocks in the middle of the Parnaíba and Paraná basins. Hatched area is the Atlantic Ocean where solid line (200 m bathymetry) denotes the edge of the continental shelf. Dates in (a) and (b) indicate magnitudes ≥ 5.3 m_b. (c) Epicentres of the uniform catalogue (open circles) and estimates of the stress field: grey bars are compressional principal stresses (S_{Hmax}); open arrows are extensional principal stresses (S_{hmin}); numbers denote main seismic zones as discussed in Section 3.3. For colour version, see Plates section.

station NAT in northeastern Brazil and general increase of seismological research in Brazil; magnitudes above 3.5 since 1980 due to the installation of several stations in various parts of the country by the University of Brasília (including the Amazon region) and the University of São Paulo. The magnitude thresholds defined above are our best estimates of the magnitude completeness of our catalogue. The geographically "uniform" catalogue (shown in Figure 3.2b) is not declustered. Both the raw (Figure 3.2a) and the uniform catalogues (Figure 3.2b) show several seismically active areas and large regions almost completely aseismic.

The stress field in Brazil is still poorly known due to the small number of well-determined focal mechanisms and few in-situ stress measurements. Data from the compilation of Assumpção (1992), other studies (Lima *et al.*, 1997; Ferreira *et al.*, 1998; Barros *et al.*, 2009, 2012; Chimpliganond *et al.*, 2010; Lima-Neto *et al.*, 2013), and recent unpublished results were combined to provide a preliminary estimate of the stress patterns in Brazil, shown in Figure 3.2c. Compressional stresses predominate with a trend of roughly east–west-oriented maximum horizontal stress (S_{Hmax}) in eastern Brazil, probably changing to a more NW–SE orientation in the Amazon region. A strong influence of the continental margin can be seen, making S_{Hmax} roughly parallel to the coast along most of the Atlantic shore. The effect of the continental margin has been explained as due to lateral density contrasts between continental and oceanic crust (e.g., Coblentz and Richardson, 1996) and flexural stresses (Assumpção, 1992, 1998).

Combining the distribution of seismicity (Figure 3.2a, b) with the estimated stress patterns (Figure 3.2c), the main active areas could be delineated as follows: (1) southern part of the Guyana shield and middle Amazon basin, the largest known event being a 5.5 m_b in 1983; (2) a north–south-trending zone along the eastern border of the Amazon craton; (3) northern part of the Borborema Province, in northeastern Brazil around the Potiguar marginal basin, with the largest known event occurring in 1980 with 5.2 m_b; (4) the Porto dos Gauchos seismic zone with activity known since the largest event of 1955 (Barros *et al.*, 2009); (5) a NE–SW zone in the Tocantins province possibly continuing towards the Pantanal basin; (6) southern Minas Gerais zone, in and around the southern tip of the São Francisco craton; and (7) the southeast offshore zone with activity concentrated along the continental slope from the Pelotas basin in the south to the Campos basin, from 33° S to 20° S (e.g., Assumpção, 1998; Assumpção *et al.*, 2011).

Remarkably, Branner (1912), on the basis of historical accounts, had already delineated the main seismic zones in Brazil (excluding the Amazon region) such as: northeastern Brazil (zone 3), Central Brazil and Mato Grosso (zone 5), and southeast Brazil (Minas Gerais and Rio de Janeiro; roughly zone 6). Most of the largest earthquakes (magnitude ~5) in the second half of the twentieth century occurred roughly in the same areas as defined by historical events in the late nineteenth and early twentieth centuries.

An interesting aspect of Brazilian seismicity is the different characteristics of the earthquake sequences in northeastern Brazil compared with other parts of the country. In northeastern Brazil, recurrent swarms and aftershock sequences, some lasting for several years, are very common (Takeya *et al.*, 1989; Ferreira *et al.*, 1998, 2008; Bezerra *et al.*, 2007), whereas large earthquakes ($\sim$5 m_b) in other parts of the country tend to occur as a single event with a short aftershock sequence.

We calculated the Gutenberg–Richter magnitude–frequency relation using the "uniform" catalogue for both continental and offshore earthquakes (Figure 3.2b), with the time-variable detectability thresholds above, using the method of Weichert (1980) for partially complete catalogues. This catalogue was declustered (reducing from 800 to 650 events) and a maximum magnitude of 7.0 was used. We obtained a b-value of 0.97 $\pm$ 0.06. The seismicity rate can best be evaluated for the cumulative frequency of earthquakes with magnitudes $\geq$5.0 m_b, for which we get a return period of 4 years, making this mid-plate area of South America one of the SCRs with the lowest seismicity rate in the world, as shown by Johnston *et al.* (1994).

3.4 Seismotectonic correlations

We now test the statistical significance of some general hypotheses and models regarding the distribution of seismicity in mid-plate South America, as seen in Figure 3.2. One must be aware, however, that the Brazilian historical and instrumental catalogue spans a very short time window and may not be representative of the long-term behaviour of intraplate seismicity, especially when events of magnitude 6+ have return periods of the order of a century. In addition, it has been suggested that long-term migration of activity in SCRs may be a common characteristic, such as in North America (Stein *et al.*, 2009; Li *et al.*, 2009) and Australia (Clark *et al.*, Chapter 2, this book). Thus, the statistical tests below should be regarded as an initial study, and interpretation of the results should consider the relatively short time span of the catalogue.

3.4.1 Lower seismicity in Precambrian cratonic provinces

A first hypothesis to be tested is the possible lower seismicity of Mid-Proterozoic and older cratonic provinces. Cratons are usually regarded as more "stable" and rigid blocks, compared with the surrounding younger fold belts. We compared the distribution of continental seismicity only (east of the catalogue limit defined by the thick solid line in Figure 3.2), excluding the offshore area. For simplicity, we considered only the two exposed large cratonic areas in the Amazon and Atlantic shields, and have not included small cratonic blocks, such as the Luis Alves craton in Southern Brazil and the Apa block near the Brazil/Paraguay border, or cratonic blocks hidden beneath sedimentary basins such as the Rio de La Plata craton in northeast Argentina or the hypothesized cratonic blocks beneath the Parnaíba and Paraná basins. Figure 3.2b shows that the exposed cratons (Amazon and São Francisco cratons) correspond to 35% of the continental area (i.e., between the "catalogue limit" in Figure 3.2 and the coastline). If the seismicity distribution had no correlation with these large cratonic areas, we would expect about the same fraction of epicentres to be located in those cratonic areas. For the "whole" catalogue (Figure 3.2a) we would expect around 390 epicentres (35% of 1,118 epicentres) but we observe only 200. Getting half the expected rate is statistically highly significant. If we use the "uniform" catalogue, we get a similar result: about 40% of the expected rate (33 observed out of 82 expected).

We can then conclude that cratonic areas in mid-plate South America are about 50% less seismic than the average mid-plate seismicity in Brazil and neighbouring areas. Including smaller cratonic blocks (with areas less than 10% of the major Amazon and São Francisco cratons) would not change this result significantly. In fact, it would probably make the result even more clear, as most of these other small blocks have almost no earthquakes. We conclude that large cratonic areas are about four times less seismic than the other younger provinces: Brasiliano fold belts ($\sim$750–540 Ma) and intraplate basins. This may not seem surprising at first. Similar results can be found in other continents such as Australia, where Clark *et al.* (this volume) showed that cratonic crust is less seismic than non-cratonic areas.

The statistically significant result for Brazil could only be obtained on a continent-wide scale. Some cratonic areas can have high earthquake activity (such as in the middle of the São Francisco craton; e.g., Chimpliganond *et al.* [2010]) and regional comparisons of cratonic versus non-cratonic areas (such as done by Assumpção *et al.* [2004] for southeastern Brazil) had not produced statistically significant differences.

3.4.2 Intraplate seismicity and cratonic roots

Recently, Mooney *et al.* (2012) analyzed the global distribution of intraplate earthquakes and concluded that intraplate seismicity tends to concentrate around cratonic edges, especially for large magnitudes. The craton edges were defined by the S-wave anomalies from global-scale tomography. The global scale of that study included both mid-continent as well as passive margin earthquakes, and earthquakes more properly characterized by continental-shelf dynamics were included as "craton-edge" events. Here we compare the distribution of continental earthquakes with S-wave anomalies. We used the S-wave anomaly at 100 km, obtained from joint inversion of surface waves and receiver function point constraints (Assumpção *et al.*, 2013), as a proxy for lithospheric depths (Figure 3.3).

The map of Figure 3.3a clearly shows a trend of higher seismicity above areas of low S-wave velocities, which is confirmed by the high frequency of epicentres above areas with S-wave anomalies between -1 and -4% (Figure 3.3c, d). This result confirms the findings of Assumpção *et al.* (2004), who showed that in Central Brazil higher seismicity is observed above areas of low P-wave velocities at lithospheric depths. This can be interpreted as stress concentrations in the upper crust due to lithospheric thinning.

Figure 3.3c, d also shows that the number of events above areas of very high S-wave velocities (anomalies $\geq +5\%$ corresponding to cratonic cores or roots) is lower than expected. This is consistent with the finding above (Section 3.4.1) that cratonic areas tend to have lower seismicity. However, an interesting feature in these histograms is the peak of high seismicity above areas with anomalies in the range $+3$ to $+5\%$. This seems to correspond to the cratonic edge effect found by Mooney *et al.* (2012) on a global scale.

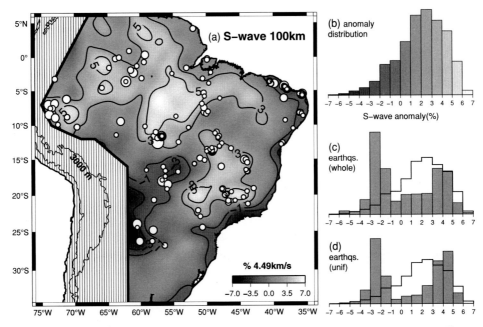

Figure 3.3 Comparison of S-wave anomalies with epicentral distribution. (a) S-wave anomalies at 100 km depth (Assumpção *et al.*, 2013) from joint inversion of surface waves and receiver function point constraints. White circles are epicentres from the "uniform" catalogue. (b) Histogram of the S-wave anomalies within the continent and east of the catalogue limit, corresponding to the grey-shaded area in (a). (c) Histogram of the S-wave anomalies beneath all continental epicentres of the "whole" catalogue. (d) Similarly for the "uniform" catalogue. The solid black lines in (c) and (d) indicate the distribution expected from (b) if the earthquakes had no correlation with lithospheric structure. For colour version, see Plates section.

Areas with "normal" lithospheric depths (anomalies between −1 and +3%) seem to be the least seismic.

3.4.3 Passive margin seismicity

Schulte and Mooney (2005) confirmed earlier findings (e.g., Sykes, 1978; Johnston and Kanter, 1990) that, on a global scale, intraplate earthquakes (M > 4.5) tend to occur preferentially in rifted crust, that is, in interior continental rifts (31%) or rifted continental margins (28%), compared with continental non-extended crust (41%). Considering that the area of non-extended crust in SCRs is several times that of rifted margins, it is clear that passive margins have a much higher seismicity rate. For magnitudes greater than 6, the preference for rifted crust is even stronger. Here we test if the extended crust along the Brazilian passive margin is significantly higher than the average continental area. The whole Brazilian margin was subjected to rifting during the Pangea breakup in the Jurassic–Cretaceous (Chang *et al.*, 1992). Therefore, we do not differentiate between interior rifts

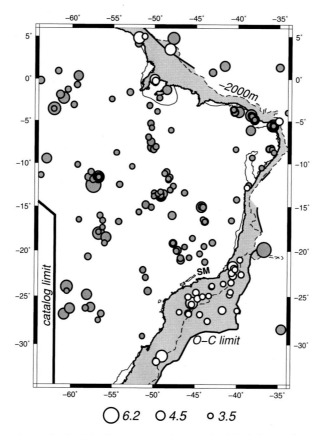

Figure 3.4 Comparison of seismicity between passive margin (shaded area, between the coastline and the oceanic–continental [O-C] limit) and interior SCR. The extension of the passive margin was defined using the mapped O-C limit, if available (thick solid line) offshore, or the 2,000 m bathymetric contour (dashed line) used as a proxy for the O-C limit. Thin solid lines near the coast indicate onshore marginal basins, not included in our definition of "passive margin". Open circles are epicentres in the passive margin, grey circles in the continental area (from the uniform catalogue of Figure 3.2b). "SM" in southeast Brazil denotes the presently aseismic "Serra do Mar" coastal ranges.

and non-extended crust along this margin, as this would require more extensive work to better define the area of influence of interior rifts.

A common definition of a "rifted continental margin" is the region of extended crust in the transition from normal continental to oceanic crust, in areas where extension led to seafloor spreading (e.g., Schulte and Mooney, 2005). For our statistical test, here we define "passive margin" as the area between the coastline and the oceanic–continental crustal limit (O-C limit), shown in Figure 3.4. The extension of the passive margin was defined using the mapped O-C limit, if available, or the 2,000 m bathymetric contour (used as a proxy for the O-C limit). Mapped O-C limits were taken from Houtz *et al.* (1977), Chang *et al.* (1992), Gomes (1992), Mohriak *et al.* (2000), Watts *et al.* (2009), and Zalan *et al.* (2011).

For simplicity, we did not include the onshore marginal basins, as different criteria could be used to define the onshore limit of the extended crust. We traded a more detailed and complex analysis for a simpler, more objective statistical test.

The continental region (Figure 3.2b) and the passive margin (Figure 3.4) have a total of 274 events with magnitudes 3.5 or greater in the uniform catalogue (238 in the continent and 36 in the passive margin). Taking into account the different sizes of both areas, we would expect only 21 events (7.8%) in the passive margin if the seismicity were uniformly distributed. The larger number of observed events (36) indicates that the Brazilian passive margin has a seismicity rate 70% higher than the average SCR seismicity, with a confidence limit better than 98%. A test using $m_b \geq 4.5$ also shows a larger number of events in the passive margin (six observed events compared to an expected number between two and three), but the small sample prevents a reliable measure of the statistical significance.

Although the Brazilian passive margin has a higher seismicity rate, on average, Figure 3.4 shows a high concentration in two areas: one in southeastern Brazil where the continental crust was subjected to an extreme degree of extension and thinning, and the other just north of the Amazon fan, both characterized by thick sedimentary sequences. Areas of short continental shelf (most of the eastern and northeastern margin), with generally thin sedimentary layers, have almost no earthquakes. This indicates that crustal weaknesses caused by high levels of extension, as well as flexural stresses, are important factors in defining seismicity rates in passive margins (e.g., Assumpção, 1998; Assumpção *et al.*, 2011). Concentration of seismicity in the continental margin is a common pattern of global seismicity, as shown by Schulte and Mooney (2005), but local conditions can cause significant variations along the margin, as shown by Sandiford and Egholm (2008) for Australia.

3.4.4 Influence of neotectonic faults

Neotectonic has been described as the period when young tectonic events have occurred and are still occurring in a given region after its last orogeny or after its last significant tectonic or stress field set up (e.g., Pavlides, 1989). The neotectonic period varies in different regions. There is no consensus on the onset of the neotectonic period in some passive margins, such as the Brazilian margin, where neotectonic studies are patchy. Therefore, we chose to analyze a preliminary neotectonic map of Brazil and show two examples of pre-existing fabric such as ductile shear zones and their relationship with the present-day seismicity.

Neotectonic activity in Brazil has been clearly observed in several areas (e.g., Riccomini and Assumpção, 1999; Bezerra and Vita-Finzi, 2000; Bezerra *et al.*, 2006, 2011), with observed geological faulting dated as recent as Holocene (e.g., Riccomini and Assumpção, 1999). A preliminary map of neotectonic faults in Brazil was presented by Saadi *et al.* (2002), with most of the mapped faults and lineaments likely to be active in the neotectonic period based on geomorphological criteria. While this compilation is still very preliminary and very few faults have geochronological dates, it is the only compilation spanning the whole country, and for this reason may be useful for some initial tests. For

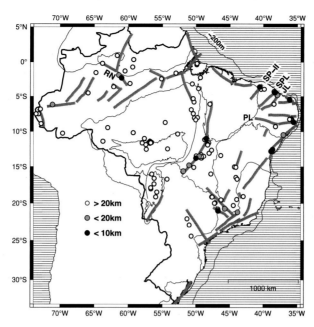

Figure 3.5 Correlation between neotectonic faults and epicentres of the "uniform" catalogue. Thick grey lines are the proposed neotectonic faults compiled by Saadi *et al.* (2002). Open circles are epicentres more than 20 km away from any fault segment; grey and solid circles denote epicentres in the range 20–10 km from any fault, and closer than 10 km, respectively. Thin solid lines are the limits of the major geological provinces as in Figure 3.2. Faults discussed in the text: RN, Rio Negro; PL, Pernambuco Lineament; SP-II, Sobral Pedro-II Lineament; SPL, Senador Pompeu Lineament; JL, Jaguaribe Lineament.

example, it is natural to wonder if the present seismicity tends to be close to the mapped neotectonic structures. Figure 3.5 shows a map of the neotectonic faults and lineaments as compiled by Saadi *et al.* (2002). We now test if the epicentres of the Brazilian catalogue (Figure 3.2) tend to lie close to the proposed neotectonic features. For these tests, we excluded the Samambaia fault (fault BR-38 in Saadi's compilation) because it had been defined only on the basis of the earthquake activity near João Câmara, northeast Brazil (5.5° S 35.5° W in Figure 3.2); the corresponding earthquakes were also removed from the database. In addition, we corrected the position of the fault BR-15 (Sobral Pedro-II in northeastern Brazil, labelled "SP-II" in Figure 3.5) because it was misplaced in Saadi's compilation.

Using the "uniform" catalogue (magnitudes above 3.5 m_b), we compared the number of observed epicentres closer than 20 km (or 10 km) to any fault segment, with the expected number assuming a random distribution of events in the continental area of Brazil. We used the criterion of 10 and 20 km distance because this is roughly the estimated epicentral accuracy of most events in the catalogue. Figure 3.5 shows the events closer than 20 km (and 10 km) from any fault segment. The chance of a random event being closer than 20 km to any fault segment is 8.8%. The uniform catalogue has 180 events, and 50 are closer than 20 km, which is about three times the expected number. For a 10 km distance, we would expect 8 events (4.3%) but observe 20 events closer than 10 km. This is highly significant (more than 99.9% confidence level). At a closer distance of 5 km, the results are similar. Both the "whole" (Figure 3.2a) and the "uniform" (Figure 3.2b) catalogues yield similar results. This indicates that areas close to the neotectonic faults mapped by Saadi *et al.* (2002) are two to three times more likely to have seismic activity, compared to areas far from the faults.

This is the case for northeastern Brazil, where seismicity is best known in the country due to many detailed studies of earthquake swarms and aftershocks using local networks. A clear example of the association of present-day seismicity with geological features is the Pernambuco Lineament, a ductile Brasiliano (~600 Ma) shear zone, reactivated in the brittle crust during the Pangea breakup in the Cretaceous. This feature was presented in the compilation of Saadi *et al.* (2002; "PL" in Figures 3.5 and 3.6a). Seismological studies indicated that this shear zone has been repeatedly reactivated in the past few decades (Ferreira *et al.*, 1998, 2008; Lopes *et al.*, 2010; Figure 3.6a). However, several other lineaments in this preliminary neotectonic map of Brazil do not have any evidence of paleoseismicity, historical or instrumental seismicity. For example, Ferreira *et al.* (1998) have shown that in northeastern Brazil most earthquakes have no direct correlation with large geological lineaments mapped at the surface, even when the epicentres are apparently very close to a major fault or shear zone. These are the cases of major shear zones shown in Figure 3.5, such as Jaguaribe Lineament (JL), Senador Pompeu Lineament (SPL), and the Sobral Pedro-II fault (SP-II).

The explanation for the reactivation of major lineaments, where others remain aseismic, is a matter of debate. For example, in northeastern Brazil, major shear zones were reactivated in the Pangea breakup and they form the boundaries of rifts (Chang *et al.*, 1992). The occurrence of pseudotachylyte in these mylonitic belts implies that they are exhumed paleoseismic zones (Kirkpatrick *et al.*, 2013). However, other continental-scale east–west-trending shear zones in northeastern Brazil (Figure 3.5) also present a long history of brittle reactivation and are also under the ~ east–west-trending maximum compression (S_{Hmax}) and ~north–south-trending extension (S_{hmin}), such as the Pernambuco shear zone (Figure 3.2c). A question remains as to why these other shear zones do not show present-day seismicity. Several processes have been proposed to explain mechanical fault weakness, such as the occurrence of low-friction material in the fault core (e.g., Morrow *et al.*, 1992) or the presence of high fluid pressure in the fault zone (e.g., Byerlee, 1990; Sibson, 1990). However, there is a paucity of evidence to support the idea that some of these processes can be responsible for the weakening of fault zones (Holdsworth *et al.*, 2001). As intraplate faults present long recurrence periods (e.g., Bezerra and Vita-Finzi, 2000), it is possible that these continental-scale features represent dormant structures. It follows that the short period of instrumental monitoring, poorly known historical seismicity, and lack of systematic paleoseismic investigations in Brazil indicate that further studies are necessary to address this problem.

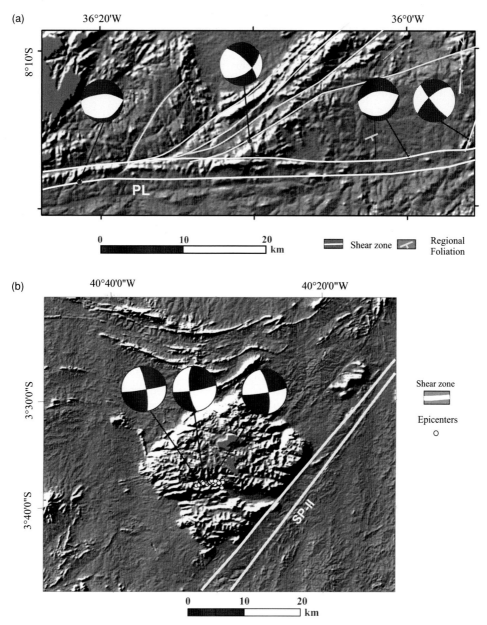

Figure 3.6 Examples of relationship of seismicity with major lineaments thought to be neotectonic features in northeastern Brazil. (a) Focal mechanisms in the present reactivation of the Pernambuco Shear zone ("PL" in this figure and in Figure 3.5). Note that events along the main E–W branch have E–W-trending normal faults, and events along the NE-trending branches are strike-slip because of the E–W compressional and N–S extensional principal stresses. Focal mechanisms from Lopes *et al.* (2010), Ferreira *et al.* (1998, 2008) and Lima-Neto *et al.* (2013). (b) Earthquake sequence in a small, E–W transcurrent fault about 10 km from the SW–NE-oriented Sobral Pedro-II Lineament ("SP-II" in Figure 3.5). The NW–SE regional maximum horizontal stress (Figure 3.2c) is not favorably oriented to reactivate the major lineament, but causes seismicity in a small previously unmapped fault nearby (Moura *et al.*, 2013).

The activity near the SW–NE-trending Jaguaribe Lineament (JL in Figure 3.5) actually occurs on an E–W-trending strike-slip seismogenic fracture ("Palhano" earthquakes in Ferreira *et al.* [1998]). The activity near the Senador Pompeu Lineament (SPL in Figure 3.5) occurs on a NW–SE-trending strike-slip fault ("Pacajus-Cascavel" activity in Ferreira *et al.*, 1998). Finally, Moura *et al.* (2013) showed that the events close to the NE–SW-trending Sobral Pedro-II shear zone actually occur on a small E–W-trending fracture, as shown in Figure 3.6b. In the Amazon basin, a similar inconsistency can be found with the "Rio Negro" fault (BR-10 of Saadi *et al.*, 2002; labelled "RN" in Figure 3.5): the event less than 10 km from the fault is the 5.1 m_b Manaus earthquake of 1963, 45 km deep, which had a well-determined reverse faulting mechanism with the two nodal planes oriented SW–NE (Assumpção and Suárez, 1988), clearly inconsistent with the SE–NW trend of the Rio Negro fault.

Our statistical tests showing higher seismicity up to about 20 km from faults interpreted as being neotectonic may, in fact, indicate a zone more prone to seismic activity, but not necessarily a direct relationship. The presence of a neotectonic feature might increase the chances of stress concentration in intersection zones involving other smaller fractures, such as proposed by Talwani (1999) and Gangopadhyay and Talwani (2003; 2007). In addition, the major lineament may not be favourably oriented with respect to the present stresses while other smaller faults may be optimally oriented, such as shown in Figure 3.6b.

3.4.5 Flexural stresses

Flexural stresses are known to contribute significantly to earthquakes in the continental shelf (e.g., Stein *et al.*, 1989; Assumpção, 1998; Assumpção *et al.*, 2011) as well as in the stable continental interior, such as in the Amazon basin (Zoback and Richardson, 1996).

Flexural stresses can reach high magnitudes, and even apparently small loads can contribute significantly to intraplate stresses as shown by Calais *et al.* (2010) by modelling the effect of Late-Pleistocene erosion in the New Madrid seismicity. In Brazil, the SW–NE-trending seismic zone in the Tocantins province (Figure 3.2) coincides with high gravity anomalies (Figure 3.7). Assumpção and Sacek (2013) showed that the uncompensated lithospheric load, which causes the high gravity anomalies, produces large compressional stresses in the upper crust (due to flexural bending of the lithosphere) and explained the seismicity distribution in Central Brazil. Here we test whether the correlation between gravity anomalies and seismicity can be extrapolated to all of the mid-plate area in South America.

We used the isostatic gravity anomaly map of South America (Sá, 2004), also used by Assumpção and Sacek (2013). Figure 3.7 shows the epicentres of the "whole" catalogue and the gravity anomalies. It can be seen that most epicentres tend to be located in areas of positive gravity anomalies, and large areas without seismicity (such as in the middle of the Paraná and Parnaíba basins, as well as in the Guaporé shield) tend to have negative gravity anomalies. The average isostatic anomaly in the continental area shown in Figure 3.7a is −13 mGal (Figure 3.7b). The distribution of gravity anomalies at the

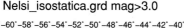

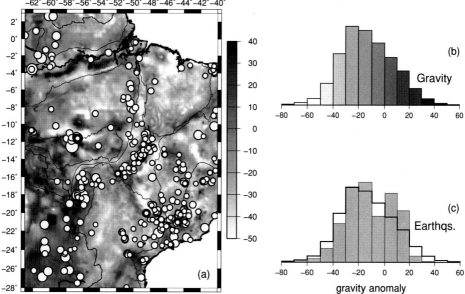

Figure 3.7 Isostatic gravity anomalies and epicentral distribution. (a) Map of epicentres of the "whole" catalogue, with magnitudes ≥ 3.0 and isostatic anomalies; scale in mGal. (b) Distribution of the gravity anomalies in the continental area shown in the map. (c) Grey bars are the gravity anomaly distribution at the epicentres. The thick solid line is the expected frequency, based on the gravity distribution, if the epicentres were uncorrelated with gravity anomalies. For colour version, see Plates section.

epicentres is shown by the grey histogram in Figure 3.7c. There are relatively more events in areas with high gravity anomalies (0 to 20 mGal) and fewer events in areas with low gravity (−50 to −30 mGal) compared to the expected numbers. This result shows that positive isostatic gravity anomalies are correlated with a higher seismicity rate, indicating that flexural stresses are an important factor in explaining intraplate seismicity in most of Brazil, and not only in the Tocantins province.

3.5 Discussion and conclusions

Mid-plate South America is less seismic than most other continental mid-plate regions such as Central and Eastern North America, Australia, India, and China. The largest known magnitude is 6.2 m_b, and the seismicity rate can be quantified by a return period of about 4 years for magnitude 5 and above. Despite this low rate of activity, it can be seen that earthquakes tend to occur in some seismic zones with large "aseismic" areas in between. No simple model can explain all the intraplate activity. Cratonic regions tend to have lower rates of seismicity, on average, but some magnitudes around 5 have been observed in old cratonic areas. In the cratons, seismicity tends to occur around deep lithospheric keels, or in craton edges, in agreement with similar findings on a global scale (Mooney *et al.*, 2012).

Areas near proposed neotectonic faults (less than 20 km) also seem to be more prone to seismic activity compared with areas farther away from these features. However, this correlation only holds on a continental scale: several neotectonic faults have no evidence of seismic activity (Figure 3.5) and no neotectonic features have been mapped yet in several seismic zones, such as the northern Parecis and the Pantanal basins. The compilation of neotectonic faults by Saadi *et al.* (2002) is a preliminary result and many more active features still remain to be mapped. Also, the sampling of neotectonic features compiled by Saadi *et al.* may have been influenced by known regions of seismicity, such as the Codajás fault (BR-05) in the Amazon basin, and the Manga fault (BR-47) in the middle of the São Francisco craton. However, despite the uneven sampling of possible neotectonic faults, the correlation with epicentral distribution is statistically highly significant. On the other hand, comparisons of the fault-plane solutions of several events close to these faults (especially in northeastern Brazil) reveal inconsistencies between the earthquake fault planes and the geologically mapped structures. We propose that the statistical correlation between epicentres and neotectonic features is due to some other mechanism, such as stress concentration near intersecting structures, as proposed by Gangopadhyay and Talwani (2003, 2007), or unfavourable orientation of the main regional fault with respect to the current intraplate stresses. At any rate, the relationship of neotectonic features and intraplate seismicity deserves further studies.

Seismic zones such as the eastern border of the Amazon craton and the northern Parecis basin (zones 2 and 4 in Figure 3.2c) could be attributed to stress concentration due both to lateral density variation in cratonic keels and to flexural stresses from intracrustal loads. Seismicity in northeastern Brazil (zone 3) and in the Pantanal basin (zone 5) could be related to stress concentration in the upper crust in areas of thin and weak lithosphere, such as proposed by Assumpção *et al.* (2004) for central and southeastern Brazil. The seismic zone in the middle of the Amazon basin (zone 1) could be influenced by flexural stresses from intracrustal loads along the Amazon rift (Figure 3.7), as proposed by Zoback and Richardson (1996), but probably other causes are also necessary as no seismicity is observed directly above the gravity high.

An interesting aspect of the seismicity distribution in southeast Brazil is the complete lack of epicentres along the Serra do Mar coastal ranges ("SM" in Figure 3.2b and Figure 3.4). A system of Cenozoic continental rifts, together with the high topography of the coastal ranges, have been subjected to several neotectonic studies (e.g., Riccomini and Assumpção, 1999; Modenesi-Gauttieri *et al.*, 2002; Cogné *et al.*, 2012), which have mapped many faults, some with Holocene reactivation (carbon-14 dating). The compilation of Saadi *et al.* (2002) includes several neotectonic faults along the southeast coastal ranges (Figure 3.5). The extensive evidence of Quaternary activity and lack of epicentres in the Brazilian catalogue led Assumpção and Riccomini (2011) to propose the possibility of long-term seismic migration in southeast Brazil.

Clearly, no simple model seems to explain the location of all seismic zones, and a combination of several factors may be necessary to explain the concentration of activity in some areas. Stress concentration mechanisms from lateral density variations (at the edge of cratonic keels, and from flexural deformation) seem to be very effective in mid-plate South America. Zones of weakness are not easily defined and have not been tested properly in this chapter. However, the higher probability of earthquakes occurring near (but not exactly on) neotectonic faults (weak zones?) may indicate that the model of stress concentrations near intersecting structures may also be applicable to Brazil.

While the statistical correlation of the present seismicity with geological and geophysical features may shed some light on the mechanics of intraplate seismogenesis, especially when similar results are found in other SCRs, one must be aware of the short time window of the Brazilian historical and instrumental catalogue. Further studies are necessary to see if long-term seismicity migration could help explain the lack of epicentres near geological features that are seismically active in other areas.

Acknowledgments

We thank all the colleagues and technicians who carried out extensive field work for several decades to study Brazilian earthquakes. Financial support from many sources such as CNPq (Brazilian National Science Council), MCT (Ministry of Science and Technology), FAPESP (São Paulo State Funding Agency), and many others is greatly appreciated. Most figures were drawn with GMT (Wessel and Smith, 1998); the statistical tests were performed using some GMT tools to select events and http://stattrek.com/online-calculator/binomial. aspx. We thank Dan Clark and Pradeep Talwani for detailed reviews and many suggestions for improvement. Work carried out under the UNESP/Petrobras project "Neotectonic Deformation of Brazil".

References

Almeida, F. F. M., Neves, B. B. B., and Carneiro, C. D. R. (2000). The origin and evolution of the South American platform. *Earth-Science Reviews*, 50, 77–111.

Assumpção, M. (1983). A regional magnitude scale for Brazil. *Bulletin of the Seismological Society of America*, 73, 237–246.

Assumpção, M. (1992). The regional intraplate stress field in South America. *Journal of Geophysical Research*, 97, 11,889–11,903.

Assumpção, M. (1998). Seismicity and stresses in the Brazilian passive margin. *Bulletin of the Seismological Society of America*, 78(1), 160–169.

Assumpção, M., and Araujo, M. (1993). Effect of the Altiplano-Puna plateau, South America, on the regional intraplate stress. *Tectonophysics*, 221, 475–496.

Assumpção, M., and Riccomini, C. (2011). Seismicity and neotectonics in the coastal ranges of SE Brazil (Serra do Mar): a case of activity migration? *Seismological Society of America Annual Meeting*, Memphis, USA, Abstract 11–148.

Assumpção, M., and Suárez, G. (1988). Source mechanisms of moderate size earthquakes and stress orientation in mid-plate South America. *Geophysics Journal*, 92, 253–267.

Assumpção, M., and Sacek, V. (2013). Intra-plate seismicity and flexural stresses in Central Brazil. *Geophysical Research Letters*, 40, 487–491, doi:10.1002/grl.50142, 2013.

Assumpção, M., Schimmel, M., Escalante, C., *et al.* (2004). Intraplate seismicity in SE Brazil: stress concentration in lithospheric thin spots. *Geophysical Journal International*, 159, 390–399.

Assumpção, M., Dourado, J. C., Ribotta, L. C., *et al.* (2011). The São Vicente earthquake of April 2008 and seismicity in the continental shelf off SE Brazil: further evidence for flexural stresses. *Geophysical Journal International*, doi: 10.1111/j.1365–246X.2011.05198.x.

Assumpção, M., Feng, M., Tassara, A., and Julià, J. (2013). Models of crustal thickness for South America from seismic refraction, receiver functions and surface wave dispersion. *Tectonophysics*, doi: 10.1016/j.tecto.2012.11.014.

Barros, L. V., Assumpção, M., Quinteros, R., and Caixeta, D. (2009). The intraplate Porto dos Gaúchos seismic zone in the Amazon craton – Brazil. *Tectonophysics*, 469, 37–47, doi: 10.1016/j.tecto.2009.01.006.

Barros, L. V., Chimpliganond, C. N., von Huelsen, M. G., *et al.* (2012). The Mara Rosa, Goiás state, Brazil, recent seismicity and its relationship with the Transbrasiliano Lineament. *Peruvian Geological Congress, Lima, extended abstracts*, September 23–26, 2012.

Berrocal, J., Assumpção, M., Antezana, R., *et al.* (1983). Seismic activity in Brazil in the period 1560–1980. *Earthquake Prediction Research*, 2, 191–208.

Berrocal, J., Assumpção, M., Antezana, R., *et al.* (1984). *Sismicidade do Brasil*. IAG/USP and Comissão Nacional de Energia Nuclear, Brazil.

Bezerra, F. H. R., and Vita-Finzi, C. (2000). How active is a passive margin? Paleoseismicity in northeastern Brazil. *Geology*, 28, 591–594.

Bezerra, F. H. R., Fonseca, V. P., Vita-Finzi, C., Lima Filho, F. P., and Saadi, (2005). Liquefaction-induced structures in Quaternary alluvial gravels and gravels sediments, NE Brazil. In *Paleoliquefaction and Appraisal of Seismic Hazards*, ed. S. F. Obermeier. *Engineering Geology*, 76, 191–208.

Bezerra, F. H. R., Ferreira, J. M., and Sousa, M. O. M. (2006). Review of the seismicity and Neogene tectonics in northeastern Brazil. *Revista de la Asociación Geológica Argentina*, 61, 525–535.

Bezerra, F. H. R., Takeya, M., Sousa, M. O. M., and do Nascimento, A. F. (2007). Coseismic reactivation of the Samambaia fault, Brazil, *Tectonophysics*, 430, 27–39.

Bezerra, F. H. R., do Nascimento, A. F., Ferreira, J. M., *et al.* (2011). Review of active faults in the Borborema Province, intraplate South America, integration of seismological and paleoseismological data. *Tectonophysics*, 510, 269–290.

Branner, J. C. (1912). Earthquakes in Brazil. *Bulletin of the Seismological Society of America*, 2(2), 105–117.

Byerlee, J. (1990). Friction, overpressure and fault normal compression. *Geophysical Research Letters*, 17, 2109–2112.

Calais, E., Freed, A. M., Van Arsdale, R., and Stein, S. (2010). Triggering of New Madrid seismicity by late-Pleistocene erosion. *Nature*, 466, 608–611.

Campbell, D. L. (1978). Investigation of the stress-concentration mechanism for intraplate earthquakes. *Geophysical Research Letters*, 5, 477– 479.

Chang, H. K., Kowsman, R., Figueiredo, A. M. F., and Bender A. A. (1992). Tectonics and stratigraphy of the east Brazil rift system: an overview. *Tectonophysics*, 213, 97–138.

Chimpliganond, C., Assumpção, M., von Huelsen, M., and França, G. S. (2010). The intracratonic Caraíbas-Itacarambi earthquake of December 09, 2007 (4.9 mb), Minas Gerais State, Brazil. *Tectonophysics*, 480, 48–56.

Cloetingh, S. A. P. L., Wortel, M. J. R., and Vlaar, N. J. (1984). Passive margin evolution, initiation of subduction and the Wilson cycle. *Tectonophysics*, 109, 147–163.

Coblentz, D. D., and Richardson, R. M. (1996). Analysis of the South American intraplate stress field. *Journal of Geophysical Research*, 101(B4), 8643–8657.

Cogné, N., Cobbold, P. R., Riccomini, C., and Gallagher, K. (2012). Tectonic setting of the Taubaté Basin (southeastern Brazil): insights from regional seismic profiles and outcrop data. *Journal of South American Earth Science*, 42, 194–204.

Ferreira, J. M., Oliveira, R. T., Takeya, M. K., and Assumpção, M. (1998). Superposition of local and regional stresses in NE Brazil: evidence from focal mechanisms around the Potiguar marginal basin. *Geophysical Journal International* 134, 341–355.

Ferreira, J. M., Bezerra, F. H. R., Sousa, M. O. L., *et al.* (2008). The role of Precambrian mylonitic belts and present-day stress field in the coseismic reactivation of the Pernambuco lineament, Brazil. *Tectonophysics*, 456, 111–126.

Gangopadhyay, A., and Talwani, P. (2003). Symptomatic features of intraplate earthquakes. *Seismological Research Letters*, 74, 863–883.

Gangopadhyay, A., and Talwani, P. (2007). Two-dimensional numerical modeling suggests that there is a preferred geometry of intersecting faults that favors intraplate earthquakes. In *Continental Intraplate Earthquakes: Science, Hazard, and Policy Issues*, ed. S. Stein, and S. Mazzotti. Geological Society of America Special Paper 425, pp. 87–99.

Gomes, B. S. (1992). Preliminary integration of marine gravimetric data of Petrobras and Leplac project: Campos, Santos and Pelotas basins. In *Proceedings of the 37th Brazilian Geological Congress*, São Paulo, SP, Brazil, 1, 559–560 (in Portuguese).

Holdsworth, R. E., Hand, M., Miller, J. A., and Buick, I. S. (2001). Continental reactivation and reworking: an introduction. In *Continental Reactivation and Reworking*, ed. J. A. Miller, R. E. Holdsworth, I. S. Buick, and M. Hand. Geological Society, London, Special Publication, 184, pp. 1–12.

Houtz, R. E., Ludwig, W. J., Milliman, J. D., and Grow, J. A. (1977). Structure of the northern Brazilian continental margin. *Geological Society of America Bulletin*, 88, 711–719.

Johnston, A. C. (1996). Seismic moment assessment of earthquakes in stable continental regions: I. Instrumental seismicity. *Geophysical Journal International*, 124, 381–414.

Johnston, A. C., and Kanter, L. R. (1990). Earthquakes in stable continental crust. *Scientific American*, 262, 68–75.

Johnston, A. C., Coppersmith, K. J., Kanter, L. R., and Cornell C. A. (1994). *The Earthquakes of Stable Continental Regions: Assessment of Large Earthquake Potential*, TR-102261, Vol. 1–5, ed. J. F. Schneider. Palo Alto, CA: Electric Power Research Institute.

Kirkpatrick, J. D., Bezerra, F. H. R., Shipton, Z. K., *et al.* (2013). Scale-dependent influence of pre-existing basement shear zones on rift faulting: a case study from NE Brazil. *Journal of the Geological Society (London)*, 170, 237–247.

Lenardic, A., Moresi, L., and Mühlhaus, H. (2000). The role of mobile belts for the longevity of deep cratonic lithosphere: the crumple zone model. *Geophysical Research Letters*, 27, 1235–1238.

Li, Q. S., Liu, M. A., and Stein, S. (2009). Spatiotemporal complexity of continental intraplate seismicity: insights from geodynamic modeling and implications for seismic hazard estimation. *Bulletin of the Seismological Society of America*, 99(1), 52–60, doi:10.1785/0120080005.

Lima, C., Nascimento, E., and Assumpção, M. (1997). Stress orientations in Brazilian sedimentary basins from breakout analysis: implications for force models in the South American plate. *Geophysical Journal International*, 130(1), 112–124.

Lima-Neto, H. C., Ferreira, J. M., Bezerra, F. H. R., *et al.* (2013). Upper crustal earthquake swarms in São Caetano: reactivation of the Pernambuco shear zone and trending branches in intraplate Brazil. *Tectonophysics*, 608, 804–811. Doi:10.1016/j.tecto2013.08001.

Lopes, A. V., Assumpção, M., do Nascimento, A. F., *et al.* (2010). Intraplate earthquake swarm in Belo Jardim, NE Brazil: reactivation of a major neoproterozoic shear zone (Pernambuco Lineament). *Geophysical Journal International*, 180, 1302–1312, doi: 10.1111/j.1365–246X.2009.04485.x.

Mazzotti, S. (2007). Geodynamic models for earthquake studies in intraplate North America. In *Continental Intraplate Earthquakes: Science, Hazard, and Policy Issues*, ed. S. Stein, and S. Mazzotti. Geological Society of America Special Paper 425, pp. 17–33.

Mazzotti, S., James, T. S., Henton, J., *et al.* (2005). GPS crustal strain, postglacial rebound, and seismic hazard in eastern North America: the Saint Lawrence valley example. *Journal of Geophysical Research*, 110, B11301, doi:10.1029/2004JB003590.

Modenesi-Gauttieri, M. C., Hiruma, S. T., and Riccomini, C. (2002). Morphotectonics of a high plateau on the northwestern flank of the Continental Rift of southeastern Brazil. *Geomorphology*, 43, 257–271.

Mohriak, W. U., Mello, M. R., Bassetto, M., Vieira, I. S., and Koutsoukos, E. A. M. (2000). Crustal architecture, sedimentation, and petroleum systems in the Sergipe–Alagoas Basin, northeastern Brazil. In *Petroleum Systems of South Atlantic Margins*, ed. M. R. Mello and B. J. Katz. AAPG Memoir 73, pp. 273–300.

Mooney, W. D., Ritsema, J., and Hwang, Y. (2012). Crustal seismicity and maximum earthquake magnitudes (Mmax) in stable continental regions (SCRs): correlation with the seismic velocity of the lithosphere. *Earth and Planetary Science Letters*, 357–358, 78–83, doi:10.1016/j.epsl.2012.08.032.

Morrow, C., Radney, B., and Byerlee, J. D. (1992). Frictional strength and the effective pressure law of montmorillonite and illite clays: fault mechanics and transport properties of rocks. In *Fault Mechanics and Transport Properties of Rocks,* ed. B. Evans and T.-F. Wong. San Diego, California: Academic Press, pp. 69–88.

Moura, A. C., Oliveira, P., Ferreira, J. M., *et al.* (2013). Seismogenic faulting in the Meruoca granite, Borborema Province, Brazil. *Annals of the Brazilian Academy of Sciences*, submitted.

Nuttli, O. W., Bollinger, G. A., and Griffiths, D. W. (1979). On the relation between Modified Mercalli intensity and body-wave magnitude. *Bulletin of the Seismological Society of America*, 69(3), 893–909.

Pavlides, S. B. (1989). Looking for a definition of neotectonics. *Terra Nova*, 1, 233–235, doi:10.1111/j.1365–3121.1989.tb00362.x.

Riccomini, C., and M. Assumpção (1999). Quaternary tectonics in Brazil. *Episodes*, 22(3), 221–225.

Rossetti, D. F., Bezerra, F. H. R., Góes, A. M., *et al.* (2011). Late Quaternary sedimentation in the Paraíba Basin, Northeastern Brazil: landform, sea level and tectonics in Eastern South America passive margin. *Palaeogeography, Palaeoclimatology, Palaeoecology*, 191–204.

Sá, N. C. (2004). O campo de gravidade, o geóide e a estrutura crustal na América do Sul: novas estratégias de representação. Tese de Livre-Docência, Universidade de São Paulo.

Saadi, A., Machette, M. N., Haller, K. M., *et al.* (2002). *Map and Database of Quaternary Faults and Lineaments in Brazil*. USGS Open-File Report 02–230.

Sandiford, M., and Egholm, D. L. (2008). Enhanced intraplate seismicity along continental margins: Some causes and consequences. *Tectonophysics*, 457, 197–208.

Schulte, S. M., and Mooney, W. D. (2005). An updated global earthquake catalogue for stable continental regions: reassessing the correlation with ancient rifts. *Geophysics Journal International*, 161, 707–721.

Sibson, R. H. (1990). Rupture nucleation of unfavorably oriented faults. *Bulletin of the Seismological Society of America*, 80, 1580–1604.

Stein, S., Cloetingh, S., Sleep, N. H., and Wortel, R. (1989). Passive margin earthquakes, stresses and rheology. In *Earthquakes at North-Atlantic Passive Margins: Neotectonics and Postglacial Rebound*, ed. S. Gregersen and P. W. Basham. Boston: Kluwer Academic, 231–259.

Stein, S., Liu, M., Calais, E., and Qingsong, L. (2009). Mid-continent earthquakes as a complex system. *Seismological Research Letters*, 80, 551–553.

Stevenson, D., Gangopadhyay, A., and Talwani, P. (2006). Booming plutons: source of microearthquakes in South Carolina. *Geophysics Research Letters*, 33, L03316, doi:10.1029/2005GL024679.

Sykes, L. (1978). Intraplate seismicity, reactivation of pre-existing zones of weakness, alkaline magmatism, and other tectonism postdating continental fragmentation. *Reviews of Geophysics and Space Physics*, 16, 621–688.

Takeya, M., Ferreira, J. M., Pearce, R. G., *et al.* (1989). The 1986–1987 intraplate earthquake sequence near João Câmara, northeast Brazil: evolution of seismicity. *Tectonophysics*, 167, 117–131.

Talwani, P. (1999). Fault geometry and earthquakes in continental interiors. *Tectonophysics* 305, 371–379.

Veloso, J. A. V. (2012). *The Earthquake that Shook Brazil*. Brasilia, Brazil: Thesaurus Press (in Portuguese).

Veloso, J. A. V. (2013). On the footprints of a major Brazilian Amazon earthquake. *Annals of the Brazilian Academy of Sciences,* in press.

Watts, A. B., Rodger, M., Peirce, C., Greenroyd, C. J., and Hobbs, R. W. (2009). Seismic structure, gravity anomalies, and flexure of the Amazon continental margin, NE Brazil. *Journal of Geophysical Research*, 114, B07103, doi:10.1029/2008JB006259.

Weichert, D. H. (1980). Estimation of the earthquake recurrence parameters for unequal observation periods for different magnitudes. *Bulletin of the Seismological Society of America*, 70(4), 1337–1346.

Wessel, P., and Smith, W. H. F. (1998). New, improved version of Generic Mapping Tools released. *Eos, Transactions, American Geophysical Union*, 79(47), 579.

Zalan, P. V., Severino, M. C. G., Rigoti, C. A., *et al.* (2011). An entirely new 3D-view of the crustal and mantle structure of a South Atlantic passive margin: Santos, Campos and Espírito Santo Basins, Brazil. *AAPG Annual Convention*, Houston Texas, 2011, extended abstracts, Search and Discovery article 30177.

Zoback, M. L., and Richardson, R. M. (1996). Stress perturbation associated with the Amazonas and other ancient continental rifts. *Journal of Geophysical Research*, B101(3), 5459–5475.

4

Earthquakes and geological structures of the St. Lawrence Rift System

MAURICE LAMONTAGNE AND GIORGIO RANALLI

Abstract

The St. Lawrence Rift System (SLRS), which includes the Ottawa–Bonnechère and Saguenay grabens, is located well inside the North American plate. Most historic and the some 350 earthquakes recorded yearly occur in three main seismically active zones, namely Charlevoix (CSZ), Western Quebec (WQSZ), and Lower St. Lawrence (LSLSZ). Outside these areas, most of the Canadian Shield and bordering regions have had a very low level of earthquake activity. In the SLRS, moderate to large earthquakes (moment magnitude, **M**, 5.5 to 7) are known to have occurred since 1663, causing landslides and damage mostly to unreinforced masonry elements of buildings located on ground capable of amplifying ground motions. Most earthquakes in these seismic zones share common characteristics such as mid- to upper-crustal focal depths, no known surface ruptures, and proximity to SLRS faults. Variations also exist such as vast seismically active regions (WQSZ and LSLSZ), the presence of a large water body (CSZ and LSLSZ), and absence of SLRS faults near concentrations of earthquakes (WQSZ). The CSZ is the best studied seismic zone; there, earthquakes occur in the Canadian Shield, mostly in a 30 × 85 km rectangle elongated along the trend of the St. Lawrence River with local variations in focal depth distribution. Faults related to the SLRS and to a meteor impact structure exist, and earthquakes occur along the SLRS faults as well as in between these faults. Overall, the SLRS faults are probably reactivated by the larger earthquakes (**M** ≥ 4.5) of the twentieth century (CSZ in 1925; WQCSZ in 1935 and 1944; Saguenay in 1988) for which we have focal mechanisms. We propose that caution be exercised when linking historical events that have uncertain epicentres with SLRS faults. Similarly, SLRS faults should not necessarily be considered to be the reactivated structures for most small to moderate

---

Intraplate Earthquakes, ed. Pradeep Talwani. Published by Cambridge University Press. © Cambridge University Press 2014.

earthquakes (**M** < 4.5). A good example of this is the earthquakes of the WQSZ, which tend to concentrate in a well-defined NW–SE alignment with no obvious geological control, except perhaps a hypothetical hotspot track. Two local factors can lead to the occurrence of SLRS earthquakes: weak faults or enhanced stress levels. We propose that local conditions, concentrated in a few seismic zones, can alter these factors and lead to the occurrence of earthquakes, especially those with **M** < 4.5. At a continent-wide scale, the correlation between the SLRS and earthquakes is appealing. We suggest, however, that pre-existing faults related to the SLRS do not explain all features of the seismicity. Seismicity is concentrated in more active areas, some with conspicuous normal faults and some with suspected weakening mechanisms, such as intense pre-fracturing (e.g., due to a meteorite impact), the passage over a hotspot, or the presence of intrusions and lateral crustal density variations.

4.1 Introduction

The St. Lawrence Rift System (SLRS), which includes the Ottawa–Bonnechère and Saguenay grabens, is located well inside the North American plate (Figure 4.1). Its earthquake activity was first described in written accounts from the early 1600s and recorded by seismograph stations since the late 1800s. Table 4.1 provides an overview of the most important earthquakes of the SLRS together with a summary of the impact on the natural and man-made environments. Some prehistoric events have also been recognized. This knowledge defines seismic zones with sustained seismic activity (Charlevoix (CSZ), Western Quebec (WQSZ), Lower St. Lawrence (LSLSZ)) and others with a lower ("background") seismic activity. Outside these areas, most of the Canadian Shield and bordering regions have had a very low level of earthquake activity for hundreds of years. Since major population centres are located in the SLRS, including the metropolitan areas of Ottawa, Montreal, and Quebec City (Figure 4.1), its seismic hazard is of great interest for the protection of the public and infrastructures critical to the Canadian economy.

Through time, there have been multiple hypotheses to explain the earthquake activity. In the 1960s, the prevalent view was that the numerous epicentres lined up with Logan's Line, which is the main boundary between the Appalachian thrust belt to the southeast and the Precambrian Shield and overlying Ordovician sedimentary rocks to the northwest (Figure 4.1; Milne and Davenport, 1969). Following the work of Leblanc *et al.* (1973, 1977) it became obvious that earthquakes occurred on faults within the Precambrian Shield. The largest of these faults were normal faults created during the opening of the Iapetus Ocean (Kumarapeli, 1985). From the concentrations of hypocentres that were dipping similarly to the normal faults in the Charlevoix Seismic Zone, this hypothesis has become one of the two seismic zoning models in Eastern Canada (Adams *et al.*, 1995). In the seismicity-rift model, it was assumed that the mild activity outside the recognized active zones was due to

Table 4.1 *Damaging earthquakes of the St. Lawrence Rift System*

Event number	Date	Time (UT)	Region	Lat.	Long.	Mag.	Landslide	Description
1	1663–02–05	22:30	Charlevoix-Kamouraska, Quebec	47.6	−70.1	7	Yes	Epicentre most likely in the Charlevoix Seismic Zone, Quebec. Some damage to masonry in Quebec City, Trois-Rivières and Montréal. Landslides reported in the Charlevoix region and along the St. Lawrence Valley. Numerous aftershocks felt in Quebec City during the following months.
2	1732–09–16	16:00	Near Montreal, Quebec	45.5	−73.6	5.8	No	Probable epicentre near Montréal. Considerable damage in the city of Montréal where hundreds of chimneys were damaged and walls cracked.
3	1791–12–06	20:00	Charlevoix-Kamouraska, Quebec	47.4	−70.5	6	No	Damage to houses and churches in Charlevoix.
4	1860–10–17	11:15	Charlevoix-Kamouraska, Quebec	47.5	−70.1	6	No	Damage in the epicentral region on both shores of the St. Lawrence River.
5	1870–10–20	16:30	Charlevoix-Kamouraska, Quebec	47.4	−70.5	6½	Yes	Considerable damage to houses in Charlevoix along the South Shore of the St. Lawrence River. Damage to chimneys reported in lower town in Quebec City. Possible rock slide along the Saguenay River.

#	Date	Time	Location	Lat.	Long.	Mag.	Damage	Description
6	1925–03–01	02:19	Charlevoix-Kamouraska, Quebec	47.8	−69.8	6.2	Yes	The earthquake caused damage to unreinforced masonry (chimneys, walls) in the epicentral region on both shores of the St. Lawrence, and in Quebec City (including damage to port facilities), and in the Trois-Rivières region. Possible liquefaction in Charlevoix. Numerous felt aftershocks followed.
7	1935–11–01	06:03	Region of Témiscaming, Quebec	46.78	−79.07	6.1	Yes	The earthquake occurred approximately 10 km east of Témiscaming, Qc. Damaged chimneys were reported.
8	1944–09–05	04:38	Cornwall, Ontario-Massena, New York	44.96	−74.77	5.6	No	Considerable damage to unreinforced masonry in both Cornwall, Ontario and Massena, New York. About 2000 chimneys were damaged in the area.
9	1988–11–25	23:46	Saguenay Region, Quebec	48.12	−71.18	5.9	Yes	Laurentides Fauna Reserve, south of Saguenay (Chicoutimi), Quebec. Damage caused to unreinforced masonry along the Saguenay River, in Charlevoix, and along the St. Lawrence river up to Montreal. Liquefaction of soft soils in the epicentral region. Eleven cases of mass movements reported.
10	1990–10–19	07:01	Mont-Laurier, Quebec	46.47	−75.59	4.5	No	Some minor damage near the epicentre (cracked chimneys, water pipes broken).
11	1997–11–6	02:34	Cap-Rouge, QC	46.80	−71.42	4.9	No	Quebec City region. Minor damage reported in a school of the region.
12	2000–01–01	11:22	Kipawa, QC	46.84	−78.92	4.7	No	Some reports of minor damage in the epicentral region.
13	2010–06–23	17:41	Val-des-Bois, QC	45.88	−75.48	5.0	Yes	Some damage and landslides occurred in the epicentral region.

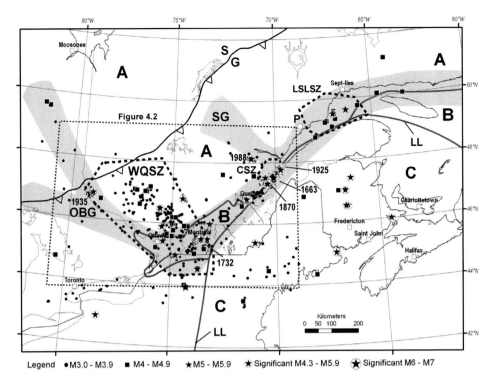

Legend   ●M3.0 - M3.9   ■ M4 - M4.9   ★M5 - M5.9   ★ Significant M4.3 - M5.9   ⬚★ Significant M6 - M7

Figure 4.1 Seismicity of the St. Lawrence Rift System together with some geological features. The shaded area is the SLRS as we define it for discussion purposes. Earthquakes of magnitude larger than 3.0 recorded between 1980 and 2012 inclusively together with significant earthquakes of Eastern Canada are shown (Lamontagne *et al.*, 2008a). Seismic zones are: WQSZ, Western Quebec Seismic Zone; CSZ, Charlevoix Seismic Zone; and LSLSZ, Lower St. Lawrence Seismic Zone, as defined by Basham *et al.* (1982). Other acronyms are: OBG, Ottawa–Bonnechère Graben; SG, Saguenay Graben; A, Canadian Shield; B, St. Lawrence Platform; C, Appalachian Orogen; LL, Logan's Line; S, Superior Geological Province; G, Grenville Geological Province; P, Pessamit. The boundaries of Figure 4.2 are shown.

the short time window of observation (i.e., since the 1600s). It was also assumed that these semi-quiescent areas could become active if longer periods, such as tens of thousands of years, were considered. The other source model assumed that the seismic zones active in the past will be the likely sites of future sizeable earthquakes.

In this chapter, we describe the current knowledge of the seismically active areas of the SLRS (Figure 4.1) and examine possible contributing factors to the earthquake activity. We are going to use the expression "St. Lawrence Rift System (SLRS)" not necessarily to express a possible relationship between the earthquakes and the rift faults. We use SLRS as a convenient way to refer to the vast region along the St. Lawrence River, including the Ottawa and Saguenay river valleys.

4.2 Historical earthquakes and their impact

Although well inside the North American plate and remote from plate boundaries, the SLRS has been subject to damaging earthquakes larger than magnitude[1] 5 (Lamontagne *et al.*, 2008a; Cassidy *et al.*, 2010; Table 4.1). For the pre-twentieth-century events, magnitudes and locations are approximate due to the variable reliability of the sources of information: felt reports, detection completeness, and geographic distribution of the population. An example of this uncertainty is the debate over the magnitude of the 1663 Charlevoix earthquake (generally estimated at M 7; see Bent [2009] for a review). More recently, Ebel (2011) rates the same earthquake as moment magnitude (**M**) 7.5 ± 0.45 based on felt reports and reported falling of masonry. This is not just an academic debate, as it may appear: the preferred magnitude impacts seismic hazard estimates. The locations of some historical events are also uncertain. An example of this is the 1732 magnitude 5.8 Montreal earthquake, located either near Montreal (Leblanc, 1981) or in northern New York State (Figure 4.2; Gouin, 2001). These uncertainties in magnitudes and locations of historical earthquakes lead to discrepancies in the various earthquake catalogues (with impact on seismotectonic interpretation and seismic hazards). An example is the correlation of the 1732 earthquake with the rift faults around Montreal (Adams and Basham, 1989). Fortunately, most instrumentally recorded twentieth-century earthquakes have better defined magnitudes and epicentres (Bent, 2009).

Historically, only a few St. Lawrence Valley earthquakes had a geological impact such as rock falls, landslides, slumps, earth flows in clay deposits, ground cracking, lateral spreading, and liquefaction. No surface rupture has ever been reported. The St. Lawrence Valley (Figure 4.2) has thick sequences of post-glacial marine clay deposits and these have proven to be highly susceptible to earth flow under ground shaking. Of all SLRS earthquakes, the M 7 1663 earthquake had the largest geotechnical impact, causing landslides in the epicentral region (Filion *et al.*, 1991), and along the Saguenay and Saint-Maurice rivers, more than 200 km away (Legget and LaSalle, 1978; Desjardins, 1980). The 1663 event may have also produced a basin collapse in the Saguenay Fjord (Syvitski and Schafer, 1996). Some submarine landslides of the Saguenay and St. Lawrence rivers are also associated with some of these large Charlevoix earthquakes. An example is a landslide near Betsiamites on the Quebec North Shore, possibly caused by the 1663 Charlevoix earthquake, which first occurred on land and continued offshore beneath the St. Lawrence River (Cauchon-Voyer *et al.*, 2007). Another notable earthquake is the 1870 M 6½ CSZ earthquake (Figure 4.2), which had some dramatic consequences, including the collapse of buildings, liquefaction, and landslides in the epicentral region, and rock falls along the Saguenay River (Lamontagne *et al.*, 2007). A landslide, most probably linked with the earthquake, killed four people five days after the mainshock (Lamontagne *et al.*, 2007). More recently, the 1988 **M** 5.9 Saguenay earthquake (Figure 4.2) exemplifies the potential impact of a moderate eastern Canadian

---

[1] M is defined as the most widely accepted magnitude value (generally the moment magnitude or the felt area magnitude for historical earthquakes) for a given event.

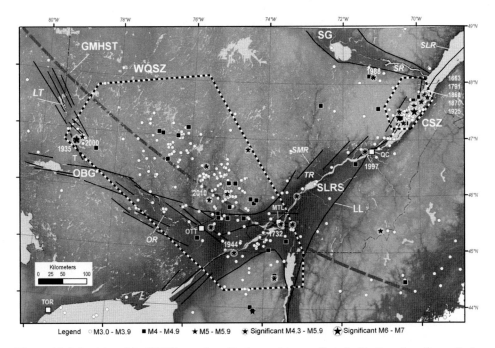

Figure 4.2 Seismicity of the SLRS in southern Quebec and eastern Ontario. Earthquakes of magnitude larger than 3.0 recorded between 1980 and 2012 inclusively together with significant earthquakes of Eastern Canada are shown (listed in Table 4.1; Lamontagne *et al.*, 2008a). Seismic zones are: WQSZ, Western Quebec Seismic Zone; CSZ, Charlevoix Seismic Zone, as defined by Basham *et al.* (1982). Other acronyms are: GMHST, Great Meteor HotSpot Track (assumed); OBG, Ottawa–Bonnechère Graben; SG, Saguenay Graben; LL, Logan's Line. Place names are: TOR, Toronto; OTT, Ottawa; MTL, Montréal; QC, Quebec City; TR, Trois-Rivières; T, Temiscaming. Rivers and lakes mentioned in the text are: LT, Lake Temiscaming; OR, Ottawa River; SMR, Saint-Maurice River; SR, Saguenay River; SLR, St. Lawrence River. For colour version, see Plates section.

earthquake. Its effects included liquefaction and rock falls in the epicentral region and landslides as far as 200 km from the epicentre. At 30 km from the epicentre, for example, liquefaction caused extensive damage to local houses (Lefebvre *et al.*, 1991; Boivin, 1992). There, evidence of older liquefaction and ground failure was found, possibly caused by older regional earthquakes (Tuttle *et al.*, 1990). Other damage was seen as far as 170 km away from the epicentre where failures of railroad embankments were reported (Mitchell *et al.*, 1990). Natural slope failures were also seen along the Saint-Maurice River (Lefebvre *et al.*, 1992). For building design, this earthquake was the first **M** 6 eastern Canadian earthquake recorded by strong ground motion instruments and showed the high-frequency content of earthquakes there (Munro and North, 1989). In general, infrastructure in the SLRS is especially at risk where marine clays are found, because even moderate earthquakes can trigger earth flows. In 2010, for example, the **M** 5.0 Val-des-Bois earthquake caused two earth flows within a few kilometres of the epicentre (Perret *et al.*, 2011).

Geological effects can provide evidence of prehistoric earthquakes. The area around Charlevoix is the site of numerous prehistoric earthquakes that caused submarine slumps and liquefaction (Doig, 1986, 1998; Tuttle *et al.*, 1990; Filion *et al.*, 1991; Ouellet, 1997; Tuttle and Atkinson, 2010; Locat, 2011). From their data, Tuttle and Atkinson (2010) could not document precise return periods. In the Ottawa River valley, landslide scars are interpreted as traces left by **M** > 6 prehistoric earthquakes (Aylsworth *et al.*, 2000; Brooks, 2013). Between Quebec City and Trois-Rivières, where current seismicity is of very low level, no such prehistoric earthquake evidence was found (Tuttle and Atkinson, 2010). In Charlevoix, in the region covered by seismic surveys, no evidence for a coseismic rupture beneath the surface of the St. Lawrence River could be found (Lamontagne, 2002).

Earthquakes of the St. Lawrence valley also damaged man-made structures. In past earthquakes, damage was more common in buildings with unreinforced masonry elements (URM; i.e., masonry without steel reinforcement) that rest on thick clay deposits (Bruneau and Lamontagne, 1994). For the 1988 Saguenay earthquake, 95% of all damage was associated with soft soils (53% with clay, 24% on multi-layer, and 18% on sand). It was also found that damage to buildings built on sandy foundations was restricted to 150 km epicentral distance, whereas for clay foundations damage existed up to 350 km distance (Paultre *et al.*, 1993). It is now recognized that urban and industrial developments in the SLRS have taken place on soils capable of amplifying earthquake ground motions. The presence of non-upgraded old buildings coupled with the population growth, particularly on sensitive soils, make earthquakes a significant natural hazard along the SLRS. Fortunately, despite the damage to buildings and earth movements, only two earthquake-caused deaths are known in Canada (Lamontagne, 2008).

In conclusion, the seismic risk in the SLRS is far from being negligible considering the seismic hazard, the low attenuation of Lg waves, the presence of unconsolidated deposits capable of amplifying ground motions and subject to mass movements, and a number of aging buildings built prior to modern building codes.

4.3 Seismic zones of the SLRS

Each year, approximately 350 earthquakes are recorded in the SLRS by the Canadian National Seismograph Network (CNSN). In the SLRS, the CNSN has had a detection completeness of about **M** 1.5 since around 1980. On average, of these 350 earthquakes, about one or two exceed **M** 3.5, ten will exceed **M** 3.0, and about twenty will be reported felt (**M** $\geq$ 2.0). A decade will, on average, include two events greater than **M** 4.5, which is the threshold of minor damage. On a yearly basis, most earthquakes occur in three main seismic zones, namely the Charlevoix, Western Quebec, and Lower St. Lawrence (Figure 4.1 and Figure 4.2).

4.3.1 Charlevoix

Based on historical and current earthquake rates, the Charlevoix Seismic Zone (CSZ) is the zone with the highest seismic hazard in continental Eastern Canada (Figure 4.3).

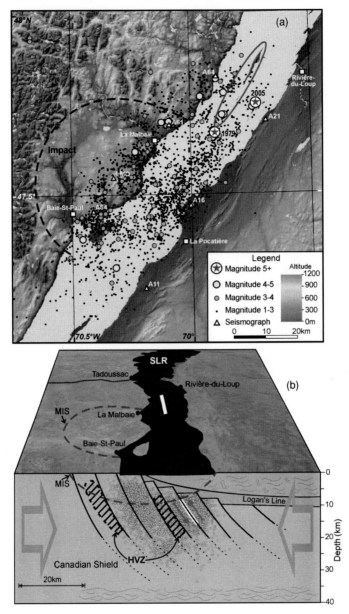

Figure 4.3 (a) Earthquakes of magnitude ≥1.0 recorded between 1978 and 2012 in the CSZ. The estimated position of the 1925 earthquake's epicentre and assumed fault plane orientation is shown as a red ellipse. The semi-circle of the Charlevoix meteorite impact crater is clearly visible on this image, where topography is shown using colours that range from blue for the lowest elevations to red for the highest ones. Since 1978, earthquakes have been monitored by a network of seven seismographs (white triangles). (b) Idealized cross-section perpendicular to the St. Lawrence River in the CSZ. The earthquake activity occurs in response to ridge push stresses (arrows) and is constrained to the Canadian Shield rocks, beneath the St. Lawrence River (SLR), St. Lawrence platform rocks (in grey), Logan's Line, and the Appalachian rocks. The 1925 earthquake ruptured along one of the large SLRS faults at focal depth of about 10 km at the extremity of the CSZ. A small proportion of earthquakes occur within the Charlevoix Meteor Impact Structure (MIS). Earthquakes tend to concentrate outside the zones of high velocities (Vlahovic *et al.*, 2003). For colour version, see Plates section.

Since the arrival of the first Europeans in the early 1600s, the CSZ has been subject to five earthquakes of magnitude 6 or larger: in 1663 (M 7), 1791 (M 6), 1860 (M 6), 1870 (M 6½), and 1925 (magnitude M_S 6.2 ± 0.3) (Lamontagne *et al.*, 2008a). The earthquake potential of the area led the Government of Canada to conduct two field surveys that defined its main seismotectonic characteristics (Leblanc *et al.*, 1973; Leblanc and Buchbinder, 1977). Earthquakes occur between the surface and 30 km depth, in the Precambrian Shield, which outcrops on the north shore of the St. Lawrence River or is found beneath Logan's Line and the Appalachian rocks (Figure 4.3). Hypocentres cluster along or between the mapped Iapetan faults (also called St. Lawrence paleo-rift faults). The largest earthquake of the twentieth century was the 1925 earthquake, and its focal mechanism has one nodal plane consistent with a reactivation of a southeast-dipping paleo-rift fault (Bent, 1992). The installation of a permanent seismograph network in 1978 (Figure 4.3) has helped to define additional characteristics of the area. The St.-Laurent fault, one of the major rift faults of the CSZ, was formed in the late Precambrian but was also active after the Devonian meteor impact (Rondot, 1979), probably during the early stages of the opening of the Atlantic Ocean in late Triassic–Jurassic times (Lemieux *et al.*, 2003). This fault is not particularly active but appears to bound concentrations of hypocentres (Lamontagne, 1999).

Due to its concentration of earthquakes, the CSZ has been the focus of various geophysical studies (Buchbinder *et al.*, 1988). Investigations of velocity structure include a seismic reflection–refraction survey (Lyons *et al.*, 1980), microearthquake surveys (Leblanc *et al.*, 1973; Leblanc and Buchbinder, 1977; Lamontagne and Ranalli, 1997), analysis of teleseismic events (Hearty *et al.*, 1977), receiver function analysis (Cassidy, 1995), shear wave splitting and anisotropy studies (Buchbinder, 1989), and focal mechanisms of microearthquakes (Lamontagne, 1998; Bent *et al.*, 2003). The concentration of earthquakes led to earthquake prediction studies in the late 1970s and early 1980s (Buchbinder *et al.*, 1988). Roughly 80% of Charlevoix earthquakes occur in the depth range 5–15 km in Grenvillian basement rocks, with some as deep as 30 km (Figure 4.3b). Comparing this depth distribution to rheological models of the region, Lamontagne and Ranalli (1996) attribute earthquakes to faulting above the brittle–ductile transition to depths of at least 25 km. The reactivation of pre-existing faults could be due to high pore-fluid pressure at temperatures below the onset of ductility for hydrated feldspar at about 350 °C and/or a low coefficient of friction, possibly related to unhealed zones of intense fracturing. The distribution of spatially clustered earthquakes within the Charlevoix seismic zone indicates that very few earthquakes have occurred on the same fractures with similar focal mechanisms, implying that these fault zones occur in highly fractured rocks, especially those within the boundaries of the Devonian impact structure (Lamontagne and Ranalli, 1997; Figure 4.3). The hypocentre-velocity simultaneous inversion of local P and S waves produced a velocity model that revealed areas of high-velocity bodies at mid-crustal depths (Vlahovic *et al.*, 2003; Figure 4.3b). These areas were interpreted to be stronger, more competent crust that separates CSZ earthquakes into two main bands elongated along the St. Lawrence River. Mazzotti and Townend (2010) noted that this seismic zone contains evidence for a local rotation of S_{Hmax} (direction of maximum horizontal stress axis) from the regionally NE–SW-oriented

S_{Hmax}. This reorientation and the maximum plausible stress difference required to cause fault reactivation can be argued to support the presence of high pore-fluid pressures along faults (Lamontagne and Ranalli, 1996). On the other hand, Hurd and Zoback (2012) believe that high pore-fluid pressures are not necessary to explain the majority of preferred focal mechanisms in eastern North America.

There are indications that the region itself may have sub-areas with different rheological properties due to the presence of the meteor crater and its faults. Larger events concentrate at both ends of the seismic zone outside the meteor impact (Figure 4.3; Stevens, 1980; Lamontagne, 1999). Recently, geomechanical modelling showed that the weakening of the rift faults produces a stress increase in the region of the crater bounded by faults, leading to low-magnitude events within the crater and large events outside it (Baird *et al.*, 2009; 2010). If this hypothesis is true, the local seismicity would be caused by local conditions rather than by more regional conditions.

4.3.2 Lower St. Lawrence

Located in the estuary of the St. Lawrence River, the Lower St. Lawrence Seismic Zone (LSLSZ) experiences about 60 events with magnitude ≥2.0 yearly but, unlike the CSZ, only two earthquakes were near magnitude 5.0 in the last 100 years (in 1944 and in 1999; Lamontagne *et al.*, 2003; Figure 4.1). Most earthquakes occur under the St. Lawrence River with hypocentres in the Precambrian mid to upper crust, between 5 and 25 km depth, similar to the CSZ. Adams and Basham (1989) suggested that the seismicity was restricted to the regional normal faults, uniquely found offshore. Focal mechanisms show, however, variable fault plane orientations with most earthquakes clustering along or between the geologically mapped and geophysically inferred Iapetan faults (Lamontagne *et al.*, 2000). Mazzotti and Townend (2010) note that this seismic zone contains evidence for a local rotation of S_{Hmax} for the generally NE–SW-oriented S_{Hmax}.

4.3.3 Western Quebec

The Western Quebec Seismic Zone (WQSZ) includes the Ottawa Valley from Montreal to Temiscaming, as well as parts of the Laurentian Mountains (Basham *et al.*, 1982; Figure 4.2). This seismic zone includes historical earthquakes as large as **M** 6.2 and frequent low-level seismic activity. Earthquake epicentres define two sub-zones: a mildly active one along the Ottawa River, including a more active cluster in the Temiscaming region, and a more active one along the Montreal–Maniwaki axis.

The first band of seismicity, which parallels the Ottawa River, includes the epicentres of moderate earthquakes: **M** 6.2 near Temiscaming in 1935; an **M** 5.6 near Cornwall–Massena in 1944 and possibly a magnitude of about 5.8 near Montréal in 1732 (Figure 4.2; Leblanc, 1981). The diffuse group of earthquakes in the first band appear to correlate with a zone of Paleozoic or younger normal faults along the Ottawa River, called the Ottawa–Bonnechère graben (Forsyth, 1981). The Temiscaming area is more active than the rest of the Ottawa

River valley and was the site of the 1935 **M** 6.2 Temiscaming earthquake. In the year 2000, an **M** 4.7 earthquake reactivated one of the local faults (Bent *et al.*, 2002), adding support to the correlation between the earthquake epicentres and the normal faults of the Ottawa–Bonnechere graben (Adams and Vonk, 2009). These earthquakes appear to reactivate faults smaller than the conspicuous ones near Lake Temiscaming (Bent *et al.*, 2002).

The majority of WQSZ earthquakes, mostly smaller than **M** 4.5, occur in an elongated NW–SE zone within the Grenville Geological Province with most focal depths varying between 7 and 25 km (Lamontagne *et al.*, 1994; Ma and Atkinson, 2006). Although northwest-trending structural features are known, correlating these with the epicentral trend is uncertain because of the thrust sheets that make up the uppermost crust in this area of the Grenville Province. The mid-crustal hypocentral depths of many earthquakes, the east–west trend of the fault planes of some earthquakes, and variations in regional focal mechanisms all suggest reactivation of deep structural features that may not have a surface expression. It is possible that at the eastern end of this active band an anorthosite body may act as a stress concentrator (Lamontagne *et al.*, 1994).

4.3.4 Background seismicity

Outside the seismic zones described above, other areas along the SLRS are much less active (Figures 4.1 and 4.2). On a yearly basis, very few earthquakes are recorded in these areas and very few significant earthquakes are known (one notable exception being the 1988 **M** 5.9 Saguenay earthquake; Table 4.1). This earthquake occurred at lower-crustal depth (29 km) in the Canadian Shield, outside the seismic zones defined by the recurring activity (North *et al.*, 1989). Possibly due to its focal depth, very few aftershocks have been recorded and only four M ≥ 3 earthquakes were recorded there between 1988 and 2013. Despite the differences in orientation between the two nodal planes of the focal mechanism and the graben fault at the surface, the proximity of the earthquake epicentre to faults of the Saguenay graben suggested a relationship between the two (North *et al.*, 1989). Other authors warned of the difficulty in linking a lower-crustal depth earthquake with the numerous faults of the Precambrian basement that have unknown fault extensions at depth (Du Berger *et al.*, 1991). This earthquake suggests that moderate to large earthquakes can occur in areas outside the zones defined by recurring seismic activity and that many faults of the whole Precambrian Shield could be near failure. This suggestion is supported by the numerous cases of reservoir-triggered seismicity in areas almost devoid of naturally occurring earthquakes (Lamontagne *et al.*, 2008b). Another noteworthy event is the moderate 1997 **M** 4.9 Cap-Rouge earthquake, which occurred at a depth of 22 km with a focal mechanism consistent with the reactivation of a mapped SLRS fault (Table 4.1; Nadeau *et al.*, 1998).

4.4 The St. Lawrence Rift System

Kumarapeli and Saull (1966) proposed that the series of normal faults along the St. Lawrence valley indicated rifting similar to the East African rift system. Although it is now recognized that the St. Lawrence Rift System (SLRS) is not an active rift, this hypothesis provided a model that explained many of the geological characteristics of the area (including faults created in an extensional regime and carbonatite intrusions). Since that pioneering work, the normal faults of the St. Lawrence paleorift system have been described and mapped at the surface as well as in the subsurface (Du Berger *et al.*, 1991; Castonguay *et al.*, 2010; Tremblay and Roden-Tice, 2011). The SLRS is a half-graben created during the late Precambrian opening of the proto-Atlantic (Iapetus) Ocean (Kumarapeli, 1985). Many of the faults trend NE–SW and mark the boundary between the Grenville Province of the Canadian Shield to the northwest and the St. Lawrence Lowlands to the southeast. The grabens of Ottawa–Bonnechère and Saguenay River intersect the SLRS and are both interpreted as Iapetan failed arms (aulacogens; Kumarapeli, 1985; Tremblay and Roden-Tice, 2011). Faults related to the Ottawa–Bonnechère and Saguenay grabens trend mostly WNW–ESE.

The SLRS faults have been studied in the Montreal region (Rocher *et al.*, 2003) as well as in Charlevoix, Quebec City, and in the St. Lawrence estuary (Figure 4.2). SLRS faults consist of cohesive cataclastic rocks, with some major fault zones being marked by 10–20-metre-thick fault breccias, ultracataclasite, and foliated fault gouge (Tremblay and Roden-Tice, 2011). The rift system is a crustal-scale fault zone where reactivation occurred along Late Precambrian to early Paleozoic faults attributed to Iapetus rifting (Sanford, 1993). Studies of the age of formation and reactivation of these faults revealed various periods of reactivation (Rocher *et al.*, 2003; Faure *et al.*, 2006). Following the formation of the Appalachian orogen (started in the Ordovician) when these faults were under a compressive stress environment, rifting of the Atlantic Ocean–Labrador Sea during Mesozoic times possibly reactivated them (e.g., Kumarapeli and Saull, 1966; Carignan *et al.*, 1997). Apatite fission-track age discontinuities between the two sides of these faults are interpreted as the result of normal faulting at *c.* 200 Ma (Jurassic) followed by tectonic inversion about 150 Ma ago (Tremblay and Roden-Tice, 2011). The latter study provides support for Atlantic-related, extensional and compressive deformation within the interior of the Canadian Shield, more than 500 km west of the axis of the Mesozoic rift basins. The faults were most recently active during the Late-Triassic–Jurassic period, which corresponds to the creation of the current Atlantic Ocean with the separation of North America and Africa. The emplacement of igneous rocks of the Monteregian Hills occurred approximately at the same time, i.e., in early Cretaceous times (McHone and Butler, 1984; Foland *et al.,* 1986; Pe-Piper and Jansa, 1987). In the Montreal region, the E–W and SE–NW extensional tectonics are documented by brittle normal faults.

Beneath the Paleozoic and Appalachian cover, seismic reflection profiles in the St. Lawrence Lowlands have shown the upper crustal morphology of the faulted blocks (Thériault *et al.*, 2005; Castonguay *et al.*, 2010). SLRS faults exist at depth beneath the Appalachians, and their positions are revealed by their magnetic and gravity signatures (Charlevoix: Lamontagne, 1999; Lower St. Lawrence: Lamontagne *et al.*, 2003; St. Lawrence Lowlands: Lamontagne *et al.*, 2012). Although we know the positions of some of these rift faults beneath the sedimentary cover, our knowledge is far from complete. Based on geophysical information, Wheeler (1995, 1996) defined the northwestern

Table 4.2 *Main characteristics and differences of the earthquake zones of the SLRS (see Table 4.1 for a description of the earthquakes)*

Common characteristics of SLRS active areas	Differences/anomalies along the SLRS
99% of earthquakes occur at less than 25 km depth	Some larger events occur near or deeper than 25 km (1988 **M** 5.9 Saguenay, 29 km; 1997 Cap-Rouge **M** 4.7, 22 km; 2010 Val-des-Bois **M** 5.0, 22 km)
Earthquakes occur in areas where paleo-rift faults are present (CSZ, WQSZ, LSLSZ)	Part of the earthquake activity is also present where no SLRS faults are present (WQSZ)
No surface rupture known for any of the large events	
Proximity to a large water body (CSZ, LSLSZ) with tides	No large water bodies (WQSZ)
Earthquakes occur over a vast region (WQSZ, LSLSZ)	Concentrated activity (CSZ)
Protracted aftershock sequence for moderate earthquakes	Very few aftershocks (1988 Saguenay at 29 km focal depth)
Presence of a meteor impact structure (CSZ)	

and southeastern boundaries of the intact Iapetan margin and its potentially seismogenic faults for seismic zoning purposes.

A seismotectonic model of the earthquakes of the St. Lawrence Valley must consider the similarities and differences of the seismic and weakly seismic zones (Figure 4.2). Table 4.2 presents an overview of the similarities and differences between the various seismically active regions of the SLRS. In the 1970s, focal mechanisms showed that microearthquakes represented the reverse-faulting reactivation of pre-existing faults in the Precambrian basement (Leblanc *et al.*, 1973; Leblanc and Buchbinder, 1977). Later studies with larger magnitude earthquakes confirmed this (see Hurd and Zoback [2012] for a review). The hypothesis that connects the SLRS faults and the current seismicity was mainly derived from the concentration of CSZ earthquake hypocentres along southeast-dipping trends, similar to SLRS faults (Anglin and Buchbinder, 1981; Anglin, 1984). By extension outside the CSZ, it was hypothesized that all earthquakes along the St. Lawrence valley represented reactivation of similar faults. This SLRS connection was also made at the time when all worldwide M ≥ 7.0 continental earthquakes were shown to correlate with rifted cratonic areas (also called "extended crust"; Johnston and Kanter, 1990). This rift-faults–seismicity connection was also used by Adams and Basham (1989) in discussing the seismicity of the SLRS.

Outside the CSZ, the best correlation between the normal faults of the SLRS and seismicity can be found in the Temiscaming region, where the 1935 M 6.2 earthquake occurred (Figure 4.2). Near the epicentre of the earthquake, many Ottawa–Bonnechère graben faults have been mapped and many have conspicuous topographical signatures visible in remote-sensing imagery (Bent *et al.*, 2002). The focal mechanism of the 1935 earthquake (the largest event of the seismic zone) indicates thrust faulting on a moderately dipping northwest-striking plane, an orientation similar to the graben faults of the area (Bent, 1996). Similarly, in the Lower St. Lawrence Seismic Zone, SLRS faults are present and may be reactivated by the current seismic activity.

The rift–seismicity hypothesis has seismic hazard implications. The seismic provisions in the 2005 National Building Code of Canada use the larger of the ground motions derived from two seismic source zone models. The first model assumes that the historical earthquake clusters denote areas that will continue to be active. The second model assumes a common geological framework for the seismicity clusters (i.e., passive Paleozoic rift faults) and regroups them into large source zones (Adams *et al.*, 1995). In areas with historical damaging earthquakes, the hazard derived from the historical model dominates, whereas in areas without significant activity but where rift faults are present, the hazard is dominated by the geologically derived model (Adams and Halchuk, 2003).

4.5 The rift hypothesis and the SLRS: discussion and conclusions

Since the mid 1980s, the rift model has served as the main hypothesis that relates seismicity and faults along the SLRS. Although the rift model explains many characteristics of the SLRS seismicity, some questions remain. At the global level, the rift model was based on a correlation with "extended crust" that was correct for 100% of earthquakes of magnitude larger than 7.0; 60% between M 6 and 7; and 46% for earthquakes smaller than M 6 (Johnston and Kanter, 1990). Recently, however, Schulte and Mooney (2005) concluded that on a global scale the correlation of seismicity within stable continental regions (SCR) and ancient rifts has been overestimated in the past. They note that several apparently non-rifted crust areas have experienced multiple large ($\mathbf{M} \geq 6.0$) events. The rift model defined from a global perspective was extrapolated to the SLRS where only the 1663 Charlevoix shock is thought to have exceeded magnitude 7.0.[2] We note that the historical earthquake of 1663 has an uncertain epicentre location based on limited intensity information and is consequently only assumed to be along a rift fault. The rift hypothesis was extrapolated to smaller earthquakes in the magnitude 5.5 to 6.5 range, with most having occurred prior to instrumental recording, with uncertain magnitudes, epicentres, and unknown focal mechanisms. Based on these limitations, we raise questions about the applicability of the global rift model to SLRS earthquakes, especially for $\mathbf{M} < 4.5$ earthquakes.

Globally, we note that many recent earthquakes do not fit the assumed local seismotectonics model. Therefore, the correlation with any earthquake that occurs within the loose geographic boundaries of the SLRS with the faults of that system is not necessarily correct. In the case of the 1988 $\mathbf{M}$ 5.9 Saguenay earthquake, for example, despite the focal mechanism that did not match the orientation of the most conspicuous graben faults that were

[2] Although the 1663 event was not part of the database of Johnston and Kanter (1990), which only considered the last 200 years.

dipping away from the epicentre (Du Berger *et al.*, 1991), this earthquake is now referred to a SLRS earthquake. To exemplify the difficulty in correlating faults and earthquakes, the 1982 **M** 5.8 Miramichi and the 1989 **M** 6.3 Ungava earthquakes, both shallow focus earthquakes, do not correlate with any fault of regional extent (Wetmiller *et al.*, 1984; Lamontagne and Graham, 1993). These events are reminders that faults with dimensions necessary to generate an **M** 6 earthquake (about 10 km) exist almost everywhere in the Canadian Shield, even where rift faults are not present. Consequently, we conclude that smaller magnitude earthquakes (**M** < 4.5), which represent the vast majority of earthquakes in Eastern Canada, do not necessarily represent reactivated SLRS faults.

We believe that some global studies have partaken in the confusion because they did not look at the SLRS at the local level. For example, Schulte and Mooney (2005) used a continent-wide view to suggest a spatial correlation of the SLRS seismicity with the Appalachian front to the east (which is known to be inactive). They also correlate SLRS with the epicentre of the 1732 Montreal M 5.8 (which they rate as **M** 6.3) earthquake, which Gouin (2001) locates in Northern New York State from the same intensity data. We argue that the current earthquake database does not have a sufficiently large number of earthquakes with reliable epicentres to strongly support a correlation between earthquakes and the entire SLRS. Another example of a continental-scale view of seismicity is the suggested correlation between earthquakes and low-angle thrust faults ("sutures") of the Grenville Province (Van Lanen and Mooney, 2006). We note that numerous aseismic areas of the Grenville Province have these low-angle thrust faults. Recently, Mooney *et al.* (2012) suggested a correlation between crustal seismicity and younger lithosphere surrounding the ancient cratons, with aseismic zones corresponding to areas with thick (and cold) lithosphere.

One must note that the strongest support of a SLRS–seismicity connection is based on the locations of earthquake epicentres, and hypocentres in the CSZ, where SLRS faults are present. We believe that the regional-scale model, suggesting a seismic hazard based on the sole presence of SLRS faults, may not hold if one considers the local scale of the various seismic zones. In the CSZ, for example, the hypocentre–rift-fault connection suggested by Anglin (1984) is based on about 6 years of earthquake recording by a local network. Today, after some 35 years of recording, a more complex picture emerges, with numerous sub-zones, each one with its special focal depth distribution, level of activity, and focal mechanism complexity. There is some control by the SLRS faults, but they appear to bound seismically active blocks rather than being active themselves (Lamontagne, 1999).

Another problem with the rift fault hypothesis is that active regions are separated by weakly active areas (Figure 4.1). Tuttle and Atkinson (2010) did not find any evidence of Holocene earthquake activity in the Quebec City to Trois-Rivières segment, despite the presence of SLRS faults. These authors suggest that the seismic activity at Charlevoix may be localized. This suggests that local factors lead to seismicity in recognized seismic zones, not uniquely the presence of SLRS faults. In the WQSZ as well, the seismicity–rift-fault correlation does not explain that most earthquakes locate well to the north of the graben faults in a NW–SE alignment. In general, the Ottawa–Bonnechère graben is weakly seismic except near the 1935 Temiscaming epicentre (Figure 4.2), where only a small portion of the regional graben faults are active (Bent *et al.*, 2002).

Our evidence in the SLRS suggests that the relation between earthquakes and rift faults breaks down for the lower magnitude earthquakes (probably smaller than **M** 6 and certainly less than **M** 4.5), where local stresses and/or fault weaknesses contribute to earthquake occurrences. This approach was also proposed on a global scale by Schulte and Mooney (2005) and Mooney *et al.* (2012). In the SLSR, and in the CSZ in particular, a study of focal mechanisms for magnitude 2 to 5 has shown that numerous fault orientations are reactivated in the smaller earthquakes, including mixed thrust–strike-slip events. When studied in detail, earthquakes in sub-areas of the CSZ do not occur on rift faults, or on some impact structure faults (Lamontagne and Ranalli, 1997). Baird *et al.* (2009, 2010) have emphasized the importance of local variations in stress level within the CSZ and of strength differences between the crater and the rift faults outside the crater. They showed that much of the background seismicity pattern can be explained by the intersection of weak faults of the St. Lawrence rift with the damage zone created by the Charlevoix impact. In terms of strain measurements, the first interpretations of crustal strain along the SLRS were supportive of a homogeneous stress system, but more recent results have shown that local stresses are anomalous in active areas such as the CSZ (Mazzotti and Townend, 2010). For smaller Eastern Canadian earthquakes, the direction of the maximum compressive stress varies, which suggests local perturbations in the stress field in contrast to studies that favour a regionally homogeneous eastern North American stress field (Zoback and Zoback, 1991).

Another possibility for the creation or reactivation of local faults is the role of sub-crustal processes, such as what was proposed for the NW–SE band of earthquakes of the WQSZ. Since no surface geological feature is seen along this trend, it was proposed that the seismicity could be due to an extension of the New England Seamount Chain track or the passage of this region over the Great Meteor hotspot between 140 and 120 million years ago (Figure 4.2; Sykes, 1978; Crough, 1981; Adams and Basham, 1989). Ma and Eaton (2007) have suggested that the passage over a hotspot, possibly imaged by lithospheric velocity anomalies at 200 km depth, may explain the enhanced level of seismicity either by thermal rejuvenation of pre-existing faults or by stress concentration caused by mid-crustal strength contrast between mafic and felsic rock. If this hypothesis is valid for the WQSZ, the hotspot would have generated crustal fractures (Adams and Basham, 1989) through the generation of thermo-elastic stress (e.g., Marret and Emerman, 1992). Two problems remain to be resolved before the hotspot–seismicity link is accepted in this area: the hotspot track has no geological expression (faulting, metamorphic grade changes, intrusives) in the WQSZ (Sleep, 1990) and there is very little seismic activity along most of the length of the hotspot track, including along the Monteregian Hills region.

Numerous hypotheses have been advanced to account for the local concentration of WQSZ earthquakes. For example, it was proposed that the NW–SE band of earthquakes in WQSZ represented crustal zones of weakness delineated by the drainage pattern (Goodacre *et al.*, 1993). A possible correlation was also suggested with positive aeromagnetic anomalies and the Helikian (Mid-Proterozoic) paragneisses. The weakness of upper-crustal geological formations could lead to enhanced stresses at mid-crustal levels. Another hypothesis is that areas that are currently active represent aftershock zones of large historic or prehistoric earthquakes (Ebel *et al.*, 2000).

If the SLRS connection only partially explains the St. Lawrence seismicity, what factors can render portions of the SLRS faults inherently weak and/or enhance the local stress levels?

Assuming that the level of tectonic stresses is everywhere similar in Eastern Canada, there are a few factors that can weaken the faults at depth and favour reactivation. In the brittle regime, the reactivation of a fault is controlled by the stress orientation, the friction coefficient on the fault, and the pore-fluid pressure acting on the fault interface. The reverse-faulting regime that prevails down to the mid-crust ($\sim$30 km) in Eastern Canada implies stress differences that are difficult to conceive. At 25 km, for example, a hydrostatic pore-fluid pressure ($\lambda = 0.4$) and a coefficient of friction of 0.75 imply a critical stress difference for sliding of about 1,200 MPa in an ideally oriented reactivated thrust fault. These high stress differences are about one order of magnitude larger than the upper limit of 100–200 MPa usually assumed for crustal stresses. This suggests that high pore-fluid pressure and/or low coefficient of friction must exist along reactivated faults to give rise to the SLRS mid-crustal seismicity (Lamontagne and Ranalli, 1996). Another factor is the orientation of the faults with respect to the maximum compressive stress (which is assumed to be sub-horizontal in Eastern Canada). Hence, only faults that strike nearly perpendicular to the maximum stress axis have a higher probability of being reactivated as reverse faults.

If, on the other hand, it is assumed that all faults are equally weak to the first order, then factors must be sought that can locally increase the stresses. Stress can concentrate in areas with lateral mass anomalies within the lithosphere (Goodacre and Hasegawa, 1980; Assameur and Mareschal, 1995). In the latter study, the fact that two out of three regions with the highest induced stress differences remain aseismic suggests that mass anomalies can favour but not control earthquake occurrences. Other models that have been suggested to explain the occurrence of seismicity within continental interiors include localized stress concentration around weak intrusions (Campbell, 1978), intersecting faults (Talwani, 1988; Gangopadhyay and Talwani, 2007) and ductile shear zones in the lower crust (Zoback, 1983). Some models relate seismicity to elevated temperatures at depth (Liu and Zoback, 1997). In these models, plate-driving forces are largely supported by the (seismogenic) upper crust; the lower crust is weakened as a result of higher temperatures and the total strength of the lithosphere is reduced. Other models refer to regional stress fields perturbed by forces associated with lithospheric flexure after deglaciation (Stein *et al.*, 1979; Quinlan, 1984; Grollimund and Zoback, 2001). Based on strain determination for Eastern Canada, deglaciation is also used by James and Bent (1994) and Mazzotti and Townend (2010) to explain perturbed stress environments. A concentration of post-glacial rebound stresses can occur in local zones of weakness, perhaps containing low-friction faults (Mazzotti and Townend, 2010). This model suggests local sources of weakness to reconcile the apparent reorientation of maximum compressive stresses in the CSZ and in the LSLSZ.

The past few decades have brought to light numerous geological and geophysical characteristics of the SLRS and their links with seismicity. At a continent-wide scale, the correlation between the SLRS and earthquakes is appealing. When a more local scale is looked at or when smaller magnitude earthquakes are considered, questions arise and the picture becomes more complex. We conclude that the sole presence of pre-existing normal faults related to the SLRS does not explain all the features of the seismicity of Eastern Canada. Seismicity is concentrated in more active areas, some with conspicuous normal faults and some with suspected weakening mechanisms such as intense pre-fracturing (e.g., due to a meteorite impact), passage over a hotspot, or the presence of intrusive and lateral crustal density variations. In most cases, the superposition of the tectonic stress field and the relatively modest post-glacial rebound stresses is likely to play a role. Irrespective of the presence of rift faults, earthquakes are caused by dynamic instabilities along pre-existing faults caused by the inherent weakness of the faults, by the locally enhanced stress differences, or a combination of the two. A general unifying explanation is likely to be useful only at the large scale, and a detailed understanding of seismicity requires further work concentrating on local conditions.

Acknowledgments

We thank Allison Bent, Pradeep Talwani, and Thierry Camelbeeck for their reviews of a draft of this paper. GR's participation in this work has been supported by a grant from NSERC (Natural Sciences and Engineering Research council of Canada), Earth Sciences Sector contribution no. 20130049.

References

Adams, J., and Basham, P. (1989). The seismicity and seismotectonics of Canada east of the Cordillera. *Geoscience Canada*, 16, 3–18.

Adams, J., and Halchuk, S. (2003). *Fourth Generation Seismic Hazard Maps of Canada: Values for Over 650 Canadian Localities Intended for the 2005 National Building Code of Canada*. Geological Survey of Canada, Open File Report 4459.

Adams, J., and Vonk, A. (2009). *The November 1, 1935, M 6.2 Timiskaming Earthquake, Its Aftershocks, and Subsequent Seismicity*. Geological Survey of Canada, Open File Report 6207.

Adams, J., Basham, P. W., and Halchuk, S. (1995). Northeastern North American earthquake potential: new challenges for seismic hazard mapping. *Current Research 1995-D*, Geological Survey of Canada, pp. 91–99.

Anglin, F. M. (1984). Seismicity and faulting in the Charlevoix zone of the St. Lawrence valley. *Bulletin of the Seismological Society of America*, 71, 1553–1560.

Anglin, F. M., and Buchbinder, G. G. R. (1981). Microseismicity in the mid-St. Lawrence Valley Charlevoix Zone, Québec. *Bulletin of the Seismological Society of America*, 71, 1553–1560.

Assameur, D. M., and Mareschal, J. C. (1995). Stress induced by topography and crustal density heterogeneities: implication for the seismicity of southeastern Canada. *Tectonophysics*, 241, 179–192.

Aylsworth, J. M., Lawrence, D. E., and Guertin, J. (2000). Did two massive earthquakes in the Holocene induce widespread landsliding and near-surface deformation in part of the Ottawa Valley, Canada? *Geology*, 28, 903–906.

Baird, A. F., McKinnon, S. D., and Godin, L. (2009). Stress channelling and partitioning of seismicity in the Charlevoix seismic zone, Québec, Canada. *Geophysical Journal International*, 179, 559–568.

Baird, A. F., McKinnon, S. D., and Godin, L. (2010). Relationship between structures, stress and seismicity in the Charlevoix seismic zone revealed by 3-D geomechanical models: implications for the seismotectonics of continental interiors. *Journal of Geophysical Research*, 115(B11), 2156–2202.

Basham, P. W., Weichert, D. H., Anglin, F. M., and Berry, M. J. (1982). *New Probabilistic Strong Ground Motion Maps of Canada: A Compilation of Earthquake Source Zones, Methods and Results*. Publications of the Earth Physics Branch, Open File 82–33 (Energy, Mines and Resources, Canada).

Bent, A. L. (1992). A re-examination of the 1925 Charlevoix, Québec, earthquake. *Bulletin of the Seismological Society of America*, 82, 2097–2113.

Bent, A. L. (1996). An improved mechanism for the 1935 Timiskaming, Québec earthquake from regional waveforms. *Pure and Applied Geophysics*, 146, 5–20.

Bent, A. L. (2009). *A Moment Magnitude Catalog for the 150 Largest Eastern Canadian Earthquakes*. Geological Survey of Canada, Open File 6080.

Bent, A. L., Lamontagne, M., Adams, J., *et al.* (2002). The Kipawa, Quebec, "Millennium" earthquake. *Seismological Research Letters*, 73, 285–297.

Bent, A. L., Drysdale, J., and Perry, H. K. C. (2003). Focal mechanisms of eastern Canadian earthquakes, 1994–2000. *Seismological Research Letters*, 74, 452–468.

Boivin, D. J. (1992). Analyse et cartographie des dommages du séisme survenu au Québec le 25 novembre 1988. *The Canadian Geographer/Le Géographe Canadien*, 36, 114–123.

Brooks, G. R. (2013). A massive sensitive clay landslide, Quyon Valley, southwestern Quebec, Canada, and evidence for a paleoearthquake triggering mechanism. *Quaternary Research*, 80, 425–434.

Bruneau, M., and Lamontagne, M. (1994). Damage from 20th century earthquakes in eastern Canada and seismic vulnerability of unreinforced masonry buildings. *Canadian Journal of Civil Engineering*, 21, 643–662.

Buchbinder, G. G. R. (1989). Shear-wave splitting and anisotropy in the Charlevoix seismic zone, Quebec, in 1985. *Canadian Journal of Earth Sciences*, 26, 2691–2696.

Buchbinder, G. G. R., Lambert, A., Kurtz, R. D., *et al.* (1988). Twelve years of geophysical research in the Charlevoix seismic zone. *Tectonophysics*, 156, 193–224.

Campbell, D. L. (1978). Investigation of the stress concentration mechanism for intraplate earthquakes. *Geophysical Research Letters*, 5, 477–479.

Carignan, J., Gariépi, C., and Hillaire-Marcel, C. (1997). Hydrothermal fluids during Mesozoic reactivation of the St. Lawrence rift system, Canada: C, O, Sr and Pb isotopic characterization. *Chemical Geology*, 137, 1–21.

Cassidy, J. F. (1995). A comparison of the receiver structure beneath stations of the Canadian Seismograph Network. *Canadian Journal of Earth Sciences*, 32, 938–951.

Cassidy, J. F., Rogers, G. C., Lamontagne, M., Halchuk, S., and Adams, J. (2010). Canada's earthquakes: "The good, the bad, and the ugly". *Geoscience Canada*, 37, 1–16.

Castonguay, S., Dietrich, J., Lavoie, D., and Laliberté, J.-Y. (2010). Structure and petroleum plays of the St. Lawrence Platform and Appalachians in southern Quebec: insights from interpretation of MRNQ seismic reflection data. *Bulletin of Canadian Petroleum Geology*, 58, 219–234.

Cauchon-Voyer, G., Locat, J., and St-Onge, G. (2007). Morpho-sédimentologie et mouvements de masse au large de la Rivière Betsiamites, Estuaire du Saint-Laurent. In *Cartographie géoscientifique dans l'estuaire du Saint-Laurent : Bilan de l'an I*, ed. A. Bolduc. Geological Survey of Canada, Open File Report 5686, CD-ROM.

Crough, S. T. (1981). Mesozoic hot spot epeirogeny in eastern North America. *Geology*, 9, 2–6.

Desjardins, R. (1980). Tremblements de terre et glissements de terrain; correlation entre des datations au (super 14) C et des donnees historiques a Shawinigan, Quebec (translated title: Earthquakes and landslides; correlation of C-14 dates and historical data of the Shawinigan region, Quebec). *Géographie Physique et Quaternaire*, 34, 359–362.

Doig, R. (1986). A method for determining the frequency of large-magnitude earthquakes using lake sediments, *Canadian Journal of Earth Sciences*, 23, 930–937.

Doig, R. (1998). 3000-year paleoseismological record from the region of the 1988 Saguenay, Quebec, earthquake. *Bulletin of the Seismological Society of America*, 88, 1198–1203.

Du Berger, R., Roy, D. W., Lamontagne, M., *et al.* (1991). The Saguenay (Québec) earthquake of November 25, 1988: seismological data and geological setting. *Tectonophysics*, 186, 59–74.

Ebel, J. E. (2011). New analysis of the magnitude of the February 1663 earthquake at Charlevoix, Quebec. *Bulletin of the Seismological Society of America*, 101, 1024–1038.

Ebel, J. E., Bonjer, K.-P., and Oncescu, M. C. (2000). Paleoseismicity: seismicity evidence for past large earthquakes. *Seismological Research Letters*, 71, 283–294.

Faure, S., Tremblay, A., Malo, M., and Angelier, J. (2006). Paleostress analysis of Atlantic crustal extension in the Quebec Appalachians. *The Journal of Geology*, 114, 435–448.

Filion, L., Quinty, F., and Bégin, C. (1991). A chronology of landslide activity in the valley of Rivière du Gouffre, Charlevoix, Québec. *Canadian Journal of Earth Sciences*, 28, 250–256.

Foland, K. A., Gilbert, L. A., Sebring, C. A., and Jiang-Feng, C. (1986). $^{40}Ar/^{39}Ar$ ages for plutons of the Monteregian Hills, Quebec: evidence for a single episode of Cretaceous magmatism. *Geological Society of America Bulletin*, 97, 966–974.

Forsyth, D. A. (1981). Characteristics of the Western Québec Seismic Zone. *Canadian Journal of Earth Sciences*, 18, 103–119.

Gangopadhyay, A., and Talwani, P. (2007). Two-dimensional numerical modeling suggests that there is a preferred geometry of intersecting faults that favors intraplate earthquakes. In *Continental Intraplate Earthquakes: Science, Hazard, and Policy Issues*, ed. S. Stein, and S. Mazzotti. Geological Society of America Special Paper 425, pp. 87–99.

Goodacre, A. K., and Hasegawa, H. S. (1980). Gravitationally induced stresses at structural boundaries. *Canadian Journal of Earth Sciences*, 17, 1286–1291.

Goodacre, A. K., Bonham-Carter, G. F., Agterberg, F. P., and Wright, D. F. (1993). A statistical analysis of the spatial association of seismicity with drainage patterns and magnetic anomalies in Western Québec. *Tectonophysics*, 217, 285–305.

Gouin, P. (2001). *Historical Earthquakes Felt in Quebec: From 1534 to March 1925, as Revealed by the Local Contemporary Literature*, Montréal: Guérin.

Grollimund, B., and Zoback, M. D. (2001). Did deglaciation trigger intraplate seismicity in the New Madrid seismic zone? *Geology*, 29, 175–178.

Hearty, D. J., Mereu, R. F., and Wright, C. (1977). Lateral variations in upper crustal structure below La Malbaie area from slowness, azimuth, and travel time measurements of teleseisms. *Canadian Journal of Earth Sciences*, 14, 2284–2293.

Hurd, O., and Zoback, M. D. (2012). Intraplate earthquakes, regional stress and fault mechanics in the Central and Eastern U.S. and southeastern Canada. *Tectonophysics*, 581, 182–192.

James, T. S., and Bent, A. L. (1994). A comparison of eastern North American seismic strain-rates to glacial rebound strain-rates. *Geophysical Research Letters*, 21, 2127–2130.

Johnston, A. C., and Kanter, L. R. (1990). Earthquakes in stable continental crust. *Scientific American*, 262, 68–75.

Kumarapeli, P. S. (1985). Vestiges of Iapetan rifting in the west of the northern Appalachians. *Geoscience Canada*, 12, 54–59.

Kumarapeli, P. S., and Saull, V. A. (1966). The St. Lawrence Valley system: a North American equivalent of the East African rift valley system. *Canadian Journal of Earth Sciences*, 3, 639–658.

Lamontagne, M. (1998). *New and Revised Earthquake Focal Mechanisms of the Charlevoix Seismic Zone, Canada*. GSC Open File Report 3556.

Lamontagne, M. (1999). *Rheological and Geological Constraints on the Earthquake Distribution in the Charlevoix Seismic Zone, Québec*. Geological Survey of Canada Open File Report D3778, CD-ROM.

Lamontagne, M. (2002). An overview of some significant Eastern Canadian earthquakes and their impacts on the geological environment, buildings and the public. *Natural Hazards*, 26, 55–67.

Lamontagne, M. (2008). Documenting the first direct losses of life caused by a Canadian earthquake: the Charlevoix, Quebec, earthquake of October 20, 1870. *Bulletin of the Seismological Society of America*, 98, 1602–1606.

Lamontagne, M., and Graham, D. F. (1993). Remote sensing looks at an intraplate earthquake surface rupture. *Eos, Transactions, American Geophysical Union*, 74, 353–357.

Lamontagne, M., and Ranalli, G. (1996). Thermal and rheological constraints on the earthquake depth distribution in the Charlevoix, Canada, intraplate seismic zone. *Tectonophysics*, 257, 55–69.

Lamontagne, M., and Ranalli, G. (1997). Faults and spatial clustering of earthquakes near La Malbaie, Charlevoix Seismic Zone, Canada. *Seismological Research Letters*, 68, 337–352.

Lamontagne, M., Hasegawa, H. S., Forsyth, D. A., Buchbinder, G. G. R., and Cajka, M. (1994). The Mont-Laurier, Quebec, earthquake of 19 October 1990 and its seismotectonic environment. *Bulletin of the Seismological Society of America*, 84, 1505–1522.

Lamontagne, M., Keating, P., and Toutin, T. (2000). Complex faulting confounds earthquake research in the Charlevoix Seismic Zone, Québec. *Eos, Transactions, American Geophysical Union*, 81, 289–293.

Lamontagne, M., Keating, P., and Perreault, S. (2003). Seismotectonic characteristics of the Lower St. Lawrence Seismic Zone, Quebec: insights from geology, magnetics, gravity, and seismics. *Canadian Journal of Earth Sciences*, 40, 317–336.

Lamontagne, M., Demers, D., and Savopol, F. (2007). Description et analyse du glissement de terrain meurtrier du 25 octobre 1870 dans le rang des Lahaie, Sainte-Geneviève-de-Batiscan, Québec. *Canadian Journal of Earth Sciences*, 44, 947–960.

Lamontagne, M., Halchuk, S., Cassidy, J. F., and Rogers, G. C. (2008a). Significant Canadian earthquakes of the period 1600–2006. *Seismological Research Letters*, 79, 211–223.

Lamontagne, M., Hammamji, Y., and Peci, V. (2008b). Reservoir-triggered seismicity at the Toulnustouc Hydroelectric Project, Québec North Shore, Canada. *Bulletin of the Seismological Society of America*, 98, 2543–2552.

Lamontagne, M., Keating, P., Bent, A., Peci, V., and Drysdale, J. (2012). The 23 July 2010 mN 4.1 Laurier-Station, Quebec, earthquake: a midcrustal tectonic earthquake occurrence unrelated to nearby underground natural gas storage. *Seismological Research Letters*, 83, 921–932.

Leblanc, G. (1981). A closer look at the September 1732, Montreal earthquake. *Canadian Journal of Earth Sciences*, 18, 539–550.

Leblanc, G., and Buchbinder, G. G. R. (1977). Second micro-earthquake survey of the St. Lawrence Valley near La Malbaie, Quebec. *Canadian Journal of Earth Sciences*, 14, 2778–2789.

Leblanc, G., Stevens, A. E., Wetmiller, R. J., and Du Berger, R. (1973). A micro-earthquake survey of the St. Lawrence Valley near La Malbaie, Quebec. *Canadian Journal of Earth Sciences*, 10, 42–53.

Lefebvre, G., Paultre, P., Devic, J.-P., and Côté, G. (1991). Distribution of damages and site effects during the 1988 Saguenay earthquake. *Proceedings of the 6th Canadian Conference on Earthquake Engineering*, Toronto 1991, 719–726.

Lefebvre, G., Leboeuf, D., Hornych, P., and Tanguay, L. (1992). Slope failures associated with the 1988 Saguenay earthquake, Quebec, Canada. *Canadian Geotechnical Journal*, 29, 117–130.

Legget, R. F., and LaSalle, P. (1978). Soil studies at Shipshaw, Quebec; 1941 and 1969. *Canadian Geotechnical Journal*, 15, 556–564.

Lemieux, Y., Tremblay, A., and Lavoie, D. (2003). Structural analysis of supracrustal faults in the Charlevoix area, Quebec: relation to impact cratering and the St-Laurent fault system. *Canadian Journal of Earth Sciences*, 40, 221–235.

Liu, L., and Zoback, M. D. (1997). Lithospheric strength and intraplate seismicity in the New Madrid seismic zone. *Tectonics*, 16, 585–595.

Locat, J. (2011). La localisation et la magnitude du séisme du 5 février 1663 (Charlevoix) revues à l'aide des mouvements de terrain. *Canadian Geotechnical Journal*, 48, 1266–1286.

Lyons, J. A., Forsyth, D. A., and Mair, J. A. (1980). Crustal studies in the La Malbaie Region, Quebec. *Canadian Journal of Earth Sciences*, 17, 478–490.

Ma, S., and Atkinson, G. M. (2006). Focal depths for small to moderate earthquakes (mN 2.8) in Western Quebec, Southern Ontario, and Northern New York. *Bulletin of the Seismological Society of America*, 96, 609–623.

Ma, S., and Eaton, D. W. (2007). Western Quebec Seismic Zone (Canada): clustered, midcrustal seismicity along a Mesozoic hotspot track. *Journal of Geophysical Research*, 112, B06305, doi:10.1029/2006JB004827.

Marret, R., and Emerman, S. H. (1992). The relations between faulting and mafic magmatism in the Altiplano-Puna plateau (central Andes). *Earth and Planetary Science Letters*, 112, 53–59.

Mazzotti, S., and Townend, J. (2010). State of stress in eastern and central North America seismic zones. *Lithosphere*, 2, 76–83.

McHone, J. G., and Butler, J. R. (1984). Mesozoic igneous provinces of New England and the opening of North Atlantic Ocean. *Geological Society of America Bulletin*, 95, 757–765.

Milne, W. G., and Davenport, A. G. (1969). Distribution of earthquake risk in Canada. *Bulletin of the Seismological Society of America*, 59, 729–754.

Mitchell, D., Tinawi, R., and Law, T. (1990). Damage caused by the November 25, 1988 Saguenay earthquake. *Canadian Journal of Civil Engineering*, 17, 338–365.

Mooney, W. D., Ritsema, J., and Hwang, J. K. (2012). Crustal seismicity and the earthquake catalog maximum moment magnitude (Mcmax) in stable continental regions (SCRs): correlation with the seismic velocity of the lithosphere. *Earth and Planetary Science Letters*, 357–358, 78–83.

Munro, P. S., and North, R. G. (1989). *The Saguenay Earthquake of November 25, 1988: Strong Motion Data*, Geological Survey of Canada Open File Report 1976.

Nadeau, L., Lamontagne, M., Wetmiller, R. J., *et al.* (1998). The November 5, 1998 Cap-Rouge, Quebec earthquake. *Current Research 1998-E*, Geological Survey of Canada, pp. 105–115.

North, R. G., Wetmiller, R. J., Adams, J., *et al.* (1989). Preliminary results from the November 1988, Saguenay (Quebec) earthquake. *Seismological Research Letters*, 60, 89–93.

Ouellet, M. (1997). Lake sediments and Holocene seismic hazard assessment within the St. Lawrence Valley, Quebec. *Geological Society of America Bulletin*, 109, 631–642.

Paultre, P., Lefebvre, G., Devic, J.-P., and Côté, G. (1993). Statistical analyses of damages to buildings in the 1988 Saguenay earthquake. *Canadian Journal of Civil Engineering*, 20, 988–998.

Pe-Piper, G., and Jansa, L. F. (1987). Geochemistry of Late Middle Jurassic–Early Cretaceous igneous rocks on the eastern North American margin. *Geological Society of America Bulletin*, 99, 803–813.

Perret, D., Mompin, R., Bosse, F., and Demers, D. (2011). Stop 2–5B: the Binette Road earth flow induced by the June 23, 2010 Val-des-Bois earthquake. In *Deglacial History of the Champlain Sea Basin and Implications for Urbanization*, Joint Annual Meeting GCMAC-SEG-SGA, Ottawa, ON, May 25–27, 2011, Field Guide Book, Geological Survey of Canada, Open File 6947, pp. 72–74.

Quinlan, G. (1984). Postglacial rebound and the focal mechanisms of eastern Canadian earthquakes. *Canadian Journal of Earth Sciences*, 21, 1018–1023.

Rocher, M., Tremblay, A., Lavoie, D., and Campeau, A. (2003). Brittle fault evolution of the Montréal area (St. Lawrence Lowlands, Canada): rift-related structural inheritance and tectonism approached by palaeostress analysis. *Geological Magazine*, 140, 157–172.

Rondot, J. (1979). *Reconnaissances géologiques dans Charlevoix-Saguenay*. Rapport DPV-682, Ministère des richesses naturelles du Québec.

Sanford, B. U. (1993). Geology of the St. Lawrence Lowland Platform. In *Sedimentary Cover of the Craton in Canada*, ed. D. F. Stott, and J. D. Aitken. Geological Survey of Canada, Geology of Canada 5, 723–786.

Schulte, S. M., and Mooney, W. D. (2005). An updated global earthquake catalogue for stable continental regions: reassessing the correlation with ancient rifts. *Geophysical Journal International*, 161, 707–721.

Sleep, N. H. (1990). Monteregian hotspot track: a long-lived mantle plume. *Journal of Geophysical Research*, 95, 21983–21990.

Stein, S., Sleep, N., Geller, R. J., Wang, S. C., and Kroeger, G. C. (1979). Earthquakes along the passive margin of eastern Canada. *Geophysical Research Letters*, 6, 537–540.

Stevens, A. E. (1980). Re-examination of some larger La Malbaie, Québec earthquakes (1924–1978). *Bulletin of the Seismological Society of America*, 70, 529–557.

Sykes, L. (1978). Intraplate seismicity, reactivation of preexisting zones of weakness, alkaline magmatism, and other tectonism postdating continental fragmentation. *Reviews of Geophysics*, 16, 621–688.

Syvitski, J. P. M., and Schafer, C. T. (1996). Evidence for an earthquake basin-triggered basin collapse in Saguenay Fjord, Canada. *Sedimentary Geology*, 104, 127–153.

Talwani, P. (1988). The intersection model for intraplate earthquakes. *Seismological Research Letters*, 59, 305–310.

Thériault, R., Laliberté, J.-Y., and Brisebois, D. (2005). *Empreintes des grabens d'Ottawa-Bonnechère du Saguenay dans la plate-forme du Saint-Laurent et les Appalaches: Cibles pour l'exploration des hydrocarbures.* Ministère des Ressources Naturelles, Québec. Available on line (last accessed March 26, 2013), http://sigpeg.mrnf.gouv.qc.ca/gpg/classes/collectionInterne.

Tremblay, A., and Roden-Tice, M. K. (2011). Iapetan versus Atlantic rifting history of Laurentia: constraints from field mapping and AFT dating of Precambrian basement rocks, Canada. *Geological Society of America, Southeastern Section, 59th Annual Meeting*, Abstracts with Programs, Geological Society of America, 42, Issue 1, p. 79.

Tuttle, M. P., and Atkinson, G. M. (2010). Localization of large earthquakes in the Charlevoix Seismic Zone, Quebec, Canada, during the past 10,000 years. *Seismological Research Letters*, 81, 140–147.

Tuttle, M., Law, K. T., Seeber, L., and Jacob, K. (1990). Liquefaction and ground failure induced by the 1988 Saguenay, Quebec, earthquake. *Canadian Geotechnical Journal*, 27, 580–589.

Van Lanen, X., and Mooney, W. D. (2006). Integrated geologic and geophysical studies of North American Continental intraplate seismicity. In *Continental Intraplate Earthquakes: Science, Hazard, and Policy Issues*, ed. S. Stein, and S. Mazzotti. Geological Society of America Special Paper 425.

Vlahovic, G., Powell, C., and Lamontagne, M. (2003). A three-dimensional P wave velocity model for the Charlevoix seismic zone, Quebec, Canada. *Journal of Geophysical Research*, 108(B9), 1–12.

Wetmiller, R. J., Adams, J., Anglin, F. M., Hasagawa, H. S., and Stevens, A. E. (1984). Aftershock sequence of the 1982 Miramichi, New Brunswick, earthquake. *Bulletin of the Seismological Society of America*, 74, 621–653.

Wheeler, R. L. (1995). Earthquakes and the cratonward limit of Iapetan faulting in eastern North America. *Geology*, 23, 105–108.

Wheeler, R. L. (1996). Earthquakes and the southeastern boundary of the intact Iapetan margin in eastern North America. *Seismological Research Letters*, 67, 77–83.

Zoback, M. D. (1983). Intraplate earthquakes, crustal deformation and in-situ stress. In *A Workshop on 'The 1886 Charleston, South Carolina, Earthquake and Its Implications for Today'*, ed. W. W. Hays, P. L. Gori, and C. Kitzmiller. Open-File Report, No. 83–843, 169–178, Reston, Virginia: U.S. Geological Survey.

Zoback, M .D., and Zoback, M. L. (1991). Tectonic stress field of North America and relative plate motions. In *Neotectonics of North America*, ed. D. B. Slemmons, E. R. Engdahl, M. D. Zoback, and D. D. Blackwell. Boulder, Colorado: Geological Society of America.

5

Intraplate earthquakes in North China

MIAN LIU, HUI WANG, JIYANG YE, AND CHENG JIA

Abstract

North China, or geologically the North China Block (NCB), is one of the most active intracontinental seismic regions in the world. More than 100 large (M > 6) earthquakes have occurred here since 23 BC, including the 1556 Huaxian earthquake (M 8.3), the deadliest one in human history with a death toll of 830,000, and the 1976 Tangshan earthquake (M 7.8) which killed ~250,000 people. The cause of active crustal deformation and earthquakes in North China remains uncertain. The NCB is part of the Archean Sino-Korean craton; thermal rejuvenation of the craton during the Mesozoic and early Cenozoic caused widespread extension and volcanism in the eastern part of the NCB. Today, this region is characterized by a thin lithosphere, low seismic velocity in the upper mantle, and a low and flat topography. The western part of the NCB consists of the Ordos Plateau, a relic of the craton with a thick lithosphere and little internal deformation and seismicity, and the surrounding rift zones of concentrated earthquakes. The spatial pattern of the present-day crustal strain rates based on GPS data is comparable to that of the total seismic moment release over the past 2,000 years, but the comparison breaks down when using shorter time windows for seismic moment release. The Chinese catalog shows long-distance roaming of large earthquakes between widespread fault systems, such that no M ≥ 7.0 events ruptured twice on the same fault segment during the past 2,000 years. The roaming of large earthquakes and their long sequences of aftershocks pose serious challenges to the current practice of seismic hazard assessment, and call for a fundamental paradigm shift for studies of intracontinental earthquakes.

Intraplate Earthquakes, ed. Pradeep Talwani. Published by Cambridge University Press. © Cambridge University Press 2014.

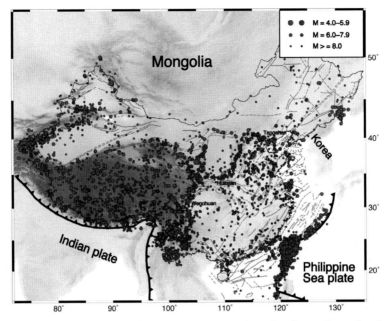

Figure 5.1 Topographic map of China and surrounding regions showing major active faults (thin lines) and seismicity (dots). Blue dots indicate epicenters of historic earthquakes before 1900; red dots indicate earthquakes during 1900–2010 (mostly instrumental records). Inset shows the scales of earthquake magnitudes. The barbed lines represent plate boundary faults. For color version, see Plates section.

5.1 Introduction

In King Jie's 10th year of the Xia Dynasty (1767 BC), an earthquake interrupted the Yi and Lo rivers and damaged houses in the capital city Zhengxuen.

In the second year of King Zhouyou (780 BC), an earthquake dried the Jin, Lo, and Wei rivers, and caused landslides in the Qi mountains.

State Records: Zhou Dynasty

These are some of the earliest written records of earthquakes in China. The Chinese catalog of historic earthquakes goes back ~3000 years, showing more than 1,000 M ≥ 6 events since 23 BC (Min *et al.*, 1995). At least 13 of these events were catastrophic (M ≥ 8). The 1556 Huaxian earthquake reportedly killed 830,000 people, making it the deadliest earthquake in human history (Min *et al.*, 1995). Modern earthquakes in China are intense and widespread. The 1976 Tangshan earthquake (Mw 7.8) killed ~242,000 people and injured millions (Chen *et al.*, 1988).

Most of these earthquakes occurred within the continental interior (Figure 5.1). In western China (approximately west of 105° E), historic earthquake records are sparse, but instrumentally recorded seismicity is intense. These earthquakes are directly related to the ongoing Indo-Asian continental collision; most of them are concentrated along the roughly E–W-trending fault systems resulting from the collision.

Seismicity in eastern China is weak today relative to western China, but the long historic records in eastern China show abundant large earthquakes, especially in the North China Block (NCB) (Figure 5.1). Here the fault systems are more complex: in the North China Plain and the coastal regions, the fault systems are mostly NE and NEE trending, owing their origin to subduction of the Pacific plate under the Eurasian plate (Deng *et al.*, 2002; Zhang *et al.*, 2003); further to the west, seismicity is concentrated in the rift fault zones around the Ordos Block.

The NCB is a geological province including the Ordos Plateau and the surrounding rift systems, the North China Plain, and the coastal regions. The cause of earthquakes in the NCB is uncertain. These earthquakes are clearly intraplate events, because the NCB is located in the interior of the Eurasian plate, within the Archean Sino-Korean craton, and thousands of kilometers away from plate boundaries. North China, being the cradle of the Chinese civilization, has the most complete historic records of earthquakes. Today, North China is one of the most densely populated regions in China with vital economic and cultural centers; hence, understanding earthquake hazards here is of great importance. In past decades, intensive geological and geophysical studies in North China have greatly refined the geological history and earth structure, and extensive Global Positioning Systems (GPS) measurements have delineated crustal kinematics. In this chapter, we briefly summarize the tectonic background of the North China region, discuss the main features of active tectonics, and describe historic and instrumentally recorded earthquakes. We highlight the complex spatiotemporal patterns of large earthquakes in North China, and discuss their implications for earthquake hazard assessment in North China and other mid-continents.

5.2 Tectonic background

The geologically defined North China Block is part of the Archean Sino-Korean craton; in China it is also referred to as the North China Craton (NCC). From west to east, the NCB includes the Ordos Plateau, the North China Plain, and the coastal regions (Figure 5.2). The Ordos Plateau is a relic of the NCC, with thick lithosphere and little internal deformation through the Cenozoic, hence is also referred to as the Ordos Block. Its margins are bounded by a system of rifts developed in the late Cenozoic, perhaps as a consequence of the Indo-Asian collision (Xu *et al.*, 1993; Zhang *et al.*, 1998). These rifts include the Weihe rift and the Shanxi rift on the southern and eastern margins of the Ordos Plateau, respectively. These rifts are structurally connected and perhaps formed together; in China they are sometimes collectively referred to as the Fenwei rift system. This is a major seismic zone in North China (Figure 5.2).

The North China Plain and the coastal regions are the part of the Sino-Korean craton where the cratonic root was destroyed by thermal rejuvenation in the Mesozoic; the process produced widespread extension and volcanism during the late Mesozoic and early

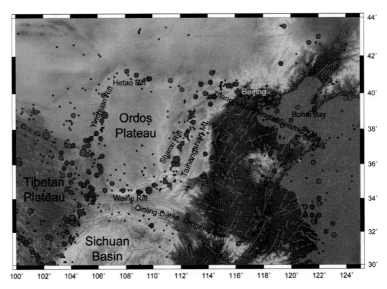

Figure 5.2 Topographic relief (colored background), faults (orange lines), and seismicity (dots for epicenters) for North China. Blue dots: historic earthquakes; red dots: instrumentally recorded earthquakes. For color version, see Plates section.

Cenozoic. Today, these regions are marked by flat, low-elevation (~50–200 m) and thin lithosphere (~60 km in places), and the upper crust includes thick Quaternary sediments covering widespread extensional basins. The fault systems are complex and widespread. The boundary between the North China Plain and the coastal regions is the northeast-trending Tanlu fault (Figure 5.2), a major strike-slip fault system developed during the Mesozoic collision between the North and South China blocks, but with little slip through the Cenozoic (Li, 1994; Yin and Nie, 1996; Wang et al., 2011).

The southern boundary of the North China Block is the east-trending, late Triassic Qinling–Dabie orogenic belt created by the collision between the North and South China blocks (Figure 5.2); the northern boundary is the Hetao rift valley along the northern rim of the Ordos Plateau, and the Yanshan–Yinshan mountain belts further to the east. Along the southwestern margin of the North China Block, the Ordos Plateau encounters the northeastern corner of the laterally expanding Tibetan Plateau, forming a fold-and-thrust belt in the late Cenozoic.

The topographic boundary between the high-standing western part of the North China Block and the low-altitude North China Plain is abrupt along the eastern flank of the Taihangshan mountain ranges (Figure 5.2). This boundary is also marked by the largest gradients of gravity in east Asia, corresponding to large gradients of crustal and lithospheric thicknesses (Ma, 1989). The upper mantle under the North China Plain and the coastal regions is characterized by low seismic velocity structures, which may be related to mantle flow above the subducted Pacific plate (Liu et al., 2004; Huang and Zhao, 2006).

5.2.1 Geological history

Much of the basement of the North China Block belongs to the Sino-Korean craton, which includes some of the oldest rocks on Earth (Liu *et al.*, 1992). During the Late Archean to Paleoproterozoic, the Western Block, which includes the Ordos Block, collided with the Eastern Block, forming the Trans-North China Orogen, which includes the basement of today's Shanxi graben and the Taihangshan mountains (Zhao *et al.*, 2005; Zhai and Santosh, 2011). Since then the North China basement has been a coherent craton and remained tectonically stable until the end of the Paleozoic. Geological evidence indicates that the lithosphere under eastern North China was more than 180 km thick in the early Mesozoic (Griffin *et al.*, 1998; Xu *et al.*, 2003).

During the Triassic to mid-Jurassic, the North China Block collided with the South China Block, resulting in the Qinling–Dabie orogenic belt with ultra-high-pressure metamorphism. The collision modified the crustal and lithospheric architecture of the North China Block, producing thick-skinned crustal thrusts in its southeastern part (Li, 1994), forming major faults, such as the Tanlu fault, which cut across the entire craton. This collision probably also initiated the thinning of the North China lithosphere (Menzies *et al.*, 2007).

Most of the lithospheric thinning in North China occurred during the late Mesozoic to early Cenozoic, accompanied by widespread extension and volcanism (Zhu *et al.*, 2012b). Petrological and geochemical probing using the upper mantle xenolith indicates that the lithosphere was thinned to ~80 km over much of the eastern NCB and less than 60 km thickness in some places (Menzies and Xu, 1988; Xu *et al.*, 2003). This is consistent with seismic data (Chen *et al.*, 2009). The cause of the removal of the cratonic root under North China remains poorly known (Zhu *et al.*, 2012a). It may have resulted from the Mesozoic collision between the North and South China blocks, which led to delamination or thermal erosion of a thickened and weakened lithosphere (Xu, 2001; Bryant *et al.*, 2004), or it may be related to the ocean-ward retreat of the western Pacific plate, which induced mantle upwelling under North China. The Cenozoic tectonism in North China may also be linked to indentation of India with Eurasia and the induced lateral mantle flow (Liu *et al.*, 2004).

5.2.2 Lithospheric structure

Seismic imaging shows that the eastern North China Block is underlain by broad low-velocity mantle structures (Figure 5.3). These low-velocity structures are limited to the upper mantle; at ~660 km depth flat subducting slabs, shown as a high-velocity layer, can be traced to the subduction zone of the Pacific plate along the eastern margins of the Eurasian plate (Huang and Zhao, 2006). The western end of the stagnant slabs extends ~1500 km inland from the active trench in the western Pacific, and can be correlated with the prominent surface topographic change between the high-standing Ordos Plateau and the Taihangshan mountains in the west, and the lowland of the North China Plain and

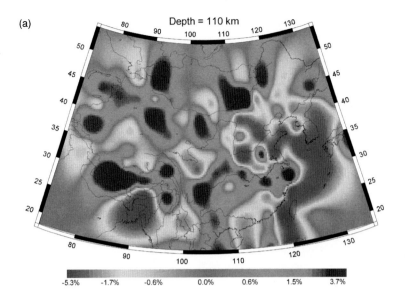

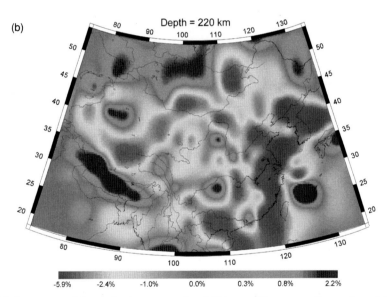

Figure 5.3 P-wave travel time seismic tomography of China and the surrounding regions. (a) P-wave velocity perturbation at 110 km depth. (b) P-wave velocity perturbation at 220 km depth. From Liu *et al.* (2004). For color version, see Plates section.

coastal regions in the east. Hence, the low-velocity mantle structures under North China may be a horizontally expanded "mantle wedge" above both the active subducting slab in the western Pacific and the stagnant slabs beneath much of the North China Plain (Huang and Zhao, 2006). This broad mantle wedge probably resulted from seaward retreat of the western Pacific trench and the sinking of the Mesozoic and Cenozoic slabs now trapped at the 660 km transition zone. Convection in this mantle wedge could have contributed to the Cenozoic lithospheric thinning, volcanism, and associated extensional basins in the eastern NCB (Liu *et al.*, 2004; Zhu *et al.*, 2009).

Accordingly, the lithosphere under the eastern NCB is abnormally thin, in some places reaching ~60–80 km (Chen *et al.*, 2009; Huang *et al.*, 2009). It thickens westward, reaching more than 200 km under the Ordos Plateau (Chen *et al.*, 2009). The thickening occurs abruptly along the foothills of the Taihangshan mountains, correlating to the western edge of the stagnant subducting slabs. Whereas the western part of the NCB avoided thermal thinning for the most part, the lithosphere under the rift systems around the Ordos Block may be thinned to 80 km (Chen *et al.*, 2009).

The variation of crustal thickness generally mimics that of the lithosphere. It is more than 40 km thick in the western NCB, and thins to 32–26 km in the eastern NCB (Ma, 1989). The Ordos Plateau has a thin Cenozoic cover; in many places the Mesozoic strata, which are a few kilometers thick, are exposed. These strata are flat-lying except near the margins of the Ordos Plateau, indicating the tectonic stability of the western NCB since at least the Mesozoic time (Zhang *et al.*, 2007). In contrast, the eastern NCB is covered by thick Cenozoic sediments, on top of widespread extensional basins, the largest one being the basin system centered around Bohai Bay (Allen *et al.*, 1998; Li *et al.*, 2012) (Figure 5.2). In these regions the upper crust is crosscut by a complex system of listric normal faults associated with these extensional basins; they change to strike-slip faults in the middle–lower crust where most of the destructive large earthquakes in the NCB initiated (Xu *et al.*, 2002b).

5.2.3 Major seismogenic faults

Seismicity in the western NCB is concentrated within the circum-Ordos rift systems (Figure 5.2). These rifts developed mainly in the Neogene, related to the Indo-Asian collision and the northeastern expansion of the Tibetan Plateau (Ye *et al.*, 1987; Zhang *et al.*, 1998). Along the northwestern edge of the Ordos Block is the Yinchuan rift, which connects with the Hetao rift along the northern side of the Ordos. On the southern side of the Ordos Block is the Weihe rift, which produced the deadly 1556 Huaxian earthquake. Along the southeastern edge of the Ordos Block, the Weihe rift connects with the northeast-trending Shanxi rift, which is a dextral shear extension zone consisting of more than 10 discontinuously distributed fault-depression basins controlled by normal strike-slip faults or normal faults (Deng *et al.*, 2003). Together, these two rifts have hosted more than 36 large (M > 6.5) earthquakes since 1303 (Liu *et al.*, 2007). Seismicity within the Ordos Plateau is minor, similar to seismicity in other stable cratons (Mooney *et al.*, 2012).

In the eastern NCB the most prominent fault is the Tanlu fault system, which was developed during the Mesozoic collision between the South and North China blocks (Li, 1994; Yin and Nie, 1996). Its Cenozoic development may be related to the opening of Bohai Bay during the Paleogene back-arc extension (Ren *et al.*, 2002). GPS surveys show that the fault is extensional, accommodating east–west extension (Shen *et al.*, 2000). Although the present-day slip rate is less than 1 mm/yr (Wang *et al.*, 2011), its rupture in 1668 caused the Ms 8.5 Tancheng earthquake.

The northern margin of the North China Plain is bounded by the Zhangjiakou–Penglai fault, a complex, NW–SE-trending fault system that is up to a few tens of kilometers wide as it extends to the coastal regions (Figure 5.2). Its intersections with a system of NE-trending faults in the North China Plain, including the Tanlu fault, are the source regions of a number of destructive earthquakes.

The North China Plain, between the Shanxi rift and the Tanlu fault, has a complex system of faults, mainly NNE-trending, which are covered by thick Quaternary sediments, with little or no surface traces (Deng *et al.*, 2003). Some of these faults, including the Xingtai–Hejian–Tangshan fault zone, were recognized only after the destructive 1966 Xingtai earthquake (M 7.2) and the 1996 Tangshan earthquake (M 7.8).

5.3 Active tectonics and crustal kinematics

North China, with its dense population and intense seismicity, is one of the best studied regions in China. A "Map of Active Tectonics of China," at a scale of 1:4 million, identified more than 200 active tectonic zones (Deng *et al.*, 2003). This map delineates the active tectonic belts of China that bound relatively aseismic blocks.

The GPS network in North China was established in 1992; it was significantly improved with the establishment of the Crustal Motion Observation Network of China (CMONOC) in 1998. About 300 CMONOC survey mode GPS stations are located in North China, covering effectively all the known regional active faults (Figure 5.4). Shen *et al.* (2000) found that regional deformation in North China is dominated by left lateral slip (~2 mm/yr) across the east-southeast-trending Zhangjiakou–Penglai seismic zone and extension (~4 mm/yr) across the north-northeast-trending Shanxi rift, which could be comparable with the ~1.0 mm/yr estimated from seismic moment data (Wesnousky *et al.*, 1984) and 0.5–1.6 mm/yr averaged over the Late Pliocene–Quaternary time (Zhang *et al.*, 1998). However, using a more complete dataset, He *et al.* (2003) found no clear signal of extension across the Shanxi rift; they attributed the discrepancy with geological extension rate to time-dependent extensional processes.

5.4 Strain rates and seismicity

Using the GPS data from China's CMONOC network, Liu and Yang (2005) calculated the scalar strain rates, defined as $\dot{E} = (\dot{\varepsilon}_{\phi\phi}\dot{\varepsilon}_{\phi\phi} + \dot{\varepsilon}_{\lambda\lambda}\dot{\varepsilon}_{\lambda\lambda} + 2\dot{\varepsilon}_{\phi\lambda}\dot{\varepsilon}_{\phi\lambda})^{1/2}$, where ϕ and λ are longitude and latitude, respectively, in North China (Figure 5.5). The higher strain rates are found in the North China Plain and around the Ordos Block. The Shanxi rift system, which has had many large earthquakes in the past 2,000 years (Figure 5.1), surprisingly shows relatively low strain rates. This is consistent with the GPS results (He *et al.*, 2003), and may be related to the seismic quiescence within the Shanxi rift in the past 300 years (Liu *et al.*, 2007, 2011).

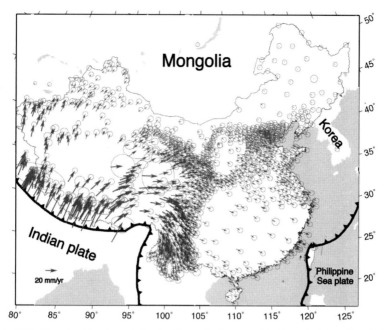

Figure 5.4 GPS site velocities in mainland China relative to stable Eurasia. Data from Zhang and Gan (2008). Error ellipses represent the 95% confidence level.

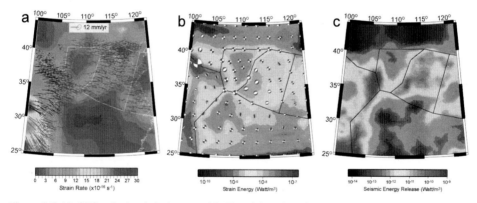

Figure 5.5 (a) GPS velocity (relative to stable Eurasia) and strain rates (background). (b) Predicted long-term strain energy (background) and stress states represented in stereographic lower-hemisphere projections. (c) Seismic strain energy released in the past ~2,000 years. Modified from Liu and Yang (2005). For color version, see Plates section.

Using the velocity boundary conditions extrapolated from the GPS data, Liu and Yang (2005) developed a 3D finite element model to calculate the long-term distribution of strain rate and strain energy (the product of scalar strain rate and stress) in North China (Figure 5.5b). The spatial pattern of strain energy is comparable with seismic moment release in the past ~2,000 years, estimated using the Chinese historic catalog (Figure 5.5c).

This indicates that the intense seismicity recorded in North China in the past 2,000 years is a close reflection of the long-term strain energy accumulation and release. One major mismatch is along the northern margin of the Ordos Plateau, where the relatively low seismic moment release contradicts the high long-term strain energy predicted by the model. A similar discrepancy was found by Wang *et al.* (2011), who derived the slip rates on major fault zones in continental China and compared the rates of moment accumulation with that released by recorded earthquakes. This inconsistency between low moment release and high strain energy may indicate a surplus of moment along the northern boundary of the NCB.

In general, however, seismic moment release, which is limited by the incomplete earthquake catalog, is incompatible with strain rates derived from GPS measurements. To illustrate the bias of inferred spatiotemporal patterns of seismicity from the incomplete earthquake record, Figure 5.6 plots the spatial distribution of moment release within a 250-year time window in North China in the past 750 years, a period of the most complete earthquake record in the Chinese catalog. Note that the spatial pattern of moment release within each time window differs from all others, and none of them is comparable with the strain rates derived from the GPS data (Figure 5.5a). These results highlight the complex spatiotemporal patterns of intracontinental earthquakes, and the problem of long recurrence time and short and incomplete records. The apparent consistency between the total seismic moment release and the strain rates based on GPS data in North China is an exception rather than the norm; this can be largely attributed to the more than 2,000-years of earthquake records in North China. The catalog is much shorter in other mid-continents. In the Central and Eastern United States, the historic catalog only extends for 200 years or so, therefore the spatial pattern of seismicity from these records may not be a reliable indicator for the long-term seismicity or what seismic activity will look like in the next few hundred years.

5.5 Seismicity

North China has the most complete earthquake records in China because it is the cradle of the Chinese civilization. The ancient Chinese regarded earthquakes and other natural hazards as the wrath of heaven; these events were faithfully recorded by the government. The earliest written record of earthquakes in North China may be traced back to the twenty-third century BC (Gu *et al.*, 1983); however, the record is likely incomplete, and most of the early records are sketchy. Recent paleoearthquake studies, mainly through trenching, have extended the earthquake history in many sites (Xu *et al.*, 2002a).

5.5.1 Paleoseismicity

Most paleoseismic studies in North China have been focused on the rift zones around the Ordos Plateau. Along the northern edge of the Ordos Block, Ran *et al.* (2003a) identified 62 paleo-earthquakes in the late Quaternary, 33 of which occurred in the Holocene. They

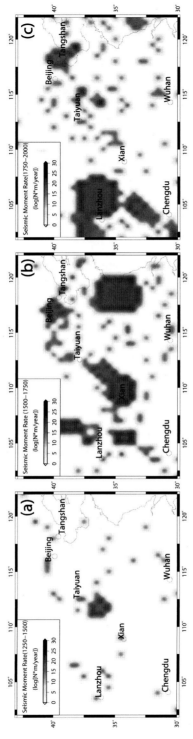

Figure 5.6 Spatial patterns of seismic moment release rate in North China averaged over different time windows: (a) 1250–1500; (b) 1500–1750; (c) 1750–2000. Moment is in Newton meter.

also found that the recurrence intervals of major earthquakes differ for individual fault segments, individual faults, and composite fault zones. One of the major faults is the 220 km long Daqingshan normal fault zone, which was initiated in the Eocene, with a total slip of more than 2.4 km since the Quaternary. Seven major paleoseismic events have occurred since 19 ka BP (thousands of years before present), with an average recurrence interval of ~2,000 ± 432 years (Ran *et al.*, 2003b).

Along the Yinchuan rift valley on the northwestern edge of the Ordos Plateau, Deng and Liao (1996) identified four large earthquakes (M 8.0) at around 8,400, 5,700, 2,600, and 256 years before present, with a recurrence interval of 2,300–3,000 years. Within the Weihe rift valley, where the Huaxian earthquake (M 8.0) occurred in 1556 (see below), trenching unveiled two more events around 2,715 and 5,610 years before present (Xu *et al.*, 1988b).

The Shanxi rift zone on the eastern side of Ordos has produced many large earthquakes (Figure 5.2). In its southern part where an M 8 event occurred in 1303 (see below), two more events occurred around 3,336–2,269 and 5,618–4,504 years before present (Bi *et al.*, 2011). On one of the faults in the central segment of the Shanxi rift, trenching shows three events around 3.06–3.53, ~5.32, and ~8.36 ka BP; the coseismic vertical slips of these events were 1.5–4.7 m, indicating magnitudes to be M 7.0 or above (Guo *et al.*, 2012). On the western branch of the northern segment of the Shanxi rift, four events occurred around 2.52, 5.68, 6.76–10.82, and 12.34 ka BP (Xie *et al.*, 2003). Trenching of various fault strands of the Shanxi rift system has revealed many other events, with recurrence times typically of a few thousand years.

Paleoseismic studies in the North China Plain, which is covered by thick Quaternary sediments, are less extensive and have been focused on where large historic earthquakes occurred. On the Tanlu fault, where the Ms 8.5 Tancheng earthquake occurred in 1668 (see below), trenching uncovered at least three more events around 3,500, 5,000–7,000, and ~10,000 years before present (Lin and Gao, 1987). The fault system bounding the northern side of the North China Block hosted the 1679 Sanhe earthquake (M 8.0); trenching studies there suggested three previous events around 10.85–9.71 ka BP, 7.390–6.68 ka BP, and 5.416–2.233 ka BP (Xu and Deng, 1996).

5.5.2 Large historic events

The Chinese catalog of historic earthquakes shows 49 large (M ≥ 6.5) events in North China since 1303 (Min *et al.*, 1995), including five catastrophic (M ≥ 8.0) events, which are described here.

1303 Hongdong earthquake (M 8.0)

The Great Hongdong earthquake occurred on September 17, 1303, near the Hongdong County within the Shanxi rift zone. The epicenter is estimated to be around 111.7° E, 36.3° N. The damaged area was about 500 km long and 250 km wide; the intensity reached

XI around the epicenter. Within the 44 km long and 18 km wide epicentral region all buildings were wiped out.

Earthquake damage was reported in 45 counties of three provinces. The death toll varies among different records, ranging from ~270,000 to ~470,000 (Yao *et al.*, 1984; Min *et al.*, 1995). Damage was widespread and extended to more than 250 km from the epicentral region. More than 100,000 houses and public buildings (mainly temples) were destroyed; fissures and sand blows were recorded in many places.

Since the 1960s, the site(s) of the Hongdong earthquake and the historic records have been studied by various groups (Su *et al.*, 2003). The 60 km long Huoshan fault was identified as the hosting fault (Xu and Deng, 1990). Xu and Deng (1990) estimated the surface rupture of the earthquake to be 45 km long with dextral and normal slips. They concluded that the maximum strike-slip displacement was about 10 m; the maximum normal slip was about 3–5 m.

1556 Huaxian earthquake (M 8.3)

The M 8.3 Huaxian earthquake occurred in the middle of the night of January 23, 1556, within the Weihe rift. The annals of the Ming Dynasty describe it this way: "Shanxi, Shannxi, and Henan provinces shook simultaneously with thundering sound. Damage was especially severe in Weinan, Huaxian, Zhouyi, Shanyuan, and Puzhou counties. Ground cracked, water gushed out, houses sank into ground, and hills were created suddenly. Rivers flooded, mountains collapsed, the death toll with named victims reached over 830,000; unnamed victims were countless." Landslide, flooding, famine, the bitter weather, and plague all contributed to the stunning death toll. This is the deadliest earthquake in human history.

Damage was reported in 101 counties in Shanxi province and four neighboring provinces, over a region of ~280,000 km^2. The maximum intensity reached XI in the epicentral region of 2,700 km^2. Historic records include detailed descriptions of surface ruptures, liquefactions, terrain changes, and flooding of the Yellow River and the Weihe River (Li, 1981).

The rupture was on the transtensional Huashan fault; the ruptured surface trace was about 200 km (Huan *et al.*, 2003). The average displacement was about 4 m (Xu and Deng, 1996). Within the epicentral region, broad subsidence occurred north of the fault and uplift south of the fault. The vertical offset reaches 10 m in Huaxian County.

1668 Tancheng earthquake (M 8.5)

The M 8.5 Tancheng earthquake occurred in the evening of July 25, 1668, on the Tanlu fault near Tancheng County (34.8° N, 118.5° E), Shandong Province. More than 50,000 people were killed, and damage was reported in 150 counties (Ma and Zhong, 2009).

This earthquake was recorded in more than 500 historic books and monuments. The annals of the Tancheng County described the earthquake as follows: "In the evening, a piercing sound suddenly came from the northwestern direction. The houses and trees moved up and down twice or three times, and then shook from side to side. The city walls, residential houses, and temples all collapsed." Near the epicenter the damage was complete, no building survived in a region of hundreds of square kilometers. The maximum intensity reached XII. Liquefaction was widespread; one village with thousands of families sank into the ground. The earthquake was followed by half a dozen large (M 6–7) aftershocks. The earthquake also caused a breach in the banks of the Yellow River, flooding a large area.

The Tancheng earthquake ruptured five segments of the Tanlu fault, with a total rupture length of 130 km. The maximum coseismic horizontal slip was about 10 m and vertical slip was about 3 m (Wang and Geng, 1996).

1679 Sanhe earthquake (M 8.0)

The M 8 Sanhe earthquake occurred on September 2, 1679, 40 km northeast of Beijing. The epicentral region includes the Sanhe and Pinggu counties, Hebei Province. The intensity reached XI in these two counties.

The Annals of the Sanhe County described the event: "... the earthquake occurred in the evening, with ground shaking from northwest to southeast. People couldn't stand on the ground, which moved like a boat tossed in a stormy sea. Nearly all houses collapsed. The ground cracked everywhere with black water swelling up for more than a month." The official death toll was more than 10,000 in Pinggu County, and 2,677 in Sanhe County. In Beijing, 485 people were killed, many buildings, including palaces in the Forbidden City and the city walls, cracked. The emperor Kangxi and the royal family stayed in tents as a precaution.

The surface rupture was estimated to be about 56 km (Xu *et al.*, 2002b), and the average coseismic vertical displacement was 1.4–3.16 m based on various drilling cores (Ran *et al.*, 1997; Jiang *et al.*, 2000). The hosting fault is part of the complex Zhangjiakou–Penglai fault system bounding the northern margin of the North China Block.

1695 Linfen earthquake (M 7.5–8.0)

On May 18, 1695, an M 7.5–8.0 earthquake occurred in the Linfen basin in the southern part of the Shanxi rift system (Wu *et al.*, 1988). According to the Annals of Linfen County, the earthquake occurred around 8:00 p.m. local time, with thundering sounds. In the epicentral region, 70–80% of buildings collapsed. Damage was reported in 125 counties of Shanxi and the neighboring provinces. The death toll varies among different records; it was ~56,200 in an official study conducted in 1875, while a Yuan Dynasty monument stated that 176,365 people lost their lives in this earthquake.

The epicenter of this earthquake is near that of the 1303 Hongdong earthquake, but on a different fault. The Linfen earthquake ruptured along a northwest-trending fault that cut the northeast-trending fault for the 1303 Hongdong earthquake. The rupture length of the Linfen earthquake was about 70 km, with a high dip angle and sinistral slip (Cheng *et al.*, 1995; Hu *et al.*, 2002).

5.5.3 Large instrumentally recorded earthquake

Several earthquakes larger than M 7.0 in the North China Plain have been recorded by instruments since 1960 (Figure 5.2). The most devastating ones include the Xingtai, Haicheng, and Tangshan earthquakes.

The 1966 Xingtai earthquake (Ms 7.2)

Between March 8 and 29, 1966, a cluster of five earthquakes larger than M 6.0 occurred in the Xingtai region of the Hebei Province. The largest one was an Ms 7.2 event which occurred on March 22; the maximum intensity reached X in the epicentral region. These earthquakes destroyed more than 5 million buildings, killed 8,064 people and injured ~38,000 (Seismological Bureau of Hebei Province, 1986).

The earthquake cluster occurred on a NNE-trending right-lateral fault along the Sulu graben (Chung and Cipar, 1983; Xu *et al.*, 1988a). The rupture length is about 50 km, and the maximum right-lateral strike-slip displacement was ~1.0 m (The Geodetic Survey Brigade for Earthquake Research, 1975). The focal depths of these events range from 9 to 25 km (Chen *et al.*, 1975), consistent with the depths of aftershocks.

The 1975 Haicheng earthquake (Ms 7.3)

The 1975 Haicheng earthquake occurred on February 4 in the Haicheng region, Northeastern China (122.83° E, 40.70° N), on the northwest-trending Jingzhou–Haicheng fault. The focal depth is 16–21 km, with a surface rupture of ~50 km length. The maximum left-lateral slip is 0.55 m (Jones *et al.*, 1982; Zhu and Wu, 1982).

The Haicheng earthquake's maximum intensity reached IX, yet the human casualties, with 1,328 killed and 16,980 injured, were much fewer than usual thanks to a successful prediction (Scholz, 1977). A major factor leading to the prediction was a long foreshock sequence: more than 500 of them were recorded within 4 days before the mainshock (Jones *et al.*, 1982). Wang *et al.* (2006) found that the role of foreshocks in this prediction was more psychological than scientific: the jolts and damage from increased seismicity in the preceding months worried earthquake workers and the general public, and the intensified foreshocks in the last day before the mainshock prompted some local officials to order an evacuation. In other places where official orders were not issued, the increased seismicity caused many residents to evacuate voluntarily.

The 1976 Tangshan earthquake (Ms 7.8)

The Great 1976 Tangshan earthquake occurred in the morning of July 28. The earthquake struck Tangshan, an industrial city 150 kilometers east of Beijing. In a brief moment, the earthquake destroyed the entire city and killed more than 242,000 people (Chen *et al.*, 1988). The maximum intensity reached XI in the epicentral region. An Ms 7.1 event followed 19 hours after the mainshock. The length of surface rupture produced by the mainshock is more than 47 km, and left-lateral slip is 1.5–2.3 m (Guo *et al.*, 2011). The seismic moment released by the mainshock was 11.7×10^{19} N m.

The earthquake occurred within the broad fault system bounding the northern margin of the North China Plain. The hosting fault for the mainshock is a northeast-trending strike-slip faulting system with a dextral bend in the middle that divides the fault into the southern and northern segments (Nábělek *et al.*, 1987; Shedlock *et al.*, 1987).

5.6 Spatiotemporal patterns of large earthquakes

The spatiotemporal pattern of seismicity for a region is essential for hazard assessment. Establishing such a pattern, however, is difficult for most intracontinental regions because of the slow loading, infrequent large earthquakes, and incomplete earthquake records. The situation is better for North China – with its relatively frequent large earthquakes and long historic records, useful insight may be obtained.

5.6.1 Long-distance roaming of large earthquakes

Here we show the spatiotemporal occurrence of large earthquakes in North China during the past 700 years. For this relatively recent period, the catalog is likely complete for M ≥ 6 events (Huang *et al.*, 1994) and it includes 49 M ≥ 6.5 events and at least four earthquakes of M ≥ 8, described in the previous section. Before the 1303 Hongdong earthquake, large earthquakes in North China were concentrated along the Weihe and Shanxi rifts and scattered over the North China Plain (Figure 5.7a). The Hongdong earthquake (M 8.0) occurred within the Shanxi rift, followed by the 1556 Huaxian earthquake (M 8.3) in the Weihe rift (Figure 5.7b). During this period, seismicity seemed to be concentrated within the Weihe–Shanxi rift systems, which would fit the model for rift zones to be the main host structures of intraplate earthquakes (Johnston and Kanter, 1990). However, the next large earthquake, the 1668 Tancheng earthquake (M 8.5), did not occur within the rift systems but more than 700 km to the east, on the Tanlu fault zone, which had little deformation through the late Cenozoic and only moderate previous seismicity. A decade later, another large event, the 1679 Sanhe earthquake (M 8.0), occurred only 40 km north of Beijing in a fault zone with limited previous seismicity and no clear surface exposure (Figure 5.7c). Then, in 1695, the Linfen earthquake (M 7.5–8.0) occurred in the Shanxi rift again, near the site of the 1303 Hongdong earthquake but on a different fault. Since then the Shanxi and Weihe rifts have been quiescent for more than 300 years, with only a few moderate earthquakes. Meanwhile, seismicity in the North China Plain apparently increased, producing three damaging earthquakes in the past century: the 1966 Xingtai earthquakes, the 1975 Haicheng earthquake, and the 1976 Tangshan earthquake (Figure 5.7d). These three large earthquakes were unexpected and occurred on previously unrecognized faults.

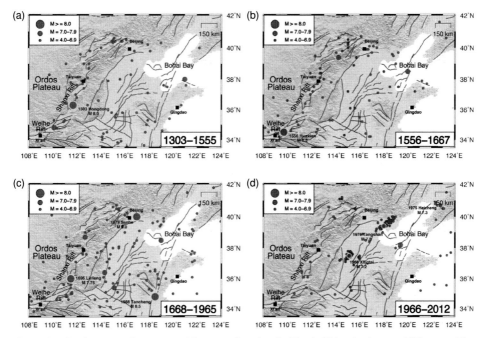

Figure 5.7 Spatiotemporal patterns of large earthquakes in North China in the past 700 years. The panels show earthquakes (epicenters indicated by dots) during various periods. Large dots with light rims are large earthquakes discussed in the text. After Liu *et al.* (2011).

5.6.2 Fault coupling and interaction

The spatiotemporal pattern of earthquakes in North China is much more complex than that at plate boundaries. Large earthquakes roamed between widespread fault systems. Is this roaming a random effect of long recurrence times on different faults? Or could these large earthquakes be related to each other through mechanical coupling and interaction between remote fault systems?

One approach is to study the stress links between these earthquakes. Wang *et al.* (1982) calculated the stress perturbations of the earthquake sequence in North China during the past 700 years and suggested that the subsequent events were linked to the previous ones. Shen *et al.* (2004) calculated the changes of Coulomb stress by the 48 M $\geq$ 6.5 events following the 1303 Hongdong earthquake, and concluded that 39 of them occurred in places where stress was elevated by previous earthquakes.

Another approach is to compare the seismic moment release on these fault systems. For a system of mechanically coupled fault zones, the total moment release rate should stay at a certain level, with the moment release rate of individual fault zones being complementary with each other (Dolan *et al.*, 2007; Luo and Liu, 2012). Liu *et al.* (2011) showed that the moment release between the Weihe and Shanxi rifts seems to be complementary to each

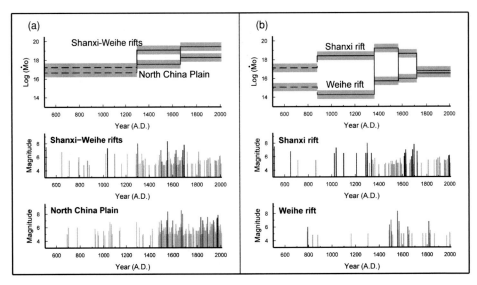

Figure 5.8 (a) Bottom two panels show occurrence and magnitude (M ≥ 4.0) of earthquakes in the North China Plain and the Shanxi–Weihe rifts. Top panel compares the time-averaged rate of seismic moment release (N m/yr) between them (tones are correlated with those in the bottom panels). Dashed lines are for older intervals, wherein some events are likely unrecorded. Gray bands indicate uncertainties associated with an estimated ±0.3 error in the magnitudes, which were based on the maximum intensity and the radius of the affected area (Ma, 1989). (b) Similar to (a) for the comparison of moment release between the Weihe and the Shanxi rifts. After Liu *et al.* (2011).

other, with increases in one corresponding to decreases in the other. Similar correlation exists between the Shanxi–Weihe rift system and the faults within the North China Plain (Figure 5.8). These results suggest that the fault systems in North China are mechanically coupled with each other.

5.6.3 A conceptual model for mid-continental earthquakes

These spatiotemporal patterns of earthquakes are not unique to North China; similar observations have been made in other mid-continents including Australia (Clark and McCue, 2003) and northwest Europe (Camelbeeck *et al.*, 2007). In the Central United States, the Meers fault in Oklahoma had a major earthquake about 1,200 years ago but is inactive today (Calais *et al.*, 2003). The New Madrid seismic zone, which has experienced several large earthquakes in the past few thousand years including at least three M ≥ 6.8 events in 1811–1812 (Johnston and Schweig, 1996; Tuttle *et al.*, 2002; Hough and Page, 2011), shows no significant surface deformation today (Calais and Stein, 2009) and its activity may be ending (Newman *et al.*, 1999; Calais and Stein, 2009; Stein and Liu, 2009).

The roaming of large earthquakes between widespread fault systems illustrates fundamental differences between earthquakes in mid-continents and at plate boundaries

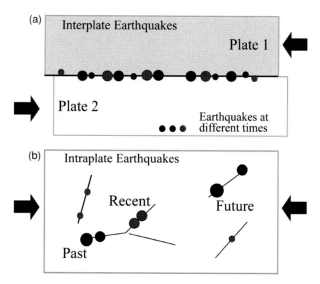

Figure 5.9 Conceptual models for the differences between interplate (a) and intraplate (b) earthquakes. For interplate earthquakes, the plate boundary fault is loaded at a constant rate by the steady relative plate motion, causing quasi-periodic earthquakes to concentrate along the plate boundary. In plate interiors, slow far-field tectonic loading is shared by a complex system of interacting faults. On each fault, the loading rate may be variable, and earthquakes may shut off on one fault and migrate to another.

(Figure 5.9). Plate boundary faults are loaded at constant rates by the steady relative plate motion. Consequently, earthquakes concentrate along the plate boundaries, and some quasi-periodic occurrences may be expected (Figure 5.9a), although the temporal patterns are often complicated (Jackson and Kagan, 2006). In contrast, in mid-continents the tectonic loading is shared by a complex system of interacting faults spread over a large region (Figure 5.9b), such that a large earthquake on one fault could affect the loading rates on remote faults (Li *et al.*, 2009). Because the slow tectonic loading is shared by many faults in mid-continents, individual faults may remain dormant for a long time and then become active for a short period, while seismicity moves to other faults.

5.7 Implications for earthquake hazards

The complex spatiotemporal patterns of earthquakes in North China pose serious challenges to the assessment of earthquake hazards. The current practice of hazard assessment is heavily influenced by the occurrence of previous large earthquakes (Figure 5.10), assuming that large earthquakes will be likely to repeat on the same faults. This line of reasoning can be traced back to Reid's elastic rebound theory (Reid, 1910), which implies cycles of energy accumulation and release on a given fault. However, the validity of this premise for intracontinental earthquakes has been questioned (Stein *et al.*, 2011, 2012). In North China,

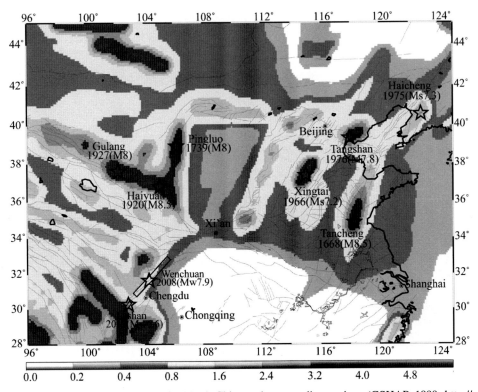

Figure 5.10 Seismic hazard map for North China and surrounding regions (GSHAP, 1999, http://www.seismo.ethz.ch/static/GSHAP). The hazard is expressed as the peak ground acceleration (PGA) on firm rocks, in m/s^2, expected to be exceeded in the next 50 years with a probability of 10%. The epicenters and rupture zones of the 2008 Wenchuan earthquake (Mw 7.9) and 2013 Lushan earthquakes (Mw 6.6) are indicated by stars and rectangles, respectively. For color version, see Plates section.

large earthquakes roamed between widespread fault systems; in the past 2,000 years, not a single M ≥ 7.0 earthquake ruptured the same fault segment twice. Such earthquake behavior raises questions about hazard maps such as that in Figure 5.10. Since the publication of this hazard map in 1999, a number of large earthquakes have occurred in this region, providing opportunities to test this assessment. Unfortunately, the map failed the test.

The devastating 2008 Wenchuan earthquake (Mw 7.9), which killed ~90,000 people, was not expected by the map maker (Figure 5.10). This earthquake ruptured the central and northern segments of the Longmenshan fault, which was assigned a moderate to low risk by the hazard map, presumably because of the low fault slip rates (<3 mm/yr) and the lack of large earthquakes in the past few centuries. The high risk zones on the map are where large earthquakes occurred in the recent past, such as along the Xianshuihe fault. Five years after the Wenchuan earthquake, the 2013 Lushan earthquake (Mw 6.6) ruptured the southern segment of the Longmenshan fault, also assigned with moderate to low risk on the hazard map (Figure 5.10). The Wenchuan and Lushan earthquakes show clearly that having no previous large earthquakes does not mean having no future large earthquakes.

Within North China, the hazard map assigns the highest risk to the Tangshan, Xingtai, and Tancheng regions, apparently because of the 1966 Xingtai and the 1976 Tangshan earthquakes, and because of the big Tancheng earthquake (M 8.5) in 1668. The Shanxi rift zone was assigned with lower risk, presumably because it has been seismically quiescent for the past 300 years. The Weihe rift zone was assigned a low risk because the 1556 Huaxian earthquake (M 8.0) was the last large earthquake within the rift.

The nearly 3,000 years of earthquake records in North China, however, indicate an earthquake behavior more complicated than that assumed in the current practice of hazard assessment. Because the large earthquakes tend to roam between widespread fault systems, previous large earthquakes may not be a good indicator of where a future large earthquake will occur; because strain can accumulate in the fault zones for thousands of years before being released by a large earthquake, low slip rates do not mean being safe; and not having previous large earthquakes does not mean no large earthquakes in the future. And finally, because it usually take thousands of years for intracontinental fault zones to accumulate enough strain energy for a large earthquake, places where large earthquakes occurred in the recent past are not necessarily more dangerous than other places.

Another challenge for earthquake hazard assessment in North China, and in other mid-continents, is the long sequences of aftershocks (Stein and Liu, 2009). Small earthquakes are often regarded as signs of stress building up towards the next big earthquake; their occurrence in source regions of previous large earthquakes therefore often causes alarm. One example is the recent sequence of moderate sized earthquakes in the Tangshan region, which includes an M 4.8 event on May 28 and an M 4.0 event on June 18, 2012. These earthquakes caused widespread concerns and heated debate in China: are they aftershocks of the great 1976 Tangshan earthquake, or are they harbingers of a new period of active seismicity in Tangshan and the rest of North China, where seismic activity seems fluctuate between highs and lows over periods of a few decades (Ma, 1989)?

Liu and Wang (2012) showed that this recent seismicity in Tangshan is likely the aftershocks of the 1976 Tangshan earthquake for the following reasons: (1) The seismicity rate in the Tangshan region has been decaying since 1976, following Omori's law, but is still clearly above the background level (Figure 5.11). (2) The seismicity rates of the Tangshan, Xingtai, and Haicheng regions for 1986–2010 (i.e., 10 years after the Tangshan earthquake and 20 years after the Xingtai earthquake) are clearly higher than the average value for the North China Plain. This indicates either that these regions are tectonically more active than the rest of the North China Plain, or the continuing influence of aftershocks of the large earthquakes decades ago. Because these regions showed no sign of abnormal tectonic activity before the large earthquakes, aftershocks are more likely the cause. (3) Strain rates calculated from GPS data are higher in the Tangshan and Xingtai regions than that of the

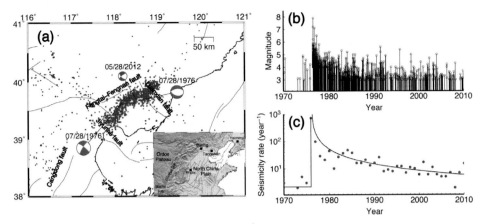

Figure 5.11 (a) Seismicity in the source region of the 1976 Tangshan earthquake. Dots are epicenters of aftershocks; circles are background seismicity (M ≥ 3.0, 1970–2011). The focal mechanism solutions are for the two mainshocks of 1976 and the May 5, 2012 event. The inset map shows the location of the Tangshan region. (b) Earthquake sequence in the source region of the great Tangshan earthquake since 1970. (c) Seismicity rates (number of events per year) of the Tangshan earthquake sequence. Solid lines are least-square fitting.

rest of the North China Plain. Again, because these regions have no evidence of unusually high tectonic activity relative to other fault zones in the North China Plain, the higher strain rates indicate that postseismic deformation continues to the present (the Haicheng region does not have sufficient GPS stations to allow a meaningful strain calculation).

The long-distance roaming of large earthquakes and their long sequences of aftershocks hence make seismic hazard assessment, a challenging task for any region (Stein *et al.*, 2011, 2012), more difficult in North China and other mid-continents. If the earthquakes roam between widespread faults, the current practice, which uses previous large earthquakes as hints to locate future earthquakes, tends to overestimate the hazard in places where previous large earthquakes occurred, and underestimate the hazard elsewhere. And if the aftershock sequences last a long time, small aftershocks may be misread as precursors.

This situation cannot be fundamentally improved by simply extending the earthquake records alone. This is because earthquakes in North China and other mid-continents do not fit the model of well-identified faults being steadily loaded. Instead, these earthquakes are the products of widespread faults interacting with each other in a complex dynamic system (Stein *et al.*, 2009). Hence, we cannot treat each individual fault or fault segment as an isolated system and expect some regular recurrence time. In a complex system where changes of any part have nonlinear impacts on all other parts (Stein *et al.*, 2009), the concepts of stress cycles, characteristic earthquakes, and recurrence time may not hold – their validity has been questioned even for interplate earthquakes (Jackson and Kagan, 2006). Recognizing the fault systems in North China and other mid-continents as complex dynamic systems does not make hazard assessment easier, but it would be a necessary paradigm shift. Geodynamic modeling will need to explore how faults interact over long distances and multiple timescales (e.g., Luo and Liu, 2012), rather than focusing on stress evolution on an isolated fault or fault segment. Paleoseismic studies should not be limited to where large earthquakes occurred recently, and detailed and careful analysis are needed to determine the age and spatial scale of past events, without trying to fit expected recurrence intervals (Xu and Deng, 1996). As for North China, the potential for repeated large earthquakes in Tangshan and Xingtai is likely lower than that generally perceived, if the 2,000-year historic record can provide any hints. In contrast, the Shanxi rift zone, which has produced more than 36 M $\geq$ 6.5 earthquakes since 1303 but has been quiescent for the past 300 years, may deserve closer monitoring, and the Weihe rift zone, the host for the 1556 Huaxian earthquake (M 8.0) and with relatively high contemporary strain rates (Figure 5.5a), may not be as safe as suggested by the hazard map.

Acknowledgements

Seth Stein inspired our interest in intraplate earthquakes and co-authored a number of papers reviewed here. We thank Roy Van Arsdale and Pradeep Talwani for their constructive reviews. ML's research in China is supported by NSF/OISE grant 0730154 and a grant from the Chinese Academy of Sciences. HW acknowledges support from the National Natural Science Foundation of China (Grants 41104058 and 41104057).

References

Allen, M. B., D. I. M. Macdonald, Z. Xun, S. J. Vincent, and M. C. Brouet (1998). Transtensional deformation in the evolution of the Bohai Basin, northern China. In *Continental Transpressional and Transtensional Tectonics*, ed. R. E. Holdsworth, R. A. Strachan, and J. F. Dewey. Geological Society, London, Special Publication, 135 215–229.

Bi, L., H. He, Y. Xu, Z. Wei, and F. Shi (2011). The extraction of knickpoint series based on the high resolution DEM data and the identification of paleoearthquakes: a case study of the Huoshan Mts. Piedmont fault (in Chinese with English abstract). *Seismology and Geology*, 33(04), 963–977.

Bryant, D. L., J. C. Ayers, S. Gao, C. F. Miller, and H. Zhang (2004). Geochemical, age, and isotopic constraints on the location of the Sino–Korean/Yangtze Suture and evolution of the Northern Dabie Complex, east central China. *Geological Society of America Bulletin*, 116(5–6), 698–717.

Calais, E., and S. Stein (2009). Time-variable deformation in the New Madrid Seismic Zone. *Science*, 323(5920), 1442.

Calais, E., M. Vergnolle, V. San'kov, *et al.* (2003). GPS measurements of crustal deformation in the Baikal-Mongolia area (1994–2002): implications for current kinematics of Asia. *Journal of Geophysical Research, Solid Earth*, 108(B10), 2501.

Camelbeeck, T., K. Vanneste, P. Alexandre, *et al.* (2007). Relevance of active faulting and seismicity studies to assess long term earthquake activity in Northwest Europe. In *Continental Intraplate Earthquakes: Science, Hazard, and Policy Issues*, ed. S. Stein, and S. Mazzotti. Geological Society of America Special Paper 425, pp. 193–224.

Chen, L., C. Cheng, and Z. Wei (2009). Seismic evidence for significant lateral variations in lithospheric thickness beneath the central and western North China Craton. *Earth and Planetary Science Letters*, 286(1–2), 171–183.

Chen, Y., B. Lin, Z. Lin, and Z.-Y. Li (1975). The focal mechanism of the 1966 Xingtai earthquake as inferred from the ground deformation observations. *Acta Geophysica Sinica*, 18(3), 164–182.

Chen, Y., K. L. Tsoi, F. B. Chen, *et al.* (1988). *The Great Tangshan Earthquake of 1976: An Anatomy of Disaster*. Oxford: Pergamon Press.

Cheng, X., Z. Su, and W. An (1995). Tectonic background of the big Linfen earthquake in 1695. *Earthquake Research in Shanxi* (in Chinese with English Abstract), Z1, 43–48.

Chung, W.-Y., and J. J. Cipar (1983). Source modeling of the Hsingtai, China earthquakes of March 1966. *Physics of the Earth and Planetary Interiors*, 33(2), 111–125.

Clark, D., and K. McCue (2003). Australian paleoseismology: towards a better basis for seismic hazard estimation. *Annals of Geophysics*, 46(5), 1087–1106.

Deng, Q., and Y. Liao (1996). Paleoseismology along the range-front fault of Helan Mountains, north central China. *Journal of Geophysical Research: Solid Earth*, 101(B3), 5873–5893.

Deng, Q. D., P. Z. Zhang, Y. K. Ran, *et al.* (2002). Basics characteristics of active tectonics of China (in Chinese). *Science in China (Series D)*, 32(12), 1020–1030.

Deng, Q., P. Zhang, Y. Ran, *et al.* (2003). Basic characteristics of active tectonics of China. *Science in China Series D: Earth Sciences*, 46(4), 356–372.

Dolan, J. F., D. D. Bowman, and C. G. Sammis (2007). Long-range and long-term fault interactions in Southern California. *Geology*, 35(9), 855–858.

Griffin, W. L., A. Zhang, S. Y. O'Reilly, and C. G. Ryan (1998). Phanerozoic evolution of the lithosphere beneath the Sino-Korean Craton. In *Mantle Dynamics and Plate Interactions in East Asia*, ed. M. F. J. Flower, S.-L. Chung, C.-H. Lo, and T.-Y. Lee. Washington, DC: American Geophysical Union, pp. 107–126.

Gu, G. X., T. H. Lin, Z. L. Shi, and Q. Li (1983). *Earthquake Catalog of China: 1831 B.C. – 1969 A.D. (in Chinese)*. Beijing: Science Publishing House.

Guo, H., W. Jiang, and X. Xie (2011). Late-Quaternary strong earthquakes on the seismogenic fault of the 1976 Ms 7.8 Tangshan earthquake, Hebei, as revealed by drilling and trenching. *Science in China Series D-Earth Sciences*, 41(7), 1009–1028.

Guo, H., W. Jiang, and X. Xie (2012). Analysis of Holocene faulting phenomena revealed in the three trenches along the northern and central Jiaocheng fault, Shanxi. *Seismology and Geology (in Chinese with English abstract)*, (01), 76–92.

He, J., M. Liu, and Y. Li (2003). Is the Shanxi rift of northern China extending? *Geophysical Research Letters*, 30, doi:10.1029/2003GL018764.

Hough, S., and M. Page (2011). Toward a consistent model for strain accrual and release for the New Madrid Seismic Zone, central United States. *Journal of Geophysical Research*, 116, doi:10.1029/2010JB007783.

Hu, X., G. Diao, J. Gao, *et al.* (2002). Application of present small earthquakes to infer the focal faults of two large historical earthquakes in Hongdong and Linfen, Shanxi Province. *Earthquake Research in China*, 18(1), 76–85.

Huan, W., Z. Shi, and S. Li (2003). The new evidence for epicenter location and seismogenic structure of the 1556 M 8 1/4 large earthquake. *Earthquake Research in China*, 19(01), 20–32.

Huang, J. L., and D. P. Zhao (2006). High-resolution mantle tomography of China and surrounding regions. *Journal of Geophysical Research, Solid Earth*, 111, B09305, doi:10.1029/2005JB004066.

Huang, W. Q., W. X. Li, and X. F. Cao (1994). Research on the completeness of earthquake data in the Chinese mainland (II). *Acta Seismologica Sinica*, 16(4), 423–432.

Huang, Z., H. Li, Y. Zheng, and Y. Peng (2009). The lithosphere of North China Craton from surface wave tomography. *Earth and Planetary Science Letters*, 288(1–2), 164–173.

Jackson, D. D., and Y. Y. Kagan (2006). The 2004 Parkfield earthquake, the 1985 prediction, and characteristic earthquakes: lessons for the future. *Bulletin of the Seismological Society of America*, 96(4B), S397–409.

Jiang, W., Z. Hou, Z. Xiao, and X. Xie (2000). Study on paleoearthquakes of Qixinzhuang trench at the Xiadian fault, Beijing plain (in Chinese with English abstract). *Seismology and Geology*, 04, 413–422.

Johnston, A. C., and L. R. Kanter (1990). Earthquakes in stable continental crust. *Scientific American*, 262(3), 68–75.

Johnston, A. C., and E. S. Schweig (1996). The enigma of the New Madrid earthquakes of 1811–1812. *Annual Review of Earth and Planetary Sciences*, 24, 339–384.

Jones, L. M., B. Wang, S. Xu, and T. J. Fitch (1982). The foreshock sequence of the February 4, 1976, Haicheng earthquake (M = 7.3). *Journal of Geophysical Research, Solid Earth*, 87(B6), 4575–4584.

Li, Q., M. Liu, and S. Stein (2009). Spatial-temporal complexity of continental intraplate seismicity: insights from geodynamic modeling and implications for seismic hazard estimation. *Bulletin of the Seismological Society of America*, 99(1), doi: 10.1785/0120080005.

Li, S. (1981). *Earthquakes in China*. Beijing: Seismological Press.

Li, S., G. Zhao, L. Dai, *et al.* (2012). Cenozoic faulting of the Bohai Bay Basin and its bearing on the destruction of the eastern North China Craton. *Journal of Asian Earth Sciences*, 47(0), 80–93.

Li, Z.-X. (1994). Collision between the North and South China blocks: a crustal-detachment model for suturing in the region east of the Tanlu fault. *Geology*, 22(8), 739–742.

Lin, W., and W. Gao (1987). The occurrence intervals of large earthquakes in the Yishu fault zone (in Chinese). *China Earthquakes*, 03, 36–42.

Liu, D. Y., A. P. Nutman, W. Compston, J. S. Wu, and Q. H. Shen (1992). Remnants of ≥3800 Ma crust in the Chinese part of the Sino-Korean craton. *Geology*, 20(4), 339–342.

Liu, M., and H. Wang (2012). Roaming earthquakes in China highlight midcontinental hazards. *Eos, Transactions, American Geophysical Union*, 93(45), 453–454.

Liu, M., and Y. Yang (2005). Contrasting seismicity between the North China and South China blocks: kinematics and geodynamics. *Geophysical Research Letters*, 32 (L12310), doi:10.1029/2005GL023048.

Liu, M., X. Cui, and F. Liu (2004). Cenozoic rifting and volcanism in eastern China: a mantle dynamic link to the Indo-Asian collision? *Tectonophysics*, 393, 29–42.

Liu, M., Y. Yang, Z. Shen, *et al.* (2007). Active tectonics and intracontinental earthquakes in China: the kinematics and geodynamics. In *Continental Intraplate Earthquakes:*

Science, Hazard, and Policy Issues, ed. S. Stein, and S. Mazzotti. Geological Society of America Special Paper 425, pp. 209–318, doi: 210.1130/2007.2425.

Liu, M., S. Stein, and H. Wang (2011). 2000 years of migrating earthquakes in North China: how earthquakes in midcontinents differ from those at plate boundaries. *Lithosphere*, 3, 128–132.

Luo, G., and M. Liu (2012). Multi-timescale mechanical coupling between the San Jacinto Fault and the San Andreas Fault, southern California. *Lithosphere*, doi: 10.1130/L1180.1131.

Ma, X. (1989). *Lithospheric Dynamics Altas of China*. Beijing: China Cartographic Publishing House.

Ma, Y., and P. Zhong (2009). Summary of the M8 1/2 Tancheng earthquake in 1668 (in Chinese). *Recent Developments in World Seismology*, 2, 9–18.

Menzies, M. A., and Y. Xu (1988). Geodynamics of the North China Craton. In *Mantle Dynamics and Plate Interactions in East Asia*, ed. M. F. J. Flower, S. L. Chung, C. H. Lo, and T. Y. Lee. Washington, DC: American Geophysical Union, 27, 155–165.

Menzies, M., Y. G. Xu, H. F. Zhang, and W. M. Fan (2007). Integration of geology, geophysics and geochemistry: a key to understanding the North China Craton. *Lithos*, 96(1–2), 1–21.

Min, Z., G. Wu, Z. Jiang, C. Liu, and Y. Yang (1995). *The Catalog of Chinese Historic Strong Earthquakes (B.C. 23–AD 1911)* (in Chinese). Beijing: Seismological Publishing House.

Mooney, W. D., J. Ritsema, and Y. K. Hwang (2012). Crustal seismicity and the earthquake catalog maximum moment magnitude (Mc_{max}) in stable continental regions (SCRs): correlation with the seismic velocity of the lithosphere. *Earth and Planetary Science Letters*, 357–358, 78–83.

Nábělek, J., W.-P. Chen, and H. Ye (1987). The Tangshan earthquake sequence and its implications for the evolution of the North China Basin. *Journal of Geophysical Research*, 92(B12), 12615–12628.

Newman, A., S. Stein, J. Weber, *et al.* (1999). Slow deformation and lower seismic hazard at the New Madrid Seismic Zone. *Science*, 284(5414), 619–621.

Ran, Y., Q. Deng, X. Yang, *et al.* (1997). Paleoearthquakes and recurrence interval on the seismogenic fault of 1679 Sanhe-pinggu M8 earthquake, Hebei and Beijing. *Seismology and Geology*, 19(3), 193–201.

Ran, Y., L. Chen, X. Yang, and Z. Han (2003a). Recurrence characteristics of late-quaternary strong earthquakes on the major active faults along the northern border of Ordos block. *Science in China Series D, Earth Sciences*, 46, 189–200.

Ran, Y., P. Zhang, and L. Chen (2003b). Late Quaternary history of paleoseismic activity along the Hohhot Segment of the Daqingshan piedmont fault in Hetao depression zone, North China. *Annals of Geophysics*, 46, 1053–1069.

Reid, H. F. (1910). *The Mechanics of the Earthquake, The California Earthquake of April 18, 1906*, Report of the State Investigation Commission Vol. 2.

Ren, J., K. Tamaki, S. Li, and J. Zhang (2002). Late Mesozoic and Cenozoic rifting and its dynamic setting in Eastern China and adjacent areas. *Tectonophysics*, 344(3–4), 175–205.

Scholz, C. H. (1977). A physical interpretation of the Haicheng earthquake prediction. *Nature*, 267, 121–124.

Seismological Bureau of Hebei Province (1986). *The 1966 Xingtai Earthquake*. Beijing: China Seismological Press.

Shedlock, K. M., J. Baranowski, X. Weiwen, and H. X. Liang (1987). The Tangshan aftershock sequence. *Journal of Geophysical Research*, 92(B3), 2791–2803.

Shen, Z. K., C. Zhao, A. Yin, and D. Jackson (2000). Contemporary crustal deformation in east Asia constrained by Global Positioning System measurements. *Journal of Geophysical Research*, 105, 5721–5734.

Shen, Z. K., Y. Wan, W. Gan, T. Li, and Y. Zeng (2004). Crustal stress evolution of the last 700 years in North China and earthquake sequence (in Chinese). *Earthquake Research in China*, 20, 211–228.

Stein, S., and M. Liu (2009). Long aftershock sequences within continents and implications for earthquake hazard assessment. *Nature*, 462(7269), 87–89.

Stein, S., M. Liu, E. Calais, and Q. Li (2009). Mid-Continent earthquakes as a complex system. *Seismological Research Letters*, 80(4), 551–553.

Stein, S., R. Geller, and M. Liu (2011). Bad assumptions or bad luck: why earthquake hazard maps need objective testing. *Seismological Research Letters*, 82(5), 623–626.

Stein, S., R. J. Geller, and M. Liu (2012). Why earthquake hazard maps often fail and what to do about it. *Tectonophysics*, 562–563, 1–25.

Su, Z., Z. Yuan, and J. Zhao (2003). A review on studies concerned with the 1303 Hongtong earthquake (in Chinese with English abstract). *Earthquake Research in Shanxi*, 03, 4–22.

The Geodetic Survey Brigade for Earthquake Research, N.S.B. (1975). Crustal deformation associated with the Hsingtai earthquake in March, 1966. *Chinese Journal of Geophysics, Chinese Edition*, 18(3), 153–163.

Tuttle, M. P., E. S. Schweig, J. D. Sims, *et al.* (2002). The earthquake potential of the New Madrid seismic zone. *Bulletin of the Seismological Society of America*, 92(6), 2080–2089.

Wang, H., and J. Geng (1996). Discussion about hypocenter parameters of the M8.5 Tancheng earthquake in 1668 (in Chinese). *Journal of Seismology*, 4, 29–35.

Wang, H., M. Liu, J. Cao, X. Shen, and G. Zhang (2011). Slip rates and seismic moment deficits on major active faults in mainland China. *Journal of Geophysical Research*, 116, B02405, doi:02410.01029/02010JB007821.

Wang, K., Q. Chen, S. Sun, and A. Wang (2006). Predicting the 1975 Haicheng earthquake. *Bulletin of the Seismological Society of America*, 96(3), 757–795.

Wang, R., S. Y. Sun, and Y. G. Cai (1982). Numerical simulation of earthquake sequences in North China over the past 700 years (in Chinese). *Science in China, B*, 8(745–753).

Wesnousky, S. G., L. M. Jones, C. H. Scholz, and Q. Deng (1984). Historical seismicity and rates of crustal deformation along the margins of the Ordos block, North China. *Bulletin of the Seismological Society of America*, 74, 1767–1783.

Wu, L., S. Qi, and R. Wang (1988). The 1303 Hongton earthquake of M = 8 and the 1695 Linfen earthquake M = 8 in Shanxi Province. In *China Special Large Earthquakes Research* (in Chinese), ed. Z.-J. Guo and Z.-J. Ma. Beijing: Seismological Press, pp. 6–35

Xie, X., W. Jiang, R. Wang, H. Wang, and X. Feng (2003). Holocene paleo-seismic activities on the Kouquan fault zone, Datong basin, Shanxi Province (in Chinese with English abstract). *Seismology and Geology* 03, 359–374.

Xu, C., L. Dong, C. Shi, Y. Li, and X. Hu (2002a). A study on annual accumulation of strain energy density significance by using GPS measurements in North China (in Chinese with English abstract). *Chinese Journal of Geophysics, Chinese Edition*, 45(4), 497–506.

Xu, J., Z. Fang, and L. Yang (1988a). Tectonic background and causative fault of 1966 Xingtai Ms7.2 earthquake (in Chinese). *Seismology and Geology*, 10(4), 51–59.

Xu, X., and Q. Deng (1990). The features of Late Quaternary activity of the piedmont fault of Mt. Huoshan, Shanxi Province and the 1303 Hongtong earthquake (M = 8) (in Chinese). *Seismology and Geology*, 12(1), 21–30.

Xu, X., and Q. Deng (1996). Nonlinear characteristics of paleoseismicity in China. *Journal of Geophysical Research, Solid Earth*, 101(B3), 6209–6231.

Xu, X., H. Zhang, and Q. Deng (1988b). The paleoearthquake traces on Huashan front fault zone in Weihe basin and its earthquake intervals (in Chinese with English abstract). *Seismology and Geology*, 04, 206.

Xu, X., X. Ma, and Q. Deng (1993). Neotectonic activity along the Shanxi rift system, China. *Tectonophysics*, 219(4), 305–325.

Xu, X. W., W. M. Wu, X. K. Zhang, *et al.* (2002b). *Neotectonics and Earthquakes in the Capital Circle Region*. Beijing: Science Press.

Xu, Y. G. (2001). Thermo-tectonic destruction of the archaean lithospheric keel beneath the Sino-Korean craton in China: evidence, timing and mechanism. *Physics and Chemistry of the Earth, Part A: Solid Earth and Geodesy*, 26(9–10), 747–757.

Xu, Y. G., M. A. Menzies, M. F. Thirlwall, *et al.* (2003). "Reactive" harzburgites from Huinan, NE China: products of the lithosphere-asthenosphere interaction during lithospheric thinning? *Geochimica et Cosmochimica Acta*, 67, 487–505.

Yao, G., Y. Jiang, and X. Yu (1984). Investigation on the 1303 Zhaocheng, Shanxi, earthquake (M = 8) and its parameters concerned (in Chinese with English abstract). *Journal of Seismological Research*, 7(13), 313–326.

Ye, H., B. Zhang, and F. Mao (1987). The Cenozoic tectonic evolution of the Great North China: two types of rifting and crustal necking in the Great North China and their tectonic implications. *Tectonophysics*, 133, 217–227.

Yin, A., and S. Nie (1996). Phanerozoic palinspastic reconstruction of China and its neighbouring regions. In *Tectonic Evolution of Aisa*, ed. A. Yin and M. Harrison. New York: Cambridge University Press, Rubey Volume IX, pp. 442–485.

Zhai, M.-G., and M. Santosh (2011). The early Precambrian odyssey of the North China Craton: a synoptic overview. *Gondwana Research*, 20(1), 6–25.

Zhang, P., and W. Gan (2008). Combined model of rigid-block motion with continuous deformation: patterns of present-day deformation in continental China. In *Investigations into the Tectonics of the Tibetan Plateau*, ed. B. C. Burchfiel and E. Wang, Geological Society of America Special Paper 444, pp. 59–71, doi:10.1130/2008.2444(1104).

Zhang, Y., J. L. Mercier, and P. Vergely (1998). Extension in the graben systems around the Ordos (China), and its contribution to the extrusion tectonics of south China with respect to Cobi-Mongolia. *Tectonophysics*, 285, 41–75.

Zhang, Y., Y. Ma, N. Yang, W. Shi, and S. Dong (2003). Cenozoic extensional stress evolution in North China. *Geodynamics*, 36, 591–613.

Zhang, Y., C. Liao, W. Shi, T. Zhang, and F. Guo (2007). Jurassic deformation in and around the Ordos Basin, North China. *Earth Science Frontiers*, 14(2), 182–196.

Zhao, G., M. Sun, S. A. Wilde, and L. Sanzhong (2005). Late Archean to Paleoproterozoic evolution of the North China Craton: key issues revisited. *Precambrian Research*, 136(2), 177–202.

Zhu, F., and G. Wu (1982). *The 1975 Haicheng Earthquake*. Beijing: China Seismological Press.

Zhu, G. Z., Y. L. Shi, S. Chen, and H. Zhang (2009). Numerical simulations on deep subduction of western pacific plate to NE China. *Chinese Journal of Geophysics (Acta Geophysica Sinica)*, 52(4), 950–957.

Zhu, R. X., Y. G. Xu, G. Zhu, *et al.* (2012a). Destruction of the North China Craton. *Science China, Earth Sciences*, 55(10), 1565–1587.

Zhu, R. X., J. H. Yang, and F. Y. Wu (2012b). Timing of destruction of the North China Craton. *Lithos*, 149, 51–60.

6

Seismogenesis of earthquakes occurring in the ancient rift basin of Kachchh, Western India

BAL KRISHNA RASTOGI, PRANTIK MANDAL,
AND SANJIB KUMAR BISWAS

Abstract

This chapter describes the mode of formation and seismogenesis of the very active Mesozoic-age intraplate 200 km × 300 km Kachchh Rift, which continues to be seismically active. As well as aftershocks of the 2001 M_w 7.7 Bhuj earthquake, seismicity has continued to occur to M_w 5.6 levels along other newly activated faults up to a distance of 240 km over the past dozen years. This ongoing activity has provided a natural laboratory for studying seismogenesis of intraplate rifts. Over the past decade, detailed investigations included intense seismicity monitoring with up to 75 broadband seismographs, ground motion detection with a GPS network of 22 stations and InSAR, active fault investigation, and subsurface mapping with various types of geophysical surveys. The results of these studies led to the detection and mapping of subsurface faults. Though the low GPS-derived horizontal deformation (2–5 mm/yr) may be adequate to trigger earthquakes along pre-existing faults away from the rupture zone, pockets of high vertical deformation (1–27 mm/yr) were detected by InSAR, indicating greater vertical deformation. The uplift was possibly aided by migration of the stress pulse due to a 20 MPa stress drop associated with the mainshock. It is inferred that tectonic inversion of the rift is causing uplift of the region. The seismicity and receiver function analysis suggest a thin crust overlying a thinned lithosphere. Tomographic studies reveal the presence of a high-velocity ~100 km wide solid mafic intrusive with embedded low-velocity zones, which suggest the presence of metamorphic fluids or volatile carbon dioxide. The thin crust and lithosphere facilitated placement of mafic intrusive and metamorphic fluids and/or volatile carbon dioxide from the underlying magma chambers. It is inferred that the high-density mafic body acts as a local stress concentrator and the low-velocity fluid-filled patches act as asperities.

Intraplate Earthquakes, ed. Pradeep Talwani. Published by Cambridge University Press. © Cambridge University Press 2014.

6.1 Introduction

Large earthquakes are continuously occurring along different faults in the Kachchh intraplate rift basin of Western India. In this chapter we describe geological features and details of the lower crust and upper mantle derived from tomography, receiver function, and shear-wave splitting; we speculate on the cause of stress accumulation; and give details of continuing seismicity triggered after the M_w 7.7 earthquake along nearby faults, possibly caused by a stress pulse, and the unusually high uplift detected by InSAR and GPS.

During the continental breakup of India from Africa in the late Triassic (210 Ma), Western India experienced crustal stretching (thinning) and the formation of three failed rifts of Kachchh (also spelled as Kutch), Cambay, and Narmada. In the Kachchh Rift (KR), magmatic intrusives were emplaced in the lower crust (possibly starting in the late Jurassic, 175 Ma). During the process, several basic intrusive rocks have come up to the surface all over Kachchh (Biswas, 2005). The xenoliths found in these intrusives may have originated in the lower lithosphere and are dated 75 Ma (Sen *et al.*, 2009). The Deccan/Réunion mantle plume during the end of the Cretaceous caused lithosphere thinning and emplaced the Deccan volcanic flood basalt. The KR and the intraplate region in or close to the Deccan Traps has been known to be active for centuries, and many earthquakes have occurred in recent times (inset of Figure 6.1): Koyna (reservoir induced seismicity along with M_w 6.3 in 1967 to M_w 5.6 in 2011) along the Kurdwadi Rift; Latur (1993, M_w 6.2), Jabalpur (1997, M_w 5.8), and Bharuch (1970, M_w 5.4) in the Narmada Rift; and Bhuj (2001, M_w 7.7) in the Kachchh Rift. The 2001 Bhuj earthquake caused 14,000 deaths and destruction in a heavily populated and industrialized region (Rastogi, 2001; Rastogi *et al.*, 2001).

Worldwide intraplate earthquakes, which mainly occur along rifts, are not well understood due to the low level of seismicity. Measuring about 200 km × 300 km, the Kachchh Rift of Gujarat is seismically one of the most active intraplate regions of the world and has experienced large earthquakes for many centuries and intense seismicity for over a decade. Its six major E–W-trending faults are being reactivated by thrusting. This detailed study of the seismogenesis of the KR has provided insights into understanding intraplate seismicity associated with rift zones.

Large earthquakes and faults in the KR are shown in Figure 6.1. These are: the 1819 M_w 7.8 earthquake along the Allahbund Fault (ABF), the 1845 M_w 6.5 Lakhpat earthquake along the western part of the Kachchh Mainland Fault (KMF), the 1956 M_w 6.0 Anjar earthquake along the Katrol Hill Fault (KHF), the 2001 Bhuj M_w 7.7 earthquake along the eastern extension of the Kachchh Mainland Fault, and the 2006 M_w 5.6 Gedi earthquake along the Gedi Fault (GF).

A seismic reflection survey detected a fault 10 km north of Anjar as part of the KHF along which the 1956 earthquake may have occurred (Sarkar *et al.*, 2007).

Aftershocks of the 2001 Bhuj earthquake of M_w up to 5 have continued for over a decade. Also, several triggered earthquakes of $M_w \leq 5.6$ occurred along different faults in Kachchh up to about 75 km from the mainshock epicenter and also up to 240 km south in Saurashtra, which is a different tectonic province.

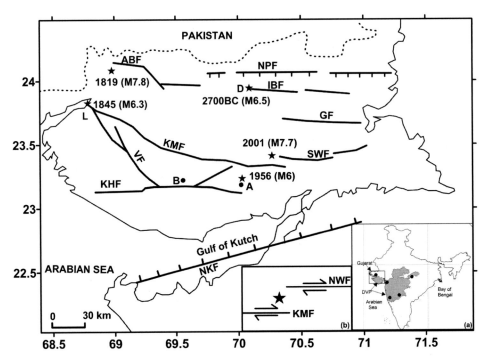

Figure 6.1 Significant earthquakes and their associated faults in Kachchh. Locations are: A, Anjar; B, Bhuj; L, Lakhpat; D, Dholavira. Major faults: ABF, Allah Bund; IBF, Island Belt; KMF, Kachchh Mainland; KHF, Katrol Hill; NPF, Nagar Parkar; NKF, North Kathiawar; VF, Vigodi; GDF, Gora Dungar; BF, Banni; GF, Gedi. The inset diagram (a) shows the Kachchh and Saurashtra area of Gujarat (small square) and the area of the Deccan Traps, along with epicenters of significant earthquakes in Peninsular India. (b) Possible stepover zone between KMF and NWF.

The seismicity over a decade has been monitored with about 75 broadband seismographs. Geodetic deformation has been detected with over 22 permanent GPS stations, 11 campaign GPS stations, and InSAR in and around the aftershock zone. Seismic, gravity, and magnetotelluric measurements have revealed the orientation of faults and information about subsurface structure (Gupta *et al.*, 2001; Sarkar *et al.*, 2007; ISR Annual Report 2012, isr.gujarat.gov.in). Geological and paleoseismological investigations indicate past activity on faults.

The availability of this large dataset has enabled detailed study for this region. In this chapter, we review the information generated on crustal and upper mantle structure in the source zone of a large paleorift to understand the rift structure and seismogenesis of the intraplate earthquakes occurring in the KR zone. A seismotectonic model of the KR area is presented based on new ideas on structural as well as seismicity patterns, fault mechanisms, and geodetic deformation.

The Indian intraplate region has negligible internal deformation (3 ± 2 mm/yr; Paul *et al.*, 2001). However, some areas such as the Narmada Rift have high *in-situ* stress (Gowd *et al.*, 1992). Moreover, the large intraplate earthquakes occurring along rifts are located in the lower crust. The presence of a magmatic intrusive body has been inferred for KR lower crust, which may accumulate stress and lead to brittle failure. Some low-velocity patches are found in and around the intrusive, which are speculated to be fluid-filled asperities.

6.2 Tectonic framework, structure, and tectonic evolution of Kachchh Rift basin

6.2.1 Structure and tectonics

The western margin of the Indian plate is the locale of three failed pericratonic rifts formed during the breakup of Gondwanaland involving stretching and thinning of the crust (Biswas, 1987, 2005): the KR formed earliest during the Late Triassic (210 Ma), followed by the Cambay Rift, which formed during the Early Cretaceous, and the Narmada Rift, which formed during the Late Cretaceous. The crust is found to be 35–37 km thick in the KR compared to 38–42 km in the surrounding region (Mandal, 2012a). Rift evolution with syn-rift sedimentation continued through the Jurassic till the Early Cretaceous as the Indian plate separated from Africa and drifted northward along an anticlockwise path. The rift expanded from north to south by successive reactivation of primordial faults of the Mid-Proterozoic Delhi fold belt. The NE–SW strike of the Delhi–Aravalli fold belt swings to E–W in the Kachchh region (Biswas, 1987). The E–W-trending rift basin has a series of E–W-trending faults (Figure 6.2). The KR is bound by the Nagar Parkar uplift on the north and the Kathiawar uplift (Saurashtra horst) on the south, lying along the sub-vertical Nagar Parkar and North Kathiawar faults (NPF and NKF). The rift is styled by three main uplifts (from north to south), Island Belt, Wagad, and Kachchh Mainland, along three intrarift faults, Island Belt (IBF), Kachchh Mainland (KMF), and South Wagad (SWF), with intervening grabens and half-grabens. The Island Belt uplift is a narrow south-tilted basement ridge, which is broken and displaced by tear faults into four separate uplifts described as "islands." The uplifts are upthrust basement blocks tilted along sub-vertical faults with initial normal separation. The structure is characterized by tilted blocks and half-grabens within a south-tilted asymmetric rift basin. The NKF is the bounding master fault along which the rift subsided most. All the faults are sub-vertical, dipping 90° to 75° towards the adjacent half-graben or graben (Biswas, 1987). Blanketing sediments over the basement drape over the tilted edges of the upthrusts as marginal flexures, which are narrow deformation zones along master faults enclosing complicated folds, locally much faulted and intruded by igneous rocks. In the western part the step-faulted uplifts are tilted to the south, with flexures draped over the faulted uplift northern edges (Biswas, 2005). In the eastern part a large uplift, Wagad, occurs between the Mainland and Island Belt uplifts. It is tilted opposite to the north with a narrow deformation zone along the faulted southern edge (Biswas, 2005). The back-slope ends against the Bela horst of the Island Belt uplift. The Mainland and Wagad uplifts occur in *en echelon* pattern. The KMF and SWF are parts of a left-stepping dextral strike-slip fault system (Biswas and Khattri, 2003). The SWF is the eastward continuation of the KMF after side-stepping with an overlap zone between Bhachau and Adhoi. Another important tectonic feature in the KR zone is a subsurface basement ridge – Median High –

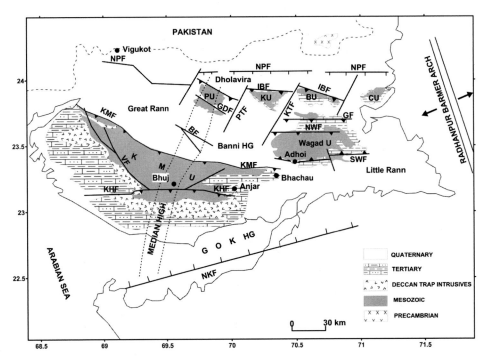

Figure 6.2 Tectonic map of Kachchh (modified after Biswas, 2005). Major faults: ABF, Allah Bund; IBF, Island Belt; KMF, Kachchh Mainland; KHF, Katrol Hill; NPF, Nagar Parkar; NKF, North Kathiawar; VF, Vigodi; GDF, Gora Dungar; BF, Banni; GF: Gedi. Uplifts: PU, Pachham Is.; KU, Khadir Is.; BU, Bela Is.; CU, Chorar Is.; Wagad U, Wagad Uplift; KMU, Kachchh Mainland Uplift. HG, Half-Graben. GOKHG, Gulf of Kutch Half-Graben. ABF is mislabeled near Vigukot.

that crosses the basin at right angles to its axis in the middle. Acting as a hinge it divides the basin into a deeper western part and a shallower and more tectonic eastern part. We also notice that the KR is terminated in the east against a transverse subsurface basement ridge, the Radhanpur arch, which is the western shoulder of the adjacent N–S oriented Cambay rift, and, to the west the KR merges with the offshore shelf.

A south to north depth section (Figure 6.3b) depicts possible dip directions of faults inferred from intensive field work by Biswas (2005), and a modified version based on new seismological and geophysical data is shown in Figure 6.3c. The dip directions of the KMF and SWF inferred by Biswas are towards the north and south, respectively, and opposite to the dip directions inferred by Rastogi. Biswas interprets that the dip directions of the two faults are towards each other, so that the low-lying Samkhiali and Banni grabens may be formed. However, Rastogi opines that the Kachchh Mainland can uplift only along a south-dipping fault while the Wagad can uplift only along a north-dipping fault. Once the uplifts are formed, the area between the two uplifts will be low-lying. Bouguer anomaly data indicate that the basement has not gone down in Samkhiali and Banni with respect to the Wagad and Kachchh Mainland uplifts. From trenching, Morino *et al.* (2008a, b) inferred

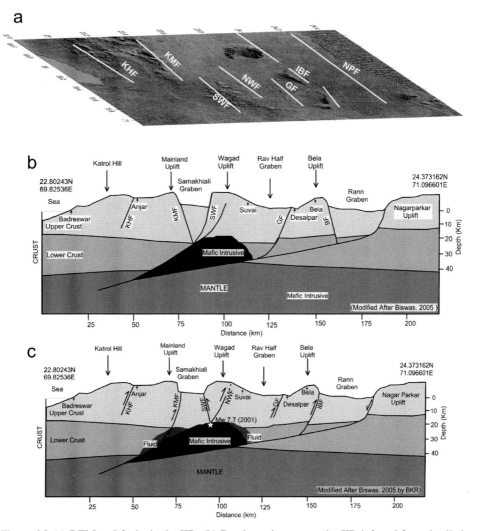

Figure 6.3 (a) DEM and faults in the KR. (b) Depth section across the KR inferred from detailed geological mapping by Biswas (2005) (lower part). (c) Depth section modified based on new geophysical surveys as well as hypocentral depths by Rastogi. The intrusive magmatic body and fluid zones around it are shown.

southward dip of the KMF. The dips given by Rastogi are based on two magnetotelluric profiles, 20 km apart, lying east and west of the epicenter of the 2001 Bhuj earthquake (23° 26.4′ N 70° 18.6′ E, about 15 km northwest of Bhachau) (Mohan *et al.*, 2013). The two faults are observed down to at least 12 km depth in the two 22 and 15 km long profiles with station spacing of 1–2 km (Figure 6.4). The surface extension of the NWF would mark the boundary between Mesozoic and Tertiary formations. From the magnetotelluric surveys, the sediments are found to be 1.5 to 2.3 km thick in the Samkhiali basin and 5–6 km in the Wagad

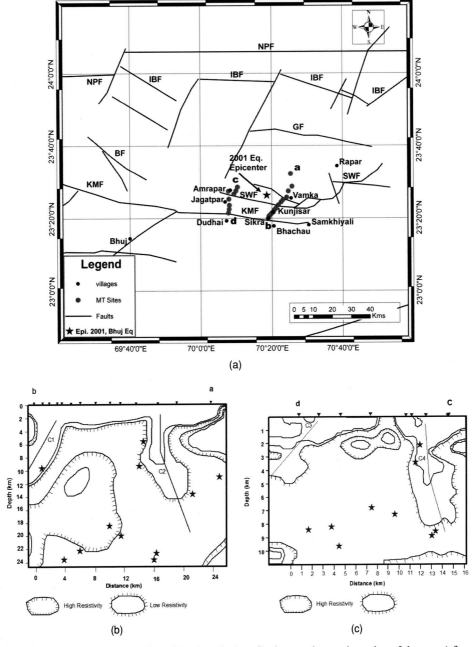

Figure 6.4 (a) Locations of MT profiles ab and cd on Geology and tectonic setting of the area (after Biswas, 2005). Profiles are east and west of the epicenter of the 2001 Bhuj earthquake (23° 26.4′ N 70° 18.6′ E (ISC), which is shown as a star 15 km northwest of Bhachau. (Epicenter by USGS is 9 km WSW.) (b) 2D geoelectric depth section (eastern profile) down to 24 km. The conductive block towards the "b" side is the Kachchh Mainland Block and the conductive block towards the "a" side (around 16 km distance) is the Wagad Block. M ≥ 4 hypocenters from 2006 to 2012 are plotted: C1, KMF; C2, SWF. (c) 2D western geoelectric depth section to 14 km depth. Epicenters of M ≥ 4 from 2006 to 2012 are plotted: C3, KMF; C4, SWF. Low resistivity zones indicate faults.

Uplift area. The Bouguer gravity anomaly also indicates larger sedimentary thickness in Wagad as well as Kachchh Mainland Uplift areas due to gravity lows as compared to high gravity value areas of the Samkhiali basin and Banni area (Singh *et al.*, 2013).

Figure 6.3c indicates that the south-dipping hidden NWF is a left step-over fault of the KMF, being separated from it by about 30 km. The step-over zone between Bhachau and Bharudia is a stressed zone of compressive stress (Figure 6.1, inset). The SWF is a conjugate fault to the NWF and intersects it at mid-crust region. The 2001 mainshock originated near this intersection. Such step-over faults have been suggested theoretically by Segall and Pollard (1980) and Sibson (1986), and through an analog sand-box model by McClay and Bonora (2001), and have been mapped in many areas (in Charleston, South Carolina, by Dura-Gomez and Talwani (2009) as well as in the New Madrid seismic zone by Russ (1982), Pratt (2012), and others referred to therein).

6.2.2 Tectono-volcanic events

Tectonic episodes were accompanied by deep crustal magmatic activity (Biswas, 2005; Ray *et al.*, 2006; Paul *et al.*, 2008; Sen *et al.*, 2009). At least two phases of magmatic activity are evident. The first activity took place during the extensional stage when ultramafic rocks intruded into the older Jurassic sediments. Presumably at this time the deep-seated magmatic body was emplaced at the site of the mantle rupture close to the basin center (Biswas, 2005). The second took place during the Late Cretaceous (65 Ma) post-rift uplift stage when plume-related alkali and tholeiitic basalts were intruded into the younger Early Cretaceous sediments and extruded as the Deccan Trap flows. The volcanic activity was associated with thinning of the sub-Kutch lithosphere. Mandal (2010) found that the lithospheric thickness varies from 62 to 63 km in the KR and from 65 to 77 km in the surrounding region. The rifting started in the Late Triassic–Early Jurassic and continued through the Deccan volcanism in the Early Cretaceous (Biswas, 1987). The bulk of the Kutch alkalic lavas came from an Indian Ridge mantle source rather than the Réunion plume. The Kutch lithosphere was composed of spinel lherzolite that was largely converted to spinel wehrlite by carbonatite metasomatism. Such carbonatite melts were generated by decompression melting of the asthenosphere during its rise to shallower levels in response to extension and thinning of the lithosphere. Mantle xenolith-bearing alkalic magmas were mainly generated from a mixture dominated by asthenospheric material and the edge of the Deccan/Réunion plume. These melts ascended, while picking up xenoliths, along pathways created by deep rift faults. Tholeiites may have come from elsewhere in the south as plumelets, from the hotter part of the Deccan plume head (Sen *et al.*, 2009).

6.2.3 Tectonic evolution and existing earthquake generation models of the Kachchh Rift zone

Rifting was aborted by the trailing edge uplift during the Late Cretaceous pre-collision stage of the Indian plate, when the leading edge of the plate was slab-pulled towards the Tethyan trench (Biswas, 2005). The uplift caused structural inversion during the rift–drift transition stage, when most of the uplifts with drape folding over the edges came into existence by upthrusting of the basement domino blocks along the master faults. This created first-order marginal flexures over the foothill uplifts. Lateral motion during the drift stage of the plate induced horizontal stress and near-vertical normal faults, which were reactivated as reverse faults during initiation of the inversion cycle, and became strike-slip faults involving divergent oblique-slip movements (Biswas, 2005). The present structural style evolved by right-lateral slip, which shifted the uplifts progressively eastward relative to each other from south to north. This resulted in the present *en echelon* positioning of the uplifts with respect to the Kutch Mainland uplift. The strike-slip related structural changes modified the linear flexures, breaking them into individual folds at the restraining and releasing bends. Narrow deformation zones complicated by second-order folds and conjugate Riedel faults formed along the master faults, modifying the initial drape folds (Biswas, 2005). Syntectonic intrusions further modified the shape and geometry of the individual second-order structures. Igneous rocks extensively intruded the Mesozoic sediments during rifting followed by post-rift hotspots related to Deccan volcanism (Sen *et al.*, 2009). Studies of the intrusive rocks and seismological data suggest the presence of mafic/ultramafic magmatic bodies close to the crust–mantle boundary (Mandal and Pujol, 2006; Mandal and Chadha, 2008; Mandal and Pandey, 2010).

Inversion continued during the post-collision compressive stress regime of the Indian plate and the KR basin became a shear zone with transpressional strike-slip movements (with thrusting at depth) along the active sub-parallel rift faults (Biswas, 2005). The same tectonic phase is continuing, as evident from neotectonic movements along these faults that are responsible for the present first-order geomorphic features and seismicity. In the current tectonic cycle, under N–S compressive stresses, the NWF and SWF are the most active faults, as evident from the concentration of aftershock hypocenters in the overlap zone (Biswas, 2005; Mandal and Horton, 2007). Pulses of movement along these faults are responsible for generation of new fault fractures within the respective deformation zones. These new fractures are propagating through the recent piedmont and scarp-fan sediments in the frontal zones of the thrusts, as seen in the trenches dug close to the KMF and KHF (Malik *et al.*, 2008; Morino *et al.*, 2008a, b) and in GPR surveys. The morphotectonic features also indicate Quaternary uplift along the above-mentioned master faults (Malik *et al.*, 2008).

During the present compressive stage, the Radhanpur arch acts as a stress barrier for eastward movements along the principal deformation zones, which is creating additional strain in this part of the basin between the arch and the Median High (Biswas, 2005). Towards the eastern end of the Mainland uplift, the right lateral KMF becomes the SWF by left-stepping with an overlap in the region between Samakhiali and Lakadiya (Biswas, 2005). This stepover zone – formed initially as the Samakhiali–Lakadiya graben – is presently a convergent transfer zone undergoing transpressional stress in the strained eastern part of the basin. This is the most strained part of the basin. Expectedly, this is the most favored site for rupture nucleation. The occurrence of the 2001 Bhuj quake (M_w 7.7) in

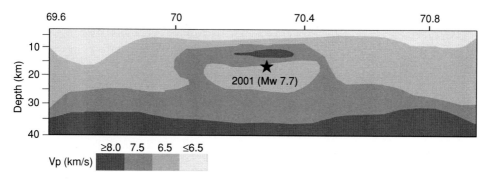

Figure 6.5 Cartoon displaying Vp distribution with depth along an E–W profile through the 2001 Bhuj earthquake epicenter. The hypocenter of the 2001 mainshock projected on this line lies in a low-velocity patch. The outer portion of the intrusive body also has low-velocity fluid-filled zones. For color version, see Plates section.

this zone and the concentration of aftershock hypocenters around it further validate this conclusion.

Aftershock data suggest a reverse slip on a fault plane (NWF) dipping 40–60° to the south as the causative fault for repeated earthquake nucleation (Mandal and Horton, 2007). Continued continental compression and repeated thrusting generate fractures related to the main fault-forming highly stressed fault zone, which seems to be the zone of repeated earthquake nucleation (Biswas, 2005). This earthquake nucleation zone is defined by several subsidiary faults branching off the main KMF. The causative fault, the NWF (North Wagad Fault), of the Bhuj 2001 earthquake seems to be one such fault that lies north of it. The occurrence of the 2001 Bhuj earthquake has been explained in terms of large local stress perturbations associated with the pre-existing fault intersection (with respect to favorably optimally oriented maximum horizontal stress direction) and mafic crustal intrusive bodies below the KR zone (Gangopadhyay and Talwani, 2003; Mandal and Pandey, 2010). The presence of faults transverse to the NWF may lead to large stress concentrations in the lower-crustal magmatic layer (at 14–34 km depth), thereby, the favored zone of earthquake nucleation (Mandal and Pandey, 2010). Presumably, the fluid released by the eclogitization of olivine-rich lower-crustal rocks aids the slippage along this causative fault and hence the occurrence of the 2001 Bhuj earthquake sequence (Mandal and Pandey, 2010; Mandal, 2011; Mandal, 2012a, b).

6.2.4 Identification of magmatic intrusive bodies

The P and S tomography results (Mandal et al., 2004a) suggest the presence of a regional high-velocity, low Poisson's ratio body around the 2001 mainshock epicenter with a head extending 60 km in the N–S direction and 40 km in the E–W direction at 10–25 km depth (Figure 6.5), which may be a mafic intrusive of the rifting stage. At deeper depths it may be larger. It has been speculated that this body is causing stress build-up. Its velocity is inferred to be high (Vp: 7.15–8.11 km/s) at 24–42 km depth and density 3.06–3.37 g/cm^3

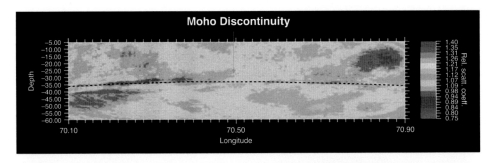

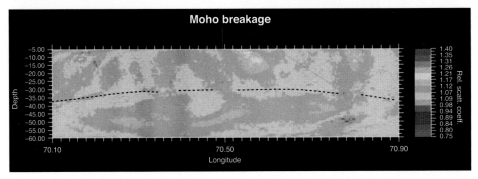

Figure 6.6 (Upper part) Intact Moho in the Kachchh region 70 km north of the epicenter of the 2001 Bhuj earthquake. (Lower part) Disturbed Moho at the 23.3° N latitude of the 2001 Bhuj earthquake. The profile length is 90 km. For color version, see Plates section.

as estimated according to the formula, ρ (g/cm^3) = 0.32 Vp (km/s) + 0.77 (Mandal and Pandey, 2010).

Due to magmatic intrusion, the Moho is likely to be broken, disturbed, and/or scattered. This is revealed by 3D Q-structure in the epicentral zone of the 2001 earthquake in Kachchh at a depth of around 40 km (Figure 6.6; Sharma *et al.*, 2008) but intact Moho north of the epicentral zone.

Such magmatic intrusives have been inferred for the Narmada Rift (Rajendran and Rajendran, 1998) and most of the worldwide rifts that act as stress concentrators and cause deepening of the brittle–ductile transition depth (Mukherjee, 1942; Mooney *et al.*, 1983; Mechie *et al.*, 1994; Prodehl *et al.*, 1994; Johnston, 1994, 1996; Mooney and Christensen, 1994; Nyblade and Langston, 1995; Liu and Zoback, 1997; Singh *et al.*, 1999; Kumar *et al.*, 2000; Deverchere *et al.*, 2001; Kruger *et al.*, 2002; Manglik and Singh, 2002; Wilson *et al.*, 2003; Gao *et al.*, 2004).

The KR has a large-scale presence of alkaline rocks and mantle xenoliths at subsurface depths in the form of plugs, cones, and sheet-like bodies containing olivine, opx, cpx, and spinel (Karmalkar *et al.*, 2000, 2008; Desai *et al.*, 2004; Sen *et al.*, 2009). Melts for such alkaline rocks are reportedly generated by the partial melting of carbon dioxide-rich lherzolite (Sen *et al.*, 2009). Eclogitization or serpentinization of these may release carbon dioxide. The availability of such sources of melts is assured by the thin crust and lithosphere.

6.2.5 Low-velocity fluid-filled zones in and around magmatic bodies in the lower crust

Mandal *et al.* (2004a) detected a low-velocity high Poisson's ratio zone within the mafic body at the hypocentral depth of the mainshock ($\sim$18–25 km), which is inferred to be a fluid-filled fractured rock mass and which might have acted as an asperity for generation of the 2001 Bhuj earthquake (Figure 6.5). From further detailed work on Vp and Vs tomography, Mandal and Chadha (2008) inferred several low-velocity patches (Figure 6.7). From 3D mapping of b-values (Mandal and Rastogi, 2005; Singh *et al.*, 2011; Nagabhushan Rao, 2012), a high b-value ($\sim$1) zone is found to be sandwiched within the maximum rupture zone at depths of 15–25 km with a low b-value (0.6 to 0.8) above and below. In the same zone Mishra and Zhao (2003) and Singh *et al.* (2012) found low Vp and Vs, high crack density, porosity, saturation rate and Poisson's ratio, and suggest the presence of a fluid-filled and fractured rock matrix.

Such a low-velocity fluid zone was inferred by Kato *et al.* (2009) from high-resolution 3D tomography around solidified intrusive bodies in the intraplate eastern margin of the Japan Sea back-arc basin, and they attributed the large earthquakes of the area to the presence of an intrusive body and fluids around it. It is speculated that the trapped aqueous fluids resulted from metamorphism or were released from degassing of mantle magmatic material or volatiles such as carbon dioxide (Zhao *et al.*, 1996; Miller *et al.*, 2004; Wang and Zhao, 2006; Mandal and Pandey, 2010).

6.2.6 Paleoseismological investigations

Based on Baker's (1846) leveling survey, Bilham (1999) inferred 11 m slip (uplift and subsidence) along the Allahbund fault due to the 1819 earthquake. The inferred fault is north-dipping with possibly listric fault geometry, with a steep dip at shallow depth but a gentle dip at larger depth. Rajendran (2000) estimated lesser uplift of 4.3 m and suggested a growing fold at shallow depths along a north-dipping thrust (Rajendran and Rajendran, 2001). From radiocarbon age data of liquefaction features they suggested occurrence of a previous earthquake 800–1000 years ago and suggested the ABF to have been previously active, as a historic event was possible in AD 893 (Rajendran and Rajendran, 2002, 2003). At Dholavira in northern Kachchh the archeological evidence indicates an earthquake during the closing decades of stage III of the settlement (2500–2200 BC) (Singh, 1996; Rajendran and Rajendran, 2003; Kovach *et al.*, 2010). Bisht (2011) suggests that this event occurred during 2100–2000 BC and identifies two earlier earthquakes also during stage II (2900 BC) and stage III (2700 BC).

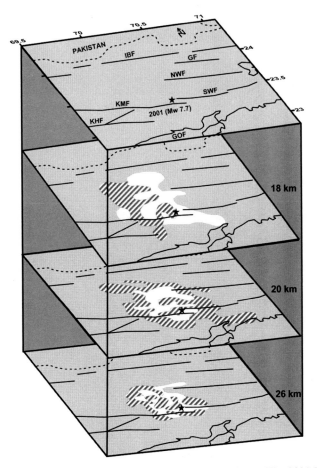

Figure 6.7 Schematic 3D model displaying Vp distribution with depth. The 2001 Bhuj earthquake hypocenter (22 km value most accepted, but ranges from 22 to 26 km) shown by a star lies on the boundary of low- and high-velocity patches. The hatched areas are low-velocity zones while the white areas mark the high-velocity zones. At a depth of 34 km (not shown) the low-velocity zone is much larger.

The ABF, KMF, SWF, IBF, and KHF are found to be active by Quaternary movements as observed in trenches (Figure 6.8; Rajendran and Rajendran, 2002, 2003; Malik *et al.*, 2008; Morino *et al.*, 2008a, b and www.isr.gujarat.gov.in/Annual Reports 2009–10, 2010–11, 2011–12). This indicates that different faults are active neotectonically. Below we describe the work of the Institute of Seismological Research (ISR), most of which was done with Malik and Morino.

The deformation zone of the 1819 Allahbund earthquake of Kachchh was mapped with a precise elevation survey in an area 80 km long and 6–7 km wide. The amount of uplift measured in 2007 is up to 5.8 m. By adding 1.8 m estimated erosion since 1819, a total uplift of 7.6 m is estimated for the 1819 earthquake. The Allahbund fault is interpreted to

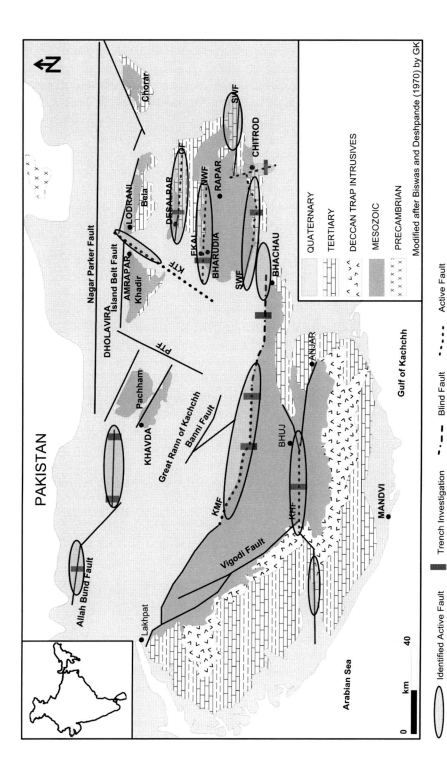

Figure 6.8 Active fault map of Kachchh by ISR prepared with data from 27 shallow trenches. The portions of faults where trenching has revealed active movements are marked with ellipses.

be north-dipping. At Vigukot, 10 km north of Allahbund, three large and three or four small events are inferred based on sand craters and disrupted human settlements found in trenches.

The presence of tidal deposits in the Sindri lake area indicates that the sea has receded by 100 km from here to the present position. In the eastern part of the ABF near Karimsahi, several trenches and uplift of paleochannels reveal uplift during the last few thousand years. Due to uplift, the paleochannels could have been formed at 3–4 ka and again uplifted by 2 m at 2 ka, as revealed by OSL dating of samples at different elevations. A trench near Dharamshala indicates three events in the last 3 ka. Along a streamcut nearby, a highly deformed zone was identified. OSL dates indicate that a 3 ka bed has thrust over a 2.7 ka bed.

The KMF is found to be south-dipping in a trench at Jhura in the central part of the KMF, where a cumulative deformation of 5 m has been assessed as due to three events. The two last events had 70 cm slip each. A rough estimate of the age of formations is 5 ka, as determined by the optically stimulated luminescence method. At Lodai, in the eastern part of the KMF, the two large events show slips of 33 cm and 40 cm during the late Pleistocene to Holocene period and older events along two older faults show a cumulative slip of 98 cm. The Mesozoic rocks override Quaternary along a south-dipping fault. Identification of a pressure ridge by remote sensing and thorough ground checks helped in identifying the activity in this part of the KMF. Further east, there is cumulative slip of 4 m in the Quaternary, and the last event at an inferred age of 2 ka indicates a 50 cm slip.

In a trench at Wandhey in the central part of the KHF, Quaternary deformation indicates three large events during the late Holocene to a few ka along three fault strands, which displaced the older terrace deposits comprising sand, silt, and gravel units along with overlying younger deposits from units 1 to 5 made of gravel, sand, and silt.

In the Wagad area, the slips noticed for events possibly in the past few ka are: the Gedi fault along a rivercut shows slip of 1 m; along the South Wagad fault two events are noticed with slips of 30 cm and 50 cm along two fault strands in a trench in the Adhoi anticline at 3–4 ka; south of the Wagad area, in the Samkhiali basin, slip of 75 cm is measured in a trench.

6.3 Seismicity of Gujarat state

Figure 6.9 shows significant faults and epicenters of about 200 mainshocks of magnitude ≥ 2 from 1684 to 2000 in Gujarat and the adjoining region bound by 20°–25.5° N and 68°–75° E. Table 6.1 shows the magnitude distribution of mainshocks during the pre- and post-2001 periods (Rastogi et al., 2013a).

About 200 km × 300 km, Kachchh is seismically one of the most active intraplate regions of the world. It has six major E–W-trending faults of the failed Mesozoic rift, which are being reactivated by thrusting. Prior to the 2001 M_w 7.7 Bhuj earthquake, Kachchh had experienced three large earthquakes: 1819 M_w 7.8 Allah Bund (24.00° N 69.00° E), 1845 M_w 6.3 Lakhpat (23.80° N 68.90° E), and 1956 M_w 6 Anjar (23.30° N 70.00° E). Some large earthquakes have been documented based on archeoseismology, such as the Dholavira earthquake (Bisht, 1997, 2011), but these earthquakes occurred along different faults (Figure 6.1). The second seismic region in Gujarat is the Narmada Rift zone, which

Table 6.1 *Magnitude distribution of mainshocks in Gujarat from 1684 to 2012*

M/Year	2–2.9	3–3.9	4–4.9	5–5.9	6–6.9	≥7	Total
1668–2000	11	79	62	27	2	1	183
2001–2012	17	26	14	6	—	1	63

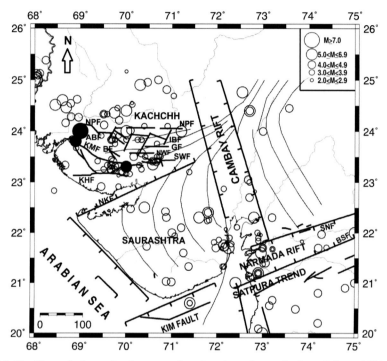

Figure 6.9 Significant faults and epicenters of earthquakes of magnitude ≥ 2 from 1684 to 2000 excluding aftershocks in Gujarat. Filled circles are the three significant events in Kachchh of 1819, 1845, and 1956. The Proterozoic-age Aravali trend branches off in three directions. In Kachchh it becomes E–W, along which faults might have formed.

experienced the severely damaging 1970 Bharuch earthquake of M_w 5.4 at its western end and larger earthquakes further east, including 1927 M_w 6.3 Son, 1938 M_w 6.5 Satpura, and 1997 M_w 5.8 Jabalpur events. The third seismic zone consists of the Cambay basin, east and southeast of Kachchh, and the Saurashtra Peninsula, south of Kachchh, which has experienced seismicity of magnitude less than 6.

Seismic networks have been operating since 1976 and earthquakes of magnitude ≥3.5 are routinely located. Most of Gujarat having been heavily populated for centuries, it is expected that for the past 200 years no earthquake of magnitude ≥4 has been missed, as such earthquakes are felt strongly over wide areas. Their locations may have an accuracy of

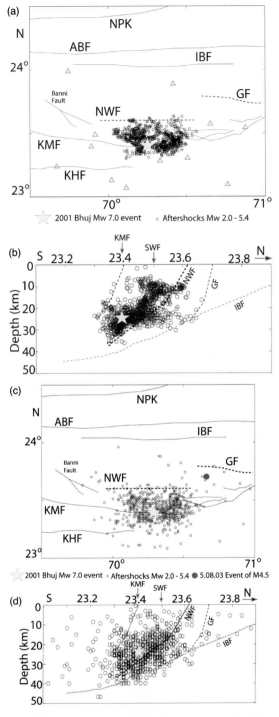

Figure 6.10 (a) Epicenters and (b) foci of Kachchh earthquakes during 2001 show that they were confined to the NWF and 40 km × 40 km area. (c) Epicenters and (d) foci of Kachchh earthquakes during the two years 2002 to 2003. A few faults near the NWF became slightly active (NGRI data): Nagar Parkar (NPF), Allah Bund (ABF), Island Belt (IBF), Gedi (GF), South Wagad (SWF), Kachchh Mainland (KMF) and Katrol Hill (KHF).

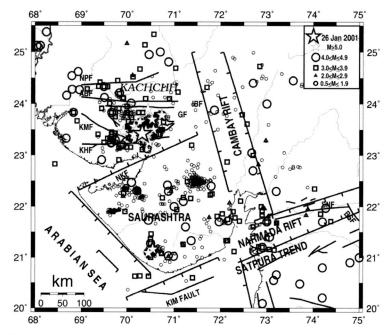

Figure 6.11 Epicenters of earthquakes in Gujarat during 2001–12. The epicenters of shocks of M < 2.0 are mostly in areas of intense local monitoring.

10–15 km. The six shocks of $M_w \geq 5.0$ and about 25 strongly felt (M_w 3 to 4.9) earthquakes that occurred in the decade after 2000 are very well located.

6.3.1 Seismicity with time in Kachchh and the nature of seismic sources

Starting in 2001, up to 75 broadband seismographs have been operating in Gujarat (500 km × 600 km), of which some 40 are in Kachchh. During 2001 to 2003, the after-shocks of the 2001 Bhuj earthquake were confined to the main rupture zone of 40 km × 40 km area north of the KMF and along an inferred hidden North Wagad fault (Figure 6.10). During 2001 to 2012 some 10,000 $M_w \geq 0.5$ shocks were located in Gujarat (Figure 6.11).

Aftershocks of the 2001 Bhuj earthquake to $M_w \sim 5$ level have continued for over a decade, including 16 $M_w \geq 5$, 250 $M_w \geq 4$, about 4000 $M_w \geq 3$ events, and some 5000 other well-located shocks of M_w 1.0 to 2.9 (Mandal and Rastogi, 2005; Mandal *et al.*, 2007; Rastogi *et al.*, 2013a, b). Up to 2003 there were 15 aftershocks of $M_w \geq 5$ and one in 2006. The latest significant aftershock was of M_w 4.9 on March 9, 2008.

6.3.2 Orientation of faults and depth of the seismogenic zone in Kachchh

The parameters of the 2001 mainshock are: 23.419° N, 70.232° E, focal depth 16 km (USGS). Some other depth estimates are 22 km or 23 km (ISC; Antolik and Dreger, 2003).

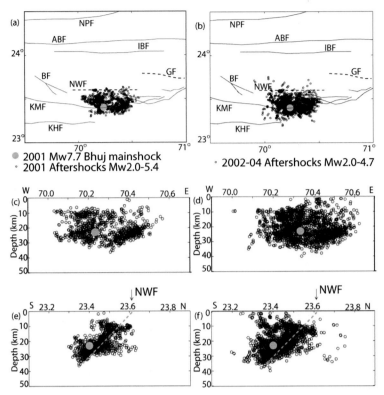

Figure 6.12 (a) HypoDD relocations of 1402 aftershocks of M_w 2.0–5.4 during 2001. Large circle marks the location of the 2001 Bhuj mainshock. (b) Epicenters of aftershocks during 2002–4. (c) E–W depth plot of relocated aftershocks during 2001. (d) E–W depth plot of relocated aftershocks during 2002–4. (e) N–S depth plot of relocated aftershocks during 2001. (f) N–S depth plot of relocated aftershocks during 2002–4.

Yagi and Kikuchi (2001, eri.u-Tokyo.ac.jp) determined source duration of 20 s, and fault dip 58°. Rupture with maximum slip of 8.5 m propagated mostly towards the west for 40 km, with a small slip towards the east for 20 km along the fault width from 8 km below the surface to 35 km depth along the NWF. If projected onto the surface, this would be about 60 km × 30 km in area between the KMF and NWF lying between 23° 20′ and 23° 34′ N, 69° 50′ and 70° 20′ E, which we take as the main rupture zone. The eastern end of the rupture zone terminates north of Bhachau. Antolik and Dreger (2003) have given a similar rupture model.

　　The epicenters of the mainshocks, foreshocks, and aftershocks during 2001 to 2009 and depth sections are shown in Figure 6.12 (Mandal and Pandey, 2010; Mandal, 2012b). The hypocenters were first determined using the Hypo71 program with up to 56 P and S phases with accuracy of better than 0.5 km horizontal and 1.0 km in depth (Rastogi *et al.*, 2013a). The subsurface orientation of seismogenic faults and depth of the seismogenic zone in the

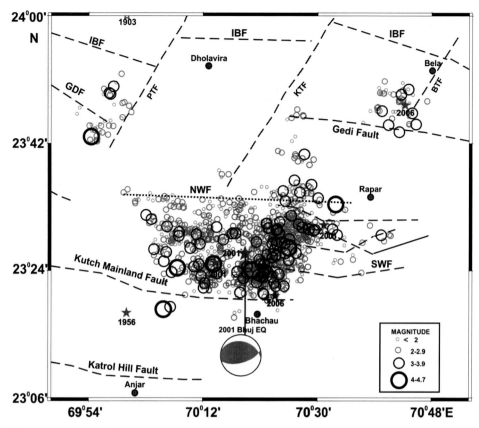

Figure 6.13 Results of HypoDD clustering analysis of the Hypo71 locations during 2007–11 in Kachchh [Kutch] (modified after Rastogi *et al.*, 2013b). Epicenters of earthquakes of M ≥ 5 are shown by stars.The North Wagad, Gedi, and Gora Dungar faults show clear concentrations. The Suvai transverse fault also shows a concentration of epicenters. Activity was seen along the Khadir Transverse Fault (KTF) along which an M_w 5.1 earthquake occurred on June 19, 2012, with a long aftershock sequence. East of it and about 20 km west of Rapar, a concentration of several shocks north of the NWF defines a N–S-trending transverse fault, as in the previous figure. A few shocks of M_w < 4.7 have occurred south of the KMF since 2006. As these are at shallow depths, they are inferred to be associated with the KMF and association with the NWF is ruled out. For color version, see Plates section.

aftershock zone of the 2001 earthquake, as well as nearby activated faults, was obtained using precise relocations by double-difference (HypoDD). The hypocentral plots depict the NWF being active from 8 to 35 km depth and the Gedi fault (GF) to 12 km depth. The HypoDD plot for the period January 2007 to May 2011 (Rastogi *et al.*, 2012) indicates clustering along the NWF and GF, and for the sequence of October 28, 2009, an M_w 4.6 earthquake along Gora Dungar Fault and some concentration along the Khadir transverse fault (Figure 6.13).

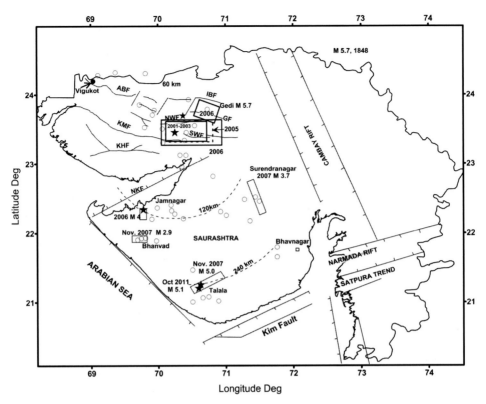

Figure 6.14 Long-time and delayed triggering of seismicity in Gujarat from 2006 to 2011. Epicentral zones for significant sequences are marked by rectangles. The seismicity migrated 120 km south to the Jamnagar and Surendranagar area and as far as 240 km in the Talala area of Saurashtra.

6.4 Long-distance delayed triggering of shocks in Gujarat after the 2001 M_w 7.7 Bhuj earthquake due to stress pulse migration

6.4.1 Triggered seismicity in Kachchh and Saurashtra

Around the aftershock zone of the M_w 7.7 Bhuj earthquake of 2001, several mainshocks of M_w 3–5.6 occurred along different faults up to distances of 75 km away during 2006–12 (Figure 6.14). Initially, the seismicity spread eastward to the SWF, but then to other faults from 2006. These faults include the KMF, IBF, GDF, SWF, and GF. The total number of such mainshocks is at least 20 (Table 6.2) in Kachchh. Many of these mainshocks are associated with foreshock–aftershock sequences, e.g., M_w 5.6 earthquake in March 2006 along the GF about 75 km northeast of the 2001 mainshock epicenter, M_w 4.4 shock of October 28, 2009, along the Gora Dungar fault, and M_w 5.1 shock of June 19, 2012, along a transverse fault north of the 2001 rupture zone (Figure 6.14 and Table 6.2). Table 6.3 compares felt and damaging earthquakes in Kachchh for the pre-2001 (200 yr) and 2001–12 periods.

Table 6.2 *Mainshocks in Kachchh after the 2001 Bhuj earthquake along different faults, and other than aftershocks along the NWF. M_w until April 2006 from NGRI and subsequently from ISR. Many of these are associated with their own foreshocks and aftershocks. USGS magnitudes are usually m_b*

SN	Y	M	D	Lat.	Long.	Dep. km	Mag. M_w	Mag. USGS	Location	Mechanism
	2001	1	26	23.44	70.31	16.0	7.7	7.7	18 km NW of Bhachau North Wagad F.	Thrust
1	2004	1	8	23.91	70.90	20.0	4.2		Gedi/Is. Belt F.	
2	2004	6	7	23.87	70.15	29.4	4.2		Dholavira/Is. Belt F.	
3	2005	3	8	23.85	69.74	11.6	4.3		Gora Dungar F.	
4	2005	10	8	23.35	70.69	24.0	4.5		South Wagad F.	
5	2005	10	9	23.74	69.93	6.6	4.3		Gora Dungar F.	
6	2006	2	03	23.92	70.44	28.0	5.0	4.5	Gedi F., foreshock	Thrust
7	2006	3	07	23.79	70.73	3.0	5.6	5.5	Tr. Bela F./Gedi F., ms	Left lateral
8	2006	4	06	23.78	70.74	3.0	4.8	5.0	Gedi F., aftershock	Thrust
9	2006	4	06	23.34	70.39	29	5.6	5.5	Lakadia, SWF	
10	2006	4	10	23.51	70.06	4.9	4.9	4.9	* Banni	
11	2006	6	12	23.88	70.43	27.3	4.4		Gedi/Is. Belt F.	
12	2007	5	13	23.44	70.42	20.4	4.7		South Wagad F.	
13	2007	5	24	23.298	70.026	9.0	4.1		Kachchh Mainland F.	Thrust
14	2007	10	8	23.295	70.075	9.6	4.7	4.5	Kachchh Mainland F.	Left lateral
15	2007	12	15	24.03	69.87	15.0	3.7		Allahbund F.	Right lateral
16	2008	3	9	23.396	70.359	18.5	4.9	4.5	South Wagad F.	Left lateral
17	2008	4	4	23.00	70.36	11.1	2.6		**Kandla	
18	2008	7	5	23.53	69.8	8.9	3.3		Banni F.	
19	2009	10	28	23.71	69.91	8.5	4.4	4.4	Gora Dungar F.	Left lateral
20	2011	1	18	23.27	70.51		3.8		Samkhiyali, SWF	
21	2011	5	17	23.55	70.57	18.2	4.2		E. of North Wagad F.	Thrust
22	2011	8	13	23.45	70.40	22.2	4.5	4.3	South Wagad F.	
23	2011	9	27	23.12	70.31	38.0	3.0		**Kandla	
24	2012	4	14	23.39	70.54	19	4.1	4.0	South Wagad F.	
25	2012	6	19	23.65	70.28	11	5.0	5.0	Khadir Tr. F.	
26	2012	12	8	23.13	70.42	21	4.5	4.1	20 km SSE of Bhachau, **Kandla	
27	2013	3	30	23.56	70.38	24	4.5		*Chobari	

* May be aftershock of 2001 Bhuj earthquake.

** May be regional shock as it is single event.

Table 6.3 *Felt and damaging earthquakes in Kachchh. Post 2001, the shocks are mostly aftershocks, but 27 earthquakes of $M_w \sim 3$ to 5.6 are independent mainshocks, many of which are associated with foreshocks and aftershocks.*

Mag.	No. of shocks in Kachchh Pre-2001 (200 yr)	No. of shocks in Kachchh 2001–2012
3.5–3.9	46	671
4.0–4.9	25	268
≥ 5.0	11	21

Moreover, seismicity (M_w 3 to 5.1) has been triggered along small faults at 20 locations up to 120 km south since 2006 and up to 240 km south since 2007 in the Saurashtra region. The three significant sequences that have continued for 5 years with hundreds of felt shocks are (i) near Jamnagar with M_w maximum 4.0; (ii) near Surendranagar with M_w maximum 3.9; and (iii) at Talala with M_w maximum 5 in 2007, M_w maximum 5.1 in 2011 and several M_w 4 to 4.8 tremors (Figure 6.14). The shocks are usually shallow (focal depth < 10 km) and are accompanied by subterranean sounds. The generation of an unusually large number of mainshocks is inferred to be due to triggering caused by migration of the stress pulse generated by the 20 MPa stress drop of the M_w 7.7 earthquake in 2001 to distances of 100–200 km and even 6–8 years after the great earthquake. The triggered seismicity may be due to an increase in Coulomb stress of up to 1 bar in Kachchh and 0.1 bar in Saurashtra, as estimated by viscoelastic modeling (Rastogi *et al.*, 2013b). Added to this may be the major stress from a stress pulse due to the 20 MPa stress drop. Scholz (1977) explained the Haicheng earthquake prediction by propagation of the deformation front at a rate of 110 km/yr. He also quoted such rates of 80 to 270 km/yr (Mogi, 1968) and explained that this may result from stress pulse migration through the mantle (Savage, 1971; Bott and Dean, 1973).

6.4.2 Coulomb stress change

Postseismic Coulomb stress changes (in bars) during 2001 at the surface due to the M_w 7.7 Bhuj 2001 earthquake have been calculated using the Yagi and Kikuchi variable slip model at 15 km depth (crust is considered elastic and mantle viscoelastic). The maximum stress change is 12 bars while a rise of 1 bar extends up to 100 km (Figure 6.15a; Mandal *et al.*, 2007; Choudhury *et al.*, 2013b).

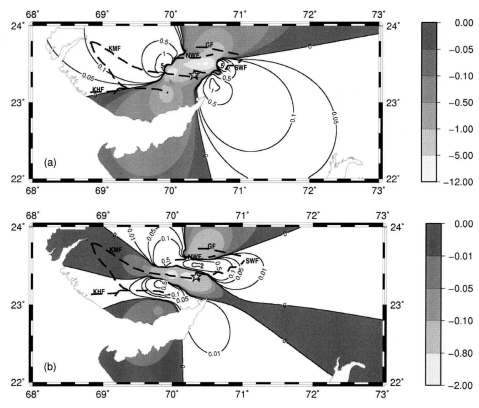

Figure 6.15 (a) Postseismic Coulomb stress changes (in bars) calculated assuming variable slip at 15 km depth due to the M_w 7.7 Bhuj 2001 earthquake soon after the earthquake (2001 mainshock epicenter is marked as a star and negative Coulomb stress areas are shaded). Maximum stress change is 12 bars while a rise of 1 bar extends up to 100 km. (b) Six years after the earthquake. Aftershocks and triggered earthquakes (during 2006–12) have occurred in the zones of positive stress change up to a distance of 75 km in Kachchh along the SWF in the east and the IBF and some other faults in the north, as well as up to 240 km south in Saurashtra. Positive Coulomb stress contours are numbered while negative Coulomb stress contours are shaded (Choudhury *et al.*, 2012).

The Coulomb stress during 2006 estimated by considering the viscoelastic process (Rastogi *et al.*, 2013b) is shown in Figure 6.15b. After 6 years the Coulomb stress remains positive towards the western part of the IBF (where a few earthquakes other than aftershocks occurred during 2006–12 along the IBF and Gora Dungar Fault) and negative along the eastern part of the IBF as well as around the KMF. It changes to positive towards the SWF and south of the KMF (maximum positive 2 bars), where seismicity that is not aftershocks was triggered from 2006 onwards. Some mainshocks, e.g., 2006 M_w 5.6 along the Gedi Fault and M_w 5.1 on June 19, 2012, north of the NWF, have occurred in the negative zone of Coulomb stress. This indicates the influence of the stress pulse in these areas.

6.4.3 Geodetic observations in Kachchh

Starting in 2006, ISR operates up to 22 permanent differential GPS stations across geological faults in Gujarat. Some 11 campaign stations in Kachchh are occupied bi-annually. Processing is done with 1 mm/yr accuracy (Choudhury *et al.*, 2013a). Local deformation has been estimated with respect to Gandhinagar station, operating more than 200 km east of the epicenter of the main earthquake of 2001. The Indian Institute of Geomagnetism operated two GPS close to two of the ISR stations during 2001–5. Combining the two datasets, the composite plot for the period 2001–9 indicates that near the epicenter the postseismic relaxation was initially large, being 12, 6, 3, and 4 mm for four consecutive 6-month periods of 2001–2, but subsequently reduced (Figure 6.16). Since 2007 all the ISR stations in Kachchh indicate the horizontal deformation to be low, i.e., of the order of 2–5 ± 1 mm/yr. Nevertheless, the horizontal deformation gives a significant strain of 0.05 µs/yr, even as late as 2008–11 (Dumka and Rastogi, 2013; Choudhury *et al.*, 2013b), triggering earthquakes along the SWF (Figure 6.16 inset). However, vertical deformation is found to be quite large, i.e., 2–13 ± 3 mm/yr, as observed by GPS measurements near the epicenter of the 2001 mainshock and up to 75 km north and northeast, being 13 mm/yr at Dholavira and 10 mm/yr at Dudhai during 2006–11 (Rastogi *et al.*, 2012) (Table 6.4).

The interferogram generated with Differential Interferometric Synthetic Aperture Radar (DInSAR) using the ENVISAT ASAR datasets of June 22, 2008 and October 25, 2009, with a baseline separation of 125 m in the area up to 75 km north of the mainshock epicenter (Figure 6.17; Rastogi *et al.*, 2012) indicates vertical deformation rates of 7–27 mm/yr.

Uplift of 16–27 mm/yr was detected along two faults (KMF and KHF) up to 50 km south of the epicenter, measurements of which were done during 2004–7 using ENVISAT ASAR datasets and during 2007–10 using ALOS PALSAR data (Choudhury *et al.*, 2012; Sreejith and Rastogi, 2013).

6.4.4 Cause of triggered earthquakes

We propose, based on the seismological data, that the viscoelastic process and rheologic changes appear to be the plausible causes of the long-distance and delayed triggering of earthquakes with diffusion rates of 5–30 km/yr, which might have also been facilitated by the migration of a stress pulse of 20 MPa stress drop caused by the 2001 M_w 7.7 earthquake. The rate of transmission of the stress pulse (Figure 6.11), considering the seismic area involved, matches a seismogenic permeability of 0.5 to 5 m^2/s (Talwani *et al.*, 2007) suggesting fluid pressure diffusion. Vertical deformation is observed to be high: measured as up to 13 mm/yr by 6 years of GPS observations and 10–27 mm/yr by 10 years of InSAR observations. We suggest that the transmission of the stress pulse into the upper crust could have resulted in vertical deformation as observed in GPS and InSAR measurements.

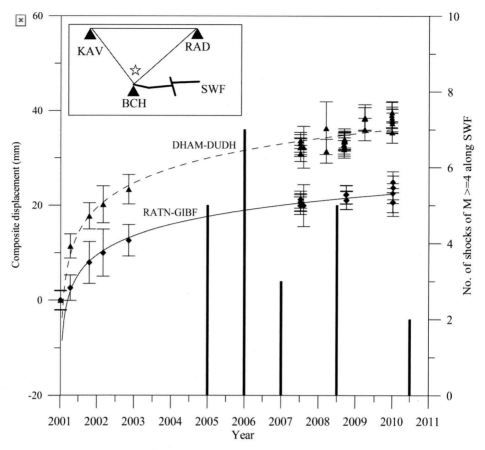

Figure 6.16 Composite map of resultant displacement of two GPS sites (DHAM-DUDH (upper curve) and RATN-GIBF (lower curve) plotted with respect to ISRG from 14 days to 8 years after the 2001 Bhuj earthquake. DHAM and RATN are two GPS stations of Reddy and Sunil (2008) co-located with DUDH and GIBF of the ISR stations, respectively. The displacement decay rate for the upper curve is $D = 7.6\ln(t) + 20.36$, for the lower curve $D = 5.8\ln(t) + 10.22$. The number of shocks of $M_w \geq 4$ along the SWF are also shown. The strain calculated in the Kachchh region during 2008–11 along the baselines, as shown in the inset, is representative of the strain in most parts of the active area. For 2005–7 the strain is likely to be greater than that for the period 2008–11 due to the larger displacements.

6.5 Focal mechanism studies

The focal mechanisms for nine significant events (M_w 4.0–5.6) by full waveform moment tensor inversion (10–20 s) reveal a dominant reverse movement with minor strike-slip component on the south-dipping NWF, indicating continued thrusting in the 2001 rupture zone. The 2006 mainshock of M_w 5.6 on the Gedi fault reveals a left-lateral strike-slip movement on an almost vertical fault, but the two aftershocks of M 4–5 indicate a greater thrust component (Mandal *et al.*, 2009; Nagabhushana Rao, 2012). It appears that the

Table 6.4 *High vertical deformations at four GPS campaign stations*

Campaign GPS station	Lat. Long.	Location	Local uplift
Dudhai	23.238N 70.145E	West of 2001 epicenter	10 ± 3 mm/yr
Gadadha	23.867N 70.373E	Close to Dholavira and 30 km north of the 2001 rupture zone	13 ± 3 mm/yr
Lilpar	23.526N 70.636E	NE of 2001 mainshock epicenter	6 ± 3 mm/yr
Fatehgadh	23.683N 70.864E	NE of 2001 mainshock epicenter	5 ± 3 mm/yr

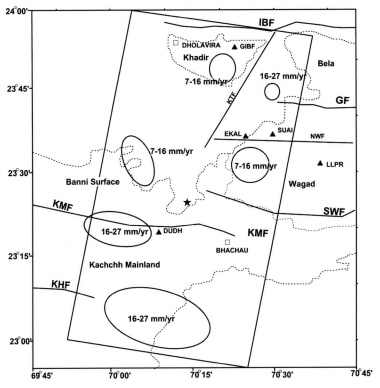

Figure 6.17 Star indicates the 2001 Bhuj earthquake. Local uplifts estimated from interferogram fringes of DInSAR data are shown north of the epicenter. ENVISAT data between two dates 1.5 yr apart (June 22, 2008–October 25, 2009) have been used. North of the epicenter the interferogram fringes were identifiable in hilly uplifted areas only, not in large low-lying areas. South of the epicenter around the KMF and KHF, ENVISAT data were used during 2004–7 and ALOS data, 2007–10 (modified after Choudhury *et al.*, 2012). Deformation could be measured for time intervals of ≥1 yr.

thrusting is along the E–W-trending Gedi Fault and strike-slip is along a northeast-trending conjugate transverse fault. The M_w 5.1 of June 19, 2012 along the Khadir transverse fault indicates left-lateral strike-slip faulting.

Moment tensor focal mechanism solutions of M > 4 earthquakes in Kachchh during 2001–10 for a few earthquakes along the Gora Dungar, Kachchh Mainland, and South Wagad faults, surrounding the 2001 rupture zone and believed to have been triggered since 2006, indicate strike-slip. The focal depths are around 3 km for the Gedi and South Wagad earthquakes, 8 km for the Gora Dungar earthquake, and 12 km for the Kachchh Mainland fault earthquake. The focal depths of earthquakes showing thrust mechanisms in the 2001 rupture zone are usually deeper than 10 km. It confirms the view that there could be strike-slip at shallow near-surface depths and thrusting at deeper than 10 km.

Stress inversion using focal mechanism data from five areas across the main rupture zone reveal a N–S orientation of maximum compressive stress with 7–32° rotation, which is attributed to the sizable local horizontal stress component associated with the crustal intrusive bodies in the main rupture zone (Mandal, 2008).

6.6 Results and discussion

The normal faults of the Mesozoic Kachchh Rift were reactivated after the continent–continent collision (40 Ma) in the form of transpressional faults (with thrusting at depth) as inferred from detailed geological field work (Biswas, 2005). New seismological and geophysical data reveal the southward dip of the Kachchh Mainland fault and northward dip of the South Wagad fault, along which large uplifted areas have been formed. Until 2000, seismicity in the region was considered low but with high hazard potential because of historical records of the destruction of Dholavira in 2700 BC due to an earthquake of estimated M_w 6.5, the 1819 Allahbund earthquake of M_w 7.8, the 1845 Lakhpat earthquake of M_w 6.3, and the 1956 Anjar earthquake of M_w 6. These earthquakes occurred along different faults. The picture of low seismicity changed after the 2001 Bhuj earthquake. The KR has been associated with over a decade of aftershock activity of $M_w \geq 5$ and triggering of ~20 earthquakes of M_w 4–5.7 along different large and small faults at up to 75 km distance in a decade. GPS measurements, which started soon after the January 2001 earthquake, revealed 12 mm horizontal displacement for the first 6-month period. It reduced to 6, 3, and 4 mm in the next three 6-month periods. Initially large horizontal displacements favor a shear deformation mode of aftershocks. However, occurrence of M_w 4–5.6 earthquakes, despite small horizontal displacements of 2–5 mm/yr across different faults, was suggested to be due to a stress pulse. The stress pulse is transferred, causing earthquakes along the pre-existing critically stressed faults. Large vertical deformation of up to 13 mm/yr was estimated by GPS for 2006–9 and 10–27 mm/yr from InSAR studies for 2008–9 north of the KMF, and during 2004–7 as well as 2007–10 south of the KMF.

We speculate that the lower crust has a mafic intrusive with high seismic wave velocities, which acts as a stress concentrator. The low-velocity patches at 14–34 km depths indicate the presence of fluids or volatile carbon dioxide, which may act as asperity zones and along which large earthquakes are associated. The presence of patches of partial melt below the lithosphere–asthenosphere boundary (LAB) suggested from receiver function analysis in the central Kachchh Rift zone may be imprints of the Deccan mantle plume (65 Ma) beneath the region (Sen *et al.*, 2009; Mandal, 2010). These may provide compressed carbon dioxide or favorable circumstances for eclogitization and consequent release of fluid, either or both of which facilitate the occurrence of earthquakes. Thin crust and lithosphere provide easy access to the magma source.

Large spatial variations in Vp (from –10 to 13%), Vs (from –11 to 13%), and Vp/Vs ratio up to 1.6 to 1.8 are seen in the main rupture zone of the 2001 Bhuj earthquake at 6–34 km depth, where the tomography model is fairly sampled by the aftershock data (Mandal and Chadha, 2008). At shallower depth (0–10 km) the low Vp, low Vs, and high Vp/Vs could indicate the presence of soft sediments with higher water content.

The aqueous fluid or volatiles containing carbon dioxide are released from the eclogitization of olivine-rich lower crustal rocks. The presence of 2 weight percent carbon dioxide fluid components is suggested in the crystalline basement rocks of the 1993 Latur earthquake (Pandey *et al.*, 2009). Interestingly, John and Schenk (2006) suggest, based on petrologic analysis of eclogites, that, once rupture begins in gabbroic crustal rocks, frictional melting can promote intermediate-to-lower crustal depth earthquakes under eclogite facies conditions and this seismic event can produce permeabilities for external fluids. Thus, the eclogitization of lower crustal olivine-rich gabbroic rocks can provide aqueous fluids or gaseous fluids (such as carbon dioxide and nitrogen) in the intermediate to lower crust. Besides, strong attenuation of seismic waves (low coda Qc) and shear-wave splitting were reported in the Bhuj aftershock zone, which could be explained in terms of the presence of fluids, cracks, or both in the fault zone (Mandal *et al.*, 2004b; Mandal, 2009; Padhy and Crampin, 2006).

Mandal (2012a) finds the crust (35 km) and lithosphere (62 to 63 km) to be thin in the central part of the KR, as compared to 40–42 km thick crust and 65–77 km thick lithosphere in the outer parts of the KR, due to updoming of the lithosphere and asthenosphere ($\sim$6–12 km). Mandal finds a decrease of Vs at 62–77 km depth over an area of 130 km $\times$ 90 km, which may indicate the presence of patches of partial melt below the LAB in the central Kachchh rift zone as a remnant of the Deccan mantle plume (65 Ma) beneath the region. Further support for this model is given by the rift axis parallel azimuthal anisotropy, with a delay of 1.6 s evaluated from the SKS splitting study, which is attributed to anisotropy induced by the rift-parallel flows within the 76 $\pm$ 6 km thick lithosphere and anisotropy associated with the rift-parallel pockets of partial melts in the asthenosphere, inherited from the plume–lithosphere interaction during the Deccan/Réunion plume episode ($\sim$65 Ma) (Mandal, 2011). Thus, there is a possibility that these patches of partial melts could provide a high input of volatiles containing carbon dioxide into the lower crust. If so, this would contribute significantly to seismo-genesis and continued aftershock activity in the intracontinental Kachchh rift zone.

6.7 Conclusions

The Kachchh rift consists of a network of conjugate faults, some of which extend to large depths. Activity on these faults has produced a series of horsts and grabens, some of which have been leveled by erosion or filled by deposition. The 2001 mainshock that occurred along a hidden North Wagad fault triggered seismicity on adjoining faults by a stress pulse.

New seismological and geophysical data have revealed the subsurface orientations of some active faults. Frequent large earthquakes up to M_w 7.8 have occurred along different faults in the past 200 years. The 2001 Bhuj earthquake has been associated with over a decade of aftershock activity of M_w up to $\sim$6. In addition to that, about 20 earthquakes of $M_w \sim$ 4–5.6 were triggered along different large and small faults up to 75 km away in Kachchh. The large vertical displacements of up to 13 mm/yr detected from GPS and 10–27 mm/yr detected from InSAR studies are contributing more to strain build-up than the 2–5 mm/yr small horizontal displacements measured from GPS network and campaign stations. There are some earthquakes that may be triggered in the zones of positive Coulomb stress but some earthquakes are in the negative Coulomb stress zone. Earthquakes 8–10 years after the mainshock, away from the rupture zone and also in the negative Coulomb stress zones indicate that the stress due to a viscoelastic process and/or rheologic process subsequent to the 2001 mainshock adds to the stress due to plate tectonics to cause triggering of earthquakes.

High-velocity mafic intrusives inferred in the lower crust may act as stress concentrators. The postulated low-velocity patches within the mafic bodies may represent fluid-filled zones that act as asperity zones and along which large earthquakes might nucleate. A velocity decrease at 62–77 km, covering an area of 120 km $\times$ 80 km, as revealed by receiver function analysis was attributed to the updoming of the asthenosphere or/and the presence of patches of partial melts beneath the central Kachchh rift zone. These may provide favorable circumstances for eclogitization and release of compressed carbon dioxide or fluids that accumulate in some pockets and facilitate occurrence of earthquakes.

We propose that large crustal stresses associated with ongoing uplift, the presence of intrusive mafic bodies, fault intersection, marked lateral variation in crustal thickness, and related sub-crustal thermal anomalies are accumulating in the heterogeneous causative fault of the 2001 mainshock. These stresses are mainly concentrated in the denser and stronger lower crust (at 14–34 km depths) beneath the Kachchh rift zone, within which most of the earthquakes nucleate along the fluid-filled fractured low-velocity zones.

Acknowledgements

A thorough review and numerous suggestions by Pradeep Talwani have tremendously improved the manuscript. Prantik Mandal is grateful to the Director, NGRI, for support and permission to publish this work. The Government of Gujarat and Ministry of Earth-Sciences, New Delhi, supported this work. Girish Kothyari drafted some of the figures.

References

Antolik, M., and Dreger, D. S. (2003). Rupture process of the 26 January 2001 M_w 7.6 Bhuj, India, earthquake from teleseismic broadband data. *Bulletin of the Seismological Society of America*, 93, 1235–1248.

Baker, W. E. (1846). Remarks on the Alla Bund and on the drainage of the Eastern part of the Sind Basin. *Transactions of the Bombay Geographical Society*, 7, 186–188.

Bilham, R. (1999). Slip parameters for the Rann of Kachchh, India, 16 June 1819, earthquake, quantified from contemporary accounts. In *Coastal Tectonics*, ed. I. S. Stewart and C. Vita-Finzi, Geological Society, London, Special Publication, 146, pp. 295–318.

Bisht, R. S. (1997). Dholavira excavations: 1990–94, Facets of Indian Civilization: Essay in Honour of Prof. B. B. Lal, ed. J. P. Joshi, New Delhi: Aryan Books, International, 1, 107–120.

Bisht, R. S. (2011). Major earthquake occurrences in archaeological strata of Harappan settlement at Dholavira (Kachchh, Gujarat). *International Symposium on the 2001 Bhuj Earthquake and Advances in Earthquake Science, AES 2011*, ISR, Gandhinagar, Gujarat, S16_IGCP-11, pp. 112–113.

Biswas, S. K. (1987). Regional framework, structure and evolution of the western marginal basins of India. *Tectonophysics*, 135, 302–327.

Biswas, S. K. (2005). A review of structure and tectonics of Kutch Basin, Western India, with special reference to earthquakes. *Current Science*, 88(10), 15.

Biswas, S. K., and Khattri, K. N. (2003). Structure and tectonics of Kutch Basin, Western India, with special reference to earthquake. *Journal of the Geological Society of India*, 61, 626–629.

Bott, M. H. P., and Dean, D. S. (1973). Stress diffusion from plate boundaries. *Nature*, 243, 339–341.

Choudhury, P., Dumka, R. K., Rastogi, B. K., Sreejith, K. M., and Majumdar, T. J. (2012). Interferometric Synthetic Aperture Radar (INSAR) Studies by ISRO and ISR. *ISR Annual Report 2011–12*, www.isr.gujarat.gov.in, pp. 30–32.

Choudhury, P., Catherine, J. K., Gahalaut, V. K., *et al.* (2013a). Post-seismic deformation associated with the 2001 Bhuj earthquake. *Natural Hazards*, 65(2), 1109–1118, doi:10.1007/s11069–012–0191–8.

Choudhury, P., Roy, K. S., and Rastogi, B. K. (2013b). Strain analysis in Kachchh and Saurashtra region using GPS data. *ISR Annual Report, 2012–13*, www.isr.gujarat.gov.in, pp. 46–48.

Desai, A. G., Markwick, A., Vaselli, O., and Downes, H. (2004). Granulite and pyroxenite xenoliths from the Deccan trap: insight into the nature and composition of the lower lithosphere beneath cratonic India. *Lithos*, 78, 263–290.

Deverchere, J., Petit, C., Gileva, N., *et al.* (2001). Depth distribution of earthquakes in the Baikal rift system and its implications for the rheology of the lithosphere. *Geophysical Journal International*, 146, 714–730.

Dumka, R. K., and Rastogi, B. K. (2013). Crustal strain in the rupture zone of 2001 Bhuj earthquake. *ISR Annual Report 2012–13*, isr.gujarat.gov.in, pp. 45–46.

Dura-Gomez, I., and Talwani, P. (2009). Finding faults in the Charleston Area, South Carolina: 1. *Seismological Data, Seismological Research Letters*, 80, 883–900.

Gangopadhyay, A., and Talwani, P. (2003). Symptomatic features of intraplate earthquakes. *Seismological Research Letters*, 74 (6), 863–883.

Gao, S. S., Kelly, H., Liu, H., and Chen, C. (2004). Significant crustal thinning beneath the Baikal rift zone: new constraints from receiver function analysis. *Geophysical Research Letters*, 31, L20610 1–4.

Gowd, T. N., Srirama Rao, S. V., and Gaur, V. K. (1992). Tectonic stress field in the Indian subcontinent. *Journal of Geophysical Research*, 97, 11,879–11,888.

Gupta, H. K., Harinarayana, T., Kousalya, M., *et al.* (2001). Bhuj earthquake of 26 January, 2001. *Journal of the Geological Society of India*, 57, 275–278.

John, T., and Schenk, V. (2006). Interrelations between intermediate-depth earthquakes and fluid flow within subducting oceanic plates: constraints from eclogite facies pseudo-tachylytes. *Geology*, 34, 557–560.

Johnston, A. C. (1994). Seismotectonic interpretations and conclusions from the stable continental regions. In *The Earthquakes of Stable Continental Regions: Vol. 1, Assessment of Large Earthquake Potential*, ed. J. F. Schneider. Palo Alto, CA: Electric Power and Research Institute, pp. 20–40.

Johnston, A. C. (1996). Seismic moment assessment of earthquakes in stable continental regions. *Geophysical Journal International*, 124, 381–414.

Karmalkar, N. R., Griffin, W. L., and O'Reilly, S. Y. (2000). Ultramafic xenoliths from Kutch, northwest India: plume related mantle samples? *International Geological Review*, 42, 416–444.

Karmalkar, N. R., Kale, M. G., Duraiswami, R. A., and Jonalgadda, M. (2008). Magma underplating and storage in the crust-building process beneath the Kutch region, NW India. *Current Science*, 94, 1582–1588.

Kato, A., Kurashimo, E., Igarashi, T., *et al.* (2009). Reactivation of ancient rift systems triggers devastating intraplate earthquakes. *Geophysical Research Letters*, 36, L05301, 1–5, doi: 10.1029/2008GL036450.

Kovach, R. L., Grijalva, K., and Nur, A. (2010). Earthquakes and civilizations of the Indus valley: a challenge for archaeoseismology. In *Ancient Earthquakes*, ed. M. Sintubin *et al.* Geological Society of America Special Paper 471, pp. 119–127.

Kruger, F., Scherbaum, F., Rosa, J. W. C., *et al.* (2002). Crustal and upper mantle structure in the Amazon region (Brazil) determined with broadband mobile stations. *Journal of Geophysical Research*, 107(B10), ESE 1711–1712.

Kumar, P., Tewari, H. C., and Khandekar, G. (2000). An anomalous high velocity layer at shallow crustal depths in the Narmada zone, India. *Geophysical Journal International*, 142, 95–107.

Liu, L., and Zoback, M. D. (1997). Lithospheric strength and intraplate seismicity in the New Madrid seismic zone. *Tectonics*, 16(4), 585–595.

Malik, J. N., Morino, M., Mishra, P., Bhuiyan, C., and Kaneko, F. (2008). First active fault exposure identified along Kachchh mainland fault: evidence from trench extraction near Lodai village, Gujarat, western India. *Journal of the Geological Society of India*, 71, 201–208.

Mandal, P. (2008). Stress rotation in the Kachchh rift zone, Gujarat, India. *Pure and Applied Geophysics*, 165, 1307–1324.

Mandal, P. (2009). Crustal shear wave splitting in the epicentral zone of the 2001 M_w 7.7 Bhuj earthquake, Gujarat, India. *Journal of Geodynamics*, 47, 246–258.

Mandal, P. (2010). Crustal and lithospheric thinning beneath the seismogenic Kachchh rift zone, Gujarat (India): its implications toward the generation of the 2001 Bhuj earthquake sequences. *Journal of Asian Earth Sciences*, doi. 10.1016/j.jseaes.2010.08.012.

Mandal, P. (2011). Upper mantle seismic anisotropy in the intra-continental Kachchh rift zone, Gujarat, India. *Tectonophysics*, 509, 81–92.

Mandal, P. (2012a). Passive-source seismic imaging of the crust and upper mantle beneath the 2001 M_w 7.7 Bhuj earthquake region, Gujarat, India. *Bulletin of the Seismological Society of America*, 102, 252–266.

Mandal, P. (2012b). Seismogenesis of the uninterrupted occurrence of the aftershock activity in the 2001 Bhuj earthquake zone, Gujarat, India, during 2001–2010. *Natural Hazards*, 65, 1063–1083.

Mandal, P., and Chadha, R. K. (2008). Three-dimensional velocity imaging of the Kachchh seismic zone, Gujarat, India. *Tectonophysics*, 452, 1–16.

Mandal, P., and Horton, S. (2007). Relocation of aftershocks, focal mechanisms and stress inversion: implications toward the seismo-tectonics of the causative fault zone of M_w 7.6 2001 Bhuj earthquake (India). *Tectonophysics*, 429, 61–78.

Mandal, P., and Pandey, O. P. (2010). Relocation of aftershocks of the 2001 Bhuj earthquake: a new insight into seismotectonics of the Kachchh seismic zone, Gujarat, India. *Journal of Geodynamics*, 49, 254–260.

Mandal, P., and Pujol, J. (2006). Seismic imaging of the aftershock zone of the 2001 M_w 7.7 Bhuj earthquake, India. *Geophysical Research Letters*, 33, L05309, 1–4.

Mandal, P., and Rastogi, B. K. (2005). Self-organized fractal seismicity and b-value of aftershocks of 2001 Bhuj earthquake in Kutch (India). *Pure and Applied Geophysics*, 162, 53–72.

Mandal, P., Rastogi, B. K., Satyanarayana, H. V. S., and Kousalya, M. (2004a). Results from local earthquake velocity tomography: implications toward the source process involved in generating the 2001 Bhuj earthquake in the lower crust beneath Kachchh (India). *Bulletin of the Seismological Society of America*, 94(2), 633–649.

Mandal, P., Jainendra, S., Joshi, S. K., Bhunia, R., and Rastogi, B. K. (2004b). Low coda-Qc in the epicentral region of the 2001 Bhuj earthquake of M_w 7.7. *Pure and Applied Geophysics*, 161, 1635–1654.

Mandal, P., Chadha, R. K., Raju, I. P., *et al.* (2007). Coulomb static stress variations in the Kachchh, Gujarat, India: implications for the occurrences of two recent earthquakes ($M_w = 5.6$) in the 2001 Bhuj earthquake region. *Geophysical Journal International*, 169, 281–285.

Mandal, P., Satyamurthy, C., and Raju, I. P. (2009). Iterative de-convolution of the local waveforms: characterization of the seismic sources in Kachchh, India. *Tectonophysics*, 478, 143–157.

Manglik, A., and Singh, R. N. (2002). Thermomechanical structure of the central Indian shield: constraints from deep crustal seismicity. *Current Science*, 82, 1151–1157.

McClay, K., and Bonora, M. (2001). Analog models of restraining stop-overs in strike-slip fault systems. *American Association of Petroleum Geologists Bulletin*, 85, 233–260.

Mechie, J., Keller, G. R., Prodehl, C., *et al.* (1994). Crustal structure beneath the Kenya rift from axial profile data. In *Crustal and Upper Mantle Structure of the Kenya Rift*, ed. C. Prodehl, G. R. Keller, and M. Khan. *Tectonophysics*, 236, 179–199.

Miller, S. A., Collettni, C., Chlaraluce, L., *et al.* (2004). Aftershocks driven by a high pressure CO_2 source at depth. *Nature*, 427, 724–727.

Mishra, O. P., and Zhao, D. (2003). Crack density, saturation rate and porosity at the 2001 Bhuj, India, earthquake hypocenter: a fluid-driven earthquake. *Earth and Planetary Science Letters*, 212, 393–405.

Mogi, K. (1968). Migration of seismic activity. *Bulletin of the Earthquake Research Institute, University of Tokyo*, 46, 53–74.

Mohan, K. (2013). Identification of Kachchh Mainland fault and South Wagad fault from magnetotellurics. *ISR Annual Report 2012–13*, pp. 31–32, www.isr.gujarat.gov.in.

Mooney, W. D., and Christensen, N. I. (1994). Composition of the crust beneath the Kenya Rift. *Tectonophysics*, 236, 391–408.

Mooney, W. D., Andrews, M. C., Ginzburg, A., Peters, D. A., and Hamilton, R. M. (1983). Crustal structure of the northern Mississippi embayment and a comparison with other continental rift zones. *Tectonophysics*, 94, 327–348.

Morino, M., Malik, J. N., Mishra, P., Bhuiyan, C., and Kaneko, F. (2008a). Active fault traces along Bhuj fault and Katrol hill fault, and trenching survey at Wandhay, Kachchh, Gujarat, India. *Journal of Earth System Science*, 117(3), 181–188.

Morino, M., Malik, J. N., Gadhavi, M. S., *et al.* (2008b). Active low-angle reverse fault and wide quaternary deformation identified in Jhura Trench across Kachchh Mainland Fault, Kachchh, Gujarat, India. *Journal of Active Fault Research, Japan*, 29, 71–77.

Mukherjee, S. M. (1942). Seismological features of the Satpura earthquake of the 14th March 1938. *Proceedings of Indian Academy of Sciences*, 16, 167–175.

Nagabhushan Rao, Ch. (2012). Moment tensor and fault mechanism solutions. *ISR Annual Report 2011–12*, pp. 10–11, www.isr.gujarat.gov.in.

Nyblade, A. A., and Langston, C. A. (1995). East African earthquakes below 20 km depth and their implications for crustal structure. *Geophysical Journal International*, 121, 49–62.

Padhy, S., and Crampin, S. (2006). High pore-fluid pressures at Bhuj, inferred from 90°-flips in shear-wave polarizations. *Geophysical Journal International*, 164, 370–376.

Pandey, O. P., Chandrakala, K., Parthasarathy, G., Reddy, P. R., and Koti Reddy, G. (2009). Upwarped high velocity mafic crust, subsurface tectonics and causes of intra plate Latur-Killari (M 6.2) and Koyna (M 6.3) earthquakes, India: a comparative study. *Journal of Asian Earth Sciences*, 34, 781–795.

Paul, J., Burgmann, R., Gaur, V. K., *et al.* (2001). The motion and active deformation of India. *Geophysical Research Letters*, 28(4), 647–651.

Paul, D. K., Ray, A., Das, B., Patil, S. K., and Biswas, S. K. (2008). Petrology, geochemistry and paleomagnetism of the earliest magmatic rocks of Deccan Volcanic Province, Kutch, Northwest India. *Lithos*, 102(1–2), 237–259.

Pratt, T. L. (2012). Kinematics of the New Madrid seismic zone, central United States, based on stepover models. *Geology*, 40, 371–374.

Prodehl, C., Keller, G. R., and Khan, M. A. (1994). Crustal and upper mantle structure of the Kenya rift. *Tectonophysics*, 236, 483.

Rajendran, C. P. (2000). Using geological data for earthquake studies: a perspective from peninsular India. *Current Science*, 79(9), 1251–1258.

Rajendaran, C. P., and Rajendaran, K. (1998). Characteristics of the 1997 Jabalpur earthquake and their bearing on its mechanism. *Current Science*, 74, 168–177.

Rajendran, C. P., and Rajendran, K. (2001). Characteristics of deformation and past seismicity associated with the 1819 Kutch earthquake, northwestern India. *Bulletin of the Seismological Society of America*, 91(3), 407–426.

Rajendran, K., and Rajendran, C. P. (2002). Historical constraints on previous seismic activity and morphologic changes near the source zone of the 1819 Rann of Kachchh earthquake: further light on the penultimate event. *Seismological Research Letters*, 73, 470–479.

Rajendran, K., and Rajendran, C. P. (2003). Seismogenesis in the stable continental regions and implications for hazard assessment: two recent examples from India. *Current Science*, 85(7), 896–903.

Rastogi, B. K. (2001). Ground deformation study of M_w 7.7 Bhuj earthquake of 2001. *Episodes*, 30(1), 160–165.

Rastogi, B. K., Gupta, H. K., Mandal, P., *et al.* (2001). The deadliest stable continental region earthquake that occurred near Bhuj on 26 January 2001. *Journal of Seismology*, 5, 609–615.

Rastogi, B. K., Choudhury, P., Dumka, R., Sreejith, K. M., and Majumdar, T. J. (2012). Stress pulse migration by viscoelastic process for long-distance delayed triggering of shocks in Gujarat, India, after the 2001 M_w 7.7 Bhuj earthquake. In *Extreme Events and Natural Hazards: The Complexity Perspective*, ed. A. S. Sharma, A. Bundle, V. P. Dimri and D. N. Baker. Geophysical Monograph Series, 196, AGU, Washington, D.C., pp. 63–73, doi:10.10.1029/GM196.

Rastogi, B. K., Kumar, S., Aggrawal, S. K. (2013a). Seismicity of Gujarat. *Natural Hazard*, 65(2), 1027–1044.

Rastogi, B. K., Aggrawal, S. K., Rao, N., and Choudhury, P. (2013b). Triggered/migrated seismicity due to the 2001 M_w 7.6 Bhuj earthquake, Western India. *Natural Hazard*, 65(2), 1085–1107.

Ray, A., Patil, S. K., Paul, D. K., *et al.* (2006). Petrology, geochemistry and magnetic properties of Sadara Sill: evidence of rift related magmatism from Kutch Basin, Northwest India. *Journal of Asian Earth Sciences*, 27, 907–921.

Reddy, C. D., and Sunil, P. S. (2008). Post seismic crustal deformation and strain rate in Bhuj region, western India, after the 2001 January 26 earthquake. *Geophysical Journal International*, 172, 593–606.

Russ, D. P. (1982). Style and significance of surface deformation in the vicinity of New Madrid, Missouri: investigations of the New Madrid, Missouri, earthquake region. In *Investigations of the New Madrid Earthquake Region*, ed. F. A. McKeown and L. C. Pakiser, U.S. Geological Survey Professional Paper 1236, pp. 95–114.

Sarkar, D., Sain, K., Reddy, P. R., Catchings, R. D., and Mooney, W. D. (2007). Seismic-reflection images of the crust beneath the 2001 M = 7.7 Kutch (Bhuj) epicentral region, western India. In *Continental Intraplate Earthquakes: Science, Hazard, and Policy Issues*, ed. S. Stein and S. Mazzotti, Geological Society of America Special Paper 425, pp. 319–327.

Savage, J. C. (1971). A theory of creep waves propagating along a transform fault. *Journal of Geophysical Research*, 76(8), 1954–1966.

Scholz, C. H. (1977). A physical interpretation of the Haicheng earthquake prediction. *Nature*, 267, 121–124.

Segall, P., and Pollard, D. D. (1980). Mechanics of discontinuous faults. *Journal of Geophysical Research*, 85, 4,337–4,350.

Sen, G., Bizimis, M., Das, R., *et al.* (2009). Deccan plume, lithospheric rifting, and volcanism in Kutch, India. *Earth and Planetary Science Letters*, 277, 101–111.

Sharma, B. S., Chopra, S., Rao, K. M., Gupta, A. K., and Rastogi, B. K. (2008). Attenuation and heterogeneity. *ISR Annual Report 2007–8*, pp. 11–14, www.isr.gujarat.gov.in.

Sibson, R. H. (1986). Rupture interaction with fault jogs. In *Earthquake Source Mechanics*, ed. S. Das, J. Boatwright, and C. H. Scholz, AGU Geophysical Monograph 37, Maurice Ewing Series 6, pp. 157–167.

Singh, B. P., ed. (1996). *Indian Archeology 1991–1992: A Review*, New Delhi: Archeological Survey of India.

Singh, S. K., Dattatrayam, R. S., Shapiro, N. M., *et al.* (1999). Crustal and upper mantle structure of Peninsular India and source parameters of the May 21, 1997, Jabalpur earthquake [M_w = 5.8]: results from a new regional broad-band network. *Bulletin of the Seismological Society of America*, 89, 1632–1641.

Singh, A. P., Mishra, O., Rastogi, B. K., and Kumar, D. (2011). 3-D seismic structure of the Kachchh, Gujarat, and its implications for the earthquake hazard mitigation. *Natural Hazards*, 57, 1–23.

Singh, A. P., Mishra, O. P., Yadav, R. B. S., and Kumar, D. (2012). New insight into crustal heterogeneity beneath the 2001 Bhuj earthquake region of Northwest India and its implications for rupture initiations. *Journal of Asian Earth Sciences*, 48, 31–42, doi: 10.1016/j.jseaes.2011.12.020.

Singh, A. P., Mishra, O. P., Rastogi, B. K., and Kumar, S. (2013). Crustal heterogeneities beneath the 2011 Talala, Saurashtra earthquake, Gujarat, India source zone: seismological evidence for neo-tectonics. *Journal of Asian Earth Sciences*, 62, 672–684.

Sreejith, K. M., and Rastogi, B. K. (2013). Interferometric Synthetic Aperture Radar (INSAR) studies in Kachchh by ISRO and ISR. *ISR Annual Report 2012–13*, pp. 49–51, www.isr.gujarat.gov.in.

Talwani, P., Chen, L., and Kalpana, G. (2007). Seismogenic permeability, Ks. *Journal of Geophysical Research*, 112, B07309, doi:10.1029/2006JB004665.

Wang, Z., and Zhao, D. (2006). Seismic evidence for the influence of fluids on the 2005 west off Fukuoka prefecture earthquake in southwest Japan. *Physics of the Earth and Planetary Interiors*, 155, 313–324.

Wilson, D., Aster, R., and the RISTRA Team (2003). Imaging crust and upper mantle seismic structure in the southwestern United States using teleseismic receiver functions. *Leading Edge*, 22, 232–237.

Zhao, D., Kanamori, H., Negishi, H., and Wiens, D. (1996). Tomography of the source area of the 1995 Kobe earthquake: evidence for fluids at the hypocenter. *Science*, 274, 1891–1894.

7

The New Madrid seismic zone of the Central United States

ROY VAN ARSDALE

Abstract

The central Mississippi River valley, within which the New Madrid seismic zone (NMSZ) lies, has undergone a history of Precambrian microplate accretion, Late Proterozoic Grenville orogenesis, Cambrian Reelfoot rifting, Late Paleozoic Ouachita/Appalachian orogenesis, Cretaceous passage above the Bermuda hotspot, and modest Cenozoic compression. Reelfoot Rift consists of northwest-striking Proterozoic and the northeast-striking Cambrian faults, which have undergone Neogene transpressive displacement that is apparently driven by the N60 °E to N80 °E S_{Hmax} of eastern North America. NMSZ earthquakes are occurring along five reactivated Reelfoot Rift faults, three of which are transpressive right-lateral strike-slip faults and two are reverse faults associated with compressional stepovers between the strike-slip faults. A number of models have been proposed for the NMSZ, but the model that best explains the Holocene onset of the seismic zone is the Mississippi River valley denudation model. In this model, Late Wisconsin entrenchment of the Mississippi River reduced the vertical stress, which reduced the normal stress across the Reelfoot Rift strike-slip faults and allowed them to slip. Quaternary displacement is not restricted to the NMSZ faults and occurs on many of the Reelfoot Rift and its outboard faults requiring an explanation for Quaternary faulting that encompasses a much larger area than the NMSZ. The denudation model appears to explain the wide distribution of low-displacement Quaternary faulting in the lower Mississippi River valley.

7.1 Introduction

The New Madrid seismic zone (NMSZ) is located in the central Mississippi River valley within the states of Missouri, Tennessee, and Arkansas (Figure 7.1) and is the most active

Intraplate Earthquakes, ed. Pradeep Talwani. Published by Cambridge University Press. © Cambridge University Press 2014.

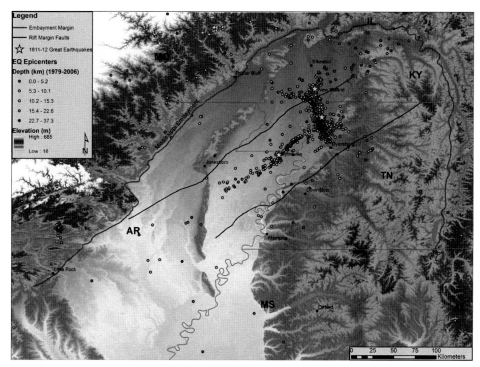

Figure 7.1 Small earthquakes recorded between 1979 and 2006 in the New Madrid seismic zone are illustrated in colored dots with the three largest 1811–1812 earthquakes marked with gold stars. Southern star, December 16, 1811; northern star, January 23, 1812; and central star, February 7, 1812. The earthquakes primarily occur within the underlying Reelfoot Rift, which is bound by the black lines (from Csontos *et al.*, 2008). For color version, see Plates section.

seismic zone of the Eastern United States. Microseismic activity (Chiu *et al.*, 1992; Mueller and Pujol, 2001) is occurring along reactivated basement faults within the Reelfoot Rift: a Cambrian aulacogen (Hildenbrand, 1985; Hildenbrand and Hendricks, 1995) (Figure 7.2). Although contemporary seismicity rarely exceeds **M** (moment magnitude) 4.0, the NMSZ has generated very large earthquakes; most recently during 1811–1812 when a minimum of four earthquakes (Hamilton and Johnston, 1990), estimated to have been in the mid **M** 7 range (Nuttli, 1973; Gomberg, 1993; Hough *et al.*, 2000; Mueller and Pujol, 2001; Hough and Page, 2011), occurred over a three-month period. The threat of this seismic zone to the United States has prompted numerous studies and expensive retrofitting, an example of which is the 276 million dollar retrofit of the Interstate 40 Bridge across the Mississippi River at Memphis. Debates continue as to the hazard posed by future New Madrid earthquakes (Stein, 2010). Many scientists argue that the late Holocene 500-year recurrence interval determined from NMSZ paleoseismic studies implies continuing hazard, whereas some of those studying GPS data argue that there is not enough strain currently accumulating to produce a very large earthquake in the near future. In this chapter I summarize the geological history of the NMSZ and its earthquakes.

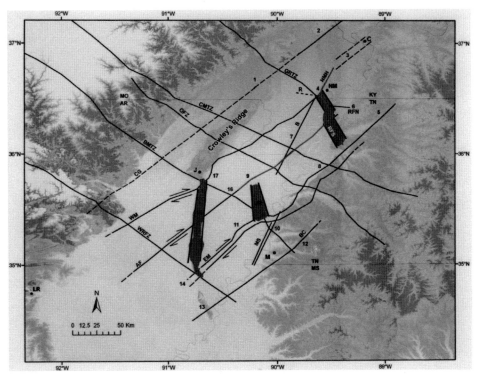

Figure 7.2 Reelfoot Rift faults and numbered locations of documented Quaternary faulting and liquefaction (from Van Arsdale and Cupples, 2013). Numbers correspond to numbers in Table 7.1.
Right-lateral shear across the Reelfoot Rift is responsible for the New Madrid seismic zone earthquakes, which occur along the Axial (northeast of number 16), Reelfoot South, and Reelfoot North
faults. Quaternary right-lateral shear on the rift faults is also causing uplift of the Lake County
uplift/Reelfoot North fault (RFN), Joiner Ridge (JR), Charleston Uplift, the southern portion of
Crowley's Ridge, and possibly the Meeman-Shelby fault. WRFZ, White River fault zone; BMTZ,
Bolivar Mansfield tectonic zone; OFZ, Osceola fault zone; CMTZ, Central Missouri tectonic zone;
GRTZ, Grand River tectonic zone; EM, Southeastern Reelfoot Rift margin faults; WM, Northwestern
Reelfoot Rift margin fault; AF, Axial fault (Cottonwood Grove fault); NMN, New Madrid North
fault; RFS, Reelfoot South fault; MS, Meeman-Shelby fault zone; CG, Commerce Geophysical linea-
ment/fault; BC, Big Creek/Ellendale fault; B, Bootheel fault; R, Risco fault (defined by seismicity);
M, Memphis, Tennessee; LR, Little Rock, Arkansas; NM, New Madrid, Missouri; C, Cairo, Illinois.

7.2 Geological history of the New Madrid seismic zone region

7.2.1 Precambrian

The NMSZ earthquakes are occurring beneath the flood plain of the central Mississippi
River valley. To more fully understand this area it is instructive to review the regional
geology and its history.

The Earth's crust beneath the Mississippi River valley has an average thickness of
42 km (Grollimund and Zoback, 2001). P-wave velocities and densities beneath the central

valley reveal ~6 km of Mississippi Embayment sediments and Reelfoot rift-fill strata overlying basement (Ginzburg *et al.*, 1983). Precambrian igneous and metamorphic rocks (6.2 km/s, 2.74 g/cm^3) of the upper crust extend from 6 to 17 km in depth. The lower crust varies in thickness and extends to depths of 39–46 km. Composition of the lower crust is speculative, but it is interpreted to be crystalline rocks of intermediate composition (6.6 km/s, 2.95 g/cm^3) (Stuart *et al.*, 1997). The variable thickness of the lower crust is due to a rift pillow at the base of the lower crust (Mooney *et al.*, 1983; Hildenbrand, 1985; Hildenbrand and Hendricks, 1995). Although not well constrained, the rift pillow rocks (7.3 km/s, 3.1 g/cm^3) are interpreted to be a quartz tholeiite intrusion or underplating that has been metamorphosed to garnet granulite. The pillow is proposed to have formed by mantle melt intrusions in the deep crust by a mantle plume during Cambrian rifting (Mooney *et al.*, 1983; Hildenbrand, 1985; Pollitz *et al.*, 2001), Appalachian orogeny (Hildenbrand and Hendricks, 1995), or during the Middle Cretaceous superplume event (Csontos, 2007). The top of the mantle (8 km/s, 3.3 g/cm^3) is at depths of 39–46 km.

Crustal rocks of the southeastern United States formed as a complex of microplates and island arcs that were accreted during the Archean and Proterozoic (Figure 7.3). The major geological events of these times were the formation of the Superior Province (3.5–2.7 Ga), Penokean Province (1.89–1.83 Ga), Yavapai Province (1.78–1.72 Ga), Mazatzal Province (1.65 Ga), Eastern Granite–Rhyolite Province (1.47 Ga), Southern Granite–Rhyolite Province (1.37 Ga), and Midcontinent Rift system (1.1 Ga), and these provinces are bordered on the east and south by the Grenville Province (1.2–1.0 Ga) (Atekwana, 1996; Van Schmus *et al.*, 1996, 2007; Tollo *et al.*, 2004; Holm *et al.*, 2007; Van Arsdale, 2009). Rocks of the Yavapai and Mazatzal Provinces (Geon 16 and 17 terranes of Figure 7.3) have a southwesterly grain and consist of metasedimentary, metaigneous, and basalt–rhyolite volcanic suites that were subsequently intruded by anorogenic granitic rocks of the Southern and Eastern Granite–Rhyolite provinces. The Mazatzal Province is bound on the east by the Eastern Granite–Rhyolite Province (EGRP), and the EGRP is generally accepted to be the basement rock of the NMSZ region. Seismic reflection data indicate widespread sub-horizontal reflectors, suggesting an interlayered sequence of rhyolites, granites, and mafic igneous rocks and/or sedimentary rocks within the EGRP (Van Schmus *et al.*, 1993). Gravity and magnetic data suggest that the Mazatzal Province rocks may continue beneath the EGRP (Atekwana, 1996), whereas isotopic data (Van Schmus *et al.*, 2007) indicate that Precambrian rocks younger than 1.6 Ga lie below the EGRP southeast of the heavy dashed line in Figure 7.3.

The Grenville Province bounds the eastern and southern flanks of the EGRP (Figure 7.3). This province formed by the collision of the North America craton first with island arcs and then with a continent to the east that is now buried beneath the Appalachian Mountains or was carried away during the late Proterozoic and early Paleozoic opening of the Iapetus Ocean (Rankin *et al.*, 1993; Atekwana, 1996; Bartholomew and Hatcher, 2010). The Grenville orogeny culminated in the formation of a huge mountain range comparable to the Himalayas of today and also in the formation of the supercontinent of Rodinia. The western boundary of the Grenville Province, the Grenville Front, has been

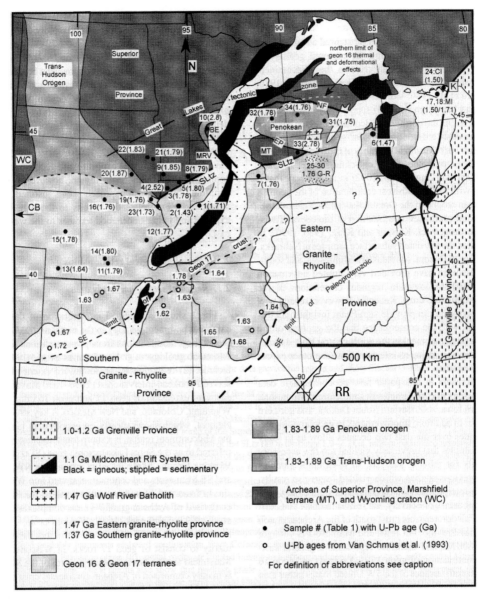

Figure 7.3 Precambrian geology of the northern Mississippi River valley (from Van Schmus *et al.*, 2007). BE, Becker Embayment; CB, Cheyenne Belt; CI, Croker Island complex; EP, Eau Pleine shear zone; G-R, granite-rhyolite province; K, Killarney magmatic complex; MI, Manitoulin Island; MRV, Minnesota River Valley terrane promontory; MT, Marshfield terrane; NF, Niagara Fault; SLtz, Spirit Lake tectonic zone; WC, Wyoming craton; RR, Reelfoot Rift.

traced primarily by gravity and magnetic data southerly through the United States, and the province's southwestern continuation has been mapped in Texas. However, it is not known where Grenville rocks exist beneath the southern United States, and in particular between southern Tennessee and its exposure in the Llano Uplift of central Texas, because there is insufficient drill-hole data in the southern United States. Culotta *et al.* (1990) have proposed that Grenville rocks underlie the NMSZ and Nelson and Zhang (1991) speculate that the western margin of the Grenville Front underlies and is responsible for the location of the Cambrian Reelfoot Rift. As the above survey of the literature indicates, the true nature of the Precambrian crust beneath the NMSZ is still an open question.

At the close of the Grenville orogeny, the Grenville Mountains extended along what is today the eastern seaboard and wrapped around the southern margin of the United States, just like the much younger 300 Ma old Appalachian–Ouachita Mountains (Van Arsdale, 2009; Bartholomew and Hatcher, 2010). Formation of the Grenville Mountains was the culmination of the collision of Laurentia (North America), Baltica (Europe), Amazonia (South America), and Kalahari (Africa) to form the supercontinent of Rodinia. Thus, during Rodinia time, the southeastern United States was a topographically high area located in the interior of Rodinia.

The Precambrian closed with the disassembly of Rodinia and the formation of a passive margin along the southern margin of the United States. More specifically, Thomas (1991, 1993) argues that the southern margin of the United States near the NMSZ area was formed by the Alabama–Oklahoma transform fault (Figure 7.4). If true, the continental shelf was sharply truncated, and immediately south of this transform fault (perhaps within as little as 25 km) was very deep water above an abyssal plain.

7.2.2 Paleozoic

Present-day NMSZ seismicity is occurring along reactivated faults of the Reelfoot Rift (Figures 7.1, 7.2, and Table 7.1). This rift formed during the Early or Middle Cambrian (520–500 Ma) as part of the Reelfoot Rift–Rough Creek graben–Rome Trough (Figure 7.4) in an abortive attempt to pull off a corner of Laurentia (United States) during the disassembly of Rodinia (Thomas, 1991). The southern end of the Reelfoot Rift was apparently the shelf edge at the Alabama–Oklahoma transform fault (Figure 7.4).

The Reelfoot Rift consists of fault-bounded blocks that were created by the intersection of Proterozoic (N~55°W) and Cambrian (N50°E) faults (Figure 7.2). North ~55° west-trending faults that displace Paleozoic strata in southeastern Missouri (Anderson, 1979) and northeastern Arkansas (Haley *et al.*, 1993) are interpreted to be reactivated faults of the Mazatzal terrane (Central Plains orogen) (McCracken, 1971). These reactivated Proterozoic faults (Cox, 1988) have been proposed to pass beneath the Mississippi River valley into western Tennessee (Figure 7.2) (Stark, 1997; Csontos *et al.*, 2008). For example, the Grand River Tectonic Zone appears to continue beneath the Mississippi River valley as the Reelfoot fault (Csontos *et al.*, 2008). During Cambrian rifting of Rodinia the N50°E-trending Reelfoot Rift formed and now underlies the Mississippi River valley in portions

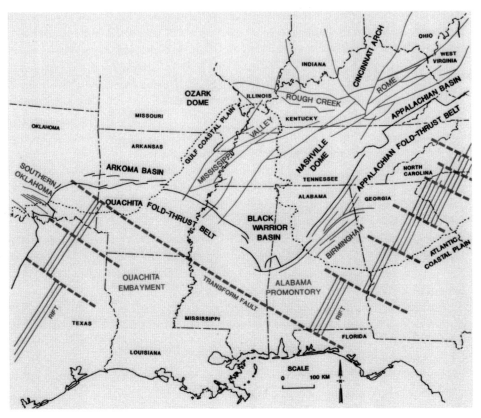

Figure 7.4 Late Precambrian–early Paleozoic transform faults (dashed lines), rifted continental margin (parallel straight lines), Cambrian basement fault systems (curved lines), and late Paleozoic orogenic belts and cratonic structures (modified from Thomas, 1988).

of Kentucky, Missouri, Arkansas, and Tennessee (Figure 7.4) (Erwin and McGinnis, 1975; Hildenbrand, 1985; Nelson and Zhang, 1991; Dart and Swolfs, 1998; Thomas, 2006; Csontos *et al.*, 2008). Superposition of the northwest- and northeast-trending faults resulted in the Reelfoot Rift consisting of sub-basin blocks.

A composite east–west COCORP deep reflection profile across the Reelfoot Rift ~50 km north of Memphis, Tennessee, reveals inward-dipping listric normal faults that form half-grabens along the southeast and northwest boundaries of the rift (Figure 7.5) (Nelson and Zhang, 1991). Inward-dipping reflections to 40 km depth indicate that extensional faulting may exist through the crust to the Moho. Extension across the rift is only ~17%; however, the amount of subsequent inversion shortening due to Appalachian/Ouachita orogeny compression is not known. Nelson and Zhang (1991) also identified northeast-trending faults within the rift: a steep east-dipping fault that projects up-section to the western margin of Crowley's Ridge, the steep west-dipping Axial fault (Cottonwood Grove

Table 7.1 *Reelfoot Rift Quaternary faulting and liquefaction locations designated with numbers in Figure 7.2 (from Van Arsdale and Cupples, 2013)*

Location/name	Structure	Deformation age	Source
1 Western Lowlands	Faulting and liquefaction	23,000–17,000 and 13,430–9,000 yr BP, AD 240–1020 and 1440–1540	Shoemaker *et al.*, 1997; Vaughn, 1994
2 Commerce fault	Faulting	60–50 ka, 35–25 ka, 5 ka, and 3,660 yr BP	Harrison *et al.*, 1999
3 Charleston Uplift	Faulting	<12 ka	Pryne *et al.*, 2013
4 New Madrid North fault	Faulting	Wisconsin	Baldwin *et al.*, 2005
5 Southeastern Reelfoot Rift margin	Faulting	Pleistocene	Cox *et al.*, 2006
6 New Madrid seismic zone	Faulting and liquefaction	2350 BC, AD 300, 900, 1450, and 1811	Kelson *et al.*, 1996; Tuttle *et al.*, 2002, 2005
7 Bootheel fault	Faulting	12.5–10.2 ka, 2.7–1.0 ka, and AD 1450	Guccione *et al.*, 2005
8 Southeastern Reelfoot Rift margin	Faulting	<20 ka	Cox *et al.*, 2006
9 Manila High	Faulting	11,500–5,400 yr BP, AD 1450 and 1811	Guccione *et al.*, 2000; Odum *et al.*, 2010
10 Southeastern Reelfoot Rift margin	Faulting	4,000–2,000 yr BP (2 events); <2000 yr BP	Cox *et al.*, 2013
11 Southeastern Reelfoot Rift margin	Faulting	Quaternary	Howe, 1985
12 Ellendale	Faulting	AD 400	Velasco *et al.*, 2005
13 Big Creek	Faulting	<27 ka	Harris & Sorrells, 2006
14 Marianna	Liquefaction	7,000–5,000 yr BP	Tuttle *et al.*, 2006
15 Crowley's Ridge	Faulting	Wisconsin	Van Arsdale *et al.*, 1995
16 Marked Tree High	Faulting	4,440–3,350 yr BP	Guccione, 2005
17 Northwestern Reelfoot Rift margin	Faulting	<19 ka	Van Arsdale *et al.*, 1995

fault) coincident with the Blytheville arch (Pratt *et al.*, 2012), and relatively minor faults. Howe (1985) interprets the Axial fault as a Cambrian half-graben fault that was inverted during the late Paleozoic Appalachian orogeny to form the Blytheville arch (Charlie's Ridge).

Although not radiometrically dated, the oldest sediments in the Reelfoot Rift are interpreted to be Early to Middle Cambrian (Johnson *et al.*, 1994). The sediments at the base of the rift consist of arkosic sandstones as much as 1 km thick overlain by shale, limestone, and dolomite with a cumulative thickness of 7 km. The deep arkosic sandstone

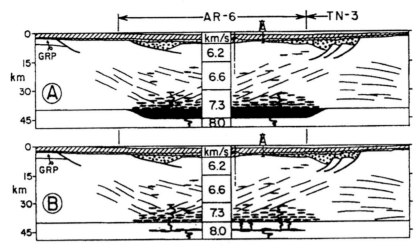

Figure 7.5 Two interpretations of a COCORP reflection line across the Reelfoot Rift located ~50 km north of Memphis, Tennessee. In (A) the 7.5 km/s velocity (black) at the base of the crust is interpreted to be a Cambrian magmatic underplate, whereas in (B) the 7.5 km/s velocity is interpreted to be magmatic sills injected into the pre-existing crust with no underplate (from Nelson and Zhang, 1991).

was deposited only within the down-dropped rift, and the detailed characteristics of these rocks remain a mystery because they have only been penetrated by a few oil exploration wells.

Subsidence of the Reelfoot Rift ceased by the end of deposition of the regionally extensive Knox Group, as revealed in the fact that the top of the Knox (470 Ma) is essentially flat across the rift margins. The NMSZ region experienced far-field effects of the Appalachian Taconic orogeny (480–450 Ma) and the Acadian orogeny (417–329 Ma) (Clendenin and Diehl, 1999) with some sediments shed westward off of the rising mountains reaching the NMSZ region. However, the Appalachian Allegheny orogeny (320–250 Ma) directly affected the NMSZ region. This orogeny resulted in the collision of North America with Europe, Africa, South America, and microcontinents to form the Appalachian–Ouachita mountain system. Appalachian Mountain thrust belt rocks continue westward beneath the Mississippi River valley and merge with the contiguous Ouachita Mountains. While formation of the Appalachian–Ouachita system was applying stress from the east and south, the Ancestral Rockies were also rising and thus applying stress from the west during the Carboniferous. The southern portion of the Reelfoot Rift was overthrust by Ouachita thrust sheets, while sediments accumulated over the rift in the Arkoma/Black Warrior foreland basin.

The Paleozoic closed with the Appalachian–Ouachita Mountain system forming a weld in the assembly of Pangea. This huge mountain system traversed the interior of Pangea resulting in drainage across the future United States flowing north and west off of the mountains (Dickinson and Gehrels, 2009).

7.2.3 Mesozoic

Pangea was torn apart during the Mesozoic, and southern North America began to separate with the initiation of the Gulf of Mexico in the Late Triassic (220 Ma). Disassembly of Pangea occurred primarily along the Appalachian–Ouachita Mountains, yet a portion of these mountains persisted and continued to control drainage across the Eastern United States until Late Cretaceous when the Mississippi Embayment formed.

The Mississippi Embayment formed as a consequence of the Central United States passing over the Bermuda hotspot during the mid-Cretaceous (Cox and Van Arsdale, 1997, 2002). Thermal heating and igneous intrusion of the crust by the hotspot caused uplift of the previously faulted Reelfoot Rift region to form the northeast-trending Mississippi Embayment arch. Erosion of the arch occurred and when the Central United States passed eastward off of the hotspot, the cooling crust subsided. Subsidence lowered the denuded crust below sea level and the Gulf of Mexico advanced up the axis of the subsided landscape, thus resulting in a regional unconformity wherein Late Cretaceous sediments locally lie on top of strata varying from Middle Cambrian shale in the northern embayment to Early Cretaceous limestone in the south. Hotspot uplift, erosion, and subsidence of the Mississippi Embayment arch included a segment of the Appalachian–Ouachita Mountains, which caused breaching of the mountains. Thus, drainage in the Central United States changed from north and west to south along the Mississippi Embayment axis to form the Mississippi River drainage system. The Bermuda hotspot is responsible for the Middle Cretaceous basalt and mafic intrusions in the NMSZ region, most of which were subsequently buried by the Late Cretaceous embayment sediments. Thus, the Reelfoot Rift and NMSZ underwent a significant thermal event during the Cretaceous.

The Early Cretaceous shelf margin is approximately coincident with the southern edge of the thick transitional crust in the northern Gulf of Mexico. This shelf margin was subsequently buried beneath a huge influx of Late Cretaceous clastic sediment. Sediment influx, primarily from the rising Rocky Mountains during the Laramide orogeny, and formation of the Mississippi River drainage system (Cox and Van Arsdale, 2002; Van Arsdale and Cox, 2007) were responsible for the seaward growth of the shoreline from southern Mississippi through eastern Texas. High Late Cretaceous sea level and subsidence of the Mississippi Embayment resulted in the Gulf of Mexico extending north into southern Illinois, thereby depositing approximately 500 m of Late Cretaceous sediments across the NMSZ. The Cretaceous ended with the Chicxulub impact occurring in the Yucatan Peninsula of Mexico immediately south of the Mississippi Embayment. What effect this impact may have had on the NMSZ is not known, but the Paleocene Clayton Formation at the K-T boundary in southeastern Missouri has been interpreted to be a megatsunami deposit (Campbell *et al.*, 2008).

7.2.4 Cenozoic

Subsidence of the Mississippi Embayment continued into Paleocene (65–55 Ma) Midway Group deposition. The overlying Paleocene Wilcox and Eocene (55–34 Ma) Claiborne

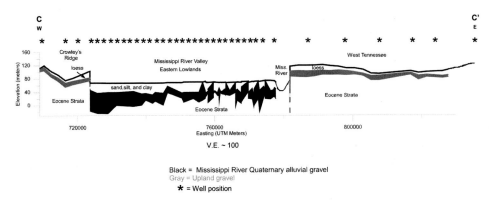

Figure 7.6 A west–east cross-section through the Mississippi River flood plain alluvium and the ancient Mississippi River flood plain sand and gravel (Upland gravel) located ~20 km north of Memphis in Figure 7.1 (from Van Arsdale *et al.*, 2007).

groups reflect alternating shallow marine, near shore, and fluvial environments as the Gulf of Mexico transgressed and regressed within the Mississippi Embayment. The last marine excursion into the Mississippi Embayment was during the Oligocene (34–24 Ma).

During early Pliocene (5.5–4.5 Ma) the ancestral Mississippi River valley looked very much like it does today in the NMSZ region. A major river system flowed south within a vast flood plain that covered portions of the same states that the Mississippi River flood plain covers today. However, the Pliocene ancestral Mississippi River flowed at a higher elevation, perhaps as much as 100 m higher than today's Mississippi River (Van Arsdale *et al.*, 2007). Evidence for this high-level Pliocene river is the sand and gravel Upland Complex (Mississippi River terrace) discontinuously preserved on drainage divides east of the modern Mississippi River from western Kentucky to southern Louisiana and west of the Mississippi River on Crowley's Ridge in Arkansas (Figure 7.6) (Autin *et al.*, 1991; Saucier, 1994). The Upland Complex varies in thickness from 1 to 100 m primarily because it has an unconformable base and top. The Upland Complex overlays Eocene and Oligocene sediments and is overlain by Pleistocene (1.8 Ma–10 ka) loess (Clark *et al.*, 1989; Markewich *et al.*, 1998). This loess consists of up to five loess units that together reach a thickness of 30 m beneath Memphis, Tennessee.

Entrenchment of the ancestral Mississippi River valley began at ~4 Ma apparently due to growth of the Antarctic and Greenland ice sheets and resulting sea level decline. The valley has experienced a complex erosional and depositional Pleistocene history tied to the advance and retreat of perhaps 20 Laurentide ice sheets (Easterbrook, 1999). During the Pleistocene, the ancestral Mississippi River flowed down the Western Lowlands and the ancestral Ohio River flowed down the Eastern Lowlands and the two rivers merged south of Helena, Arkansas (Saucier, 1994; Van Arsdale *et al.*, 2007). During the Pleistocene their point of convergence jumped northward several times until reaching Cairo, Illinois, at 10 ka.

The Mississippi River valley has entrenched 100 m through Pliocene Upland Complex strata and is inset into Eocene strata (Figure 7.6) (Van Arsdale *et al.*, 2007). This entrenchment occurred within the past ~4 Ma.

7.3 The New Madrid seismic zone

7.3.1 The 1811–1812 earthquakes

Contemporary seismicity within the NMSZ primarily consists of small earthquakes occurring between the depths of 4 and 14 km (Figure 7.1) (Csontos and Van Arsdale, 2008). The earthquakes are occurring along a northeast trend from Marked Tree, Arkansas, to the southern end of Reelfoot Lake, Tennessee, where the seismicity merges with a broad band of earthquakes that lie along a northwest trend from Dyersburg, Tennessee, to New Madrid, Missouri. The seismicity trend continues northeast from New Madrid with a less well defined trend to the west from New Madrid towards Risco, Missouri.

During the winter months of 1811–1812 three very large mainshocks and one very large aftershock rocked the central Mississippi River valley (Johnston, 1996; Johnston and Schweig, 1996; Hough and Page, 2011). These earthquakes occurred at approximately 2:15 a.m. December 16, 1811, 8:15 a.m. December 16, 1811 (aftershock), 9 a.m. January 23, 1812, and at 3:45 a.m. February 7, 1812. Hough and Page (2011) believe that the December 16 mainshock occurred on the Axial fault, the large December 16 aftershock occurred on either the northern portion of the Axial fault or on the Reelfoot South fault, the January 23 mainshock occurred on either the New Madrid North fault or on a fault in southern Illinois, and that the February 7 mainshock occurred on the Reelfoot North fault (Figure 7.2). Shaking from the four principal events was felt at a number of locations along the eastern seaboard including Charleston, South Carolina, north to Quebec, Canada, at least as far west as the Kansas–Nebraska border, and south in New Orleans, Louisiana, encompassing an area of 500,000 km^2 (Penick, 1981; Johnston and Schweig, 1996). Few Europeans were living in the epicentral region during these earthquakes, with the largest town of New Madrid (now in Missouri) having a population of ~400. Our best accounts of these earthquakes come from diaries kept by frontier people and boatmen who were on the Mississippi River at the time (Penick, 1981). Eyewitness accounts describe the banks of the Mississippi River caving, water sloshing out of the banks and returning with many downed trees, the river flowing upstream (probably a seiche), sand and water exploding 100 feet (30.5 m) into the air, ground fissures opening, landslides along the river bluffs, temporary damming of the river, temporary waterfalls (perhaps rapids), permanent land uplift (e.g., Lake County uplift) and subsidence (e.g., Reelfoot Lake), as well as severe ground shaking. As spectacular as these earthquakes must have been, there was no systematic compilation of the geological effects until Fuller (1912), who documented fissures, landslides (Jibson and Keefer, 1989), and liquefaction deposits (Obermeier, 1989) that were still evident in the landscape. One of Fuller's most compelling arguments for very large earthquakes was his work on the distribution of liquefaction deposits, mapped in detail by Tuttle *et al.*

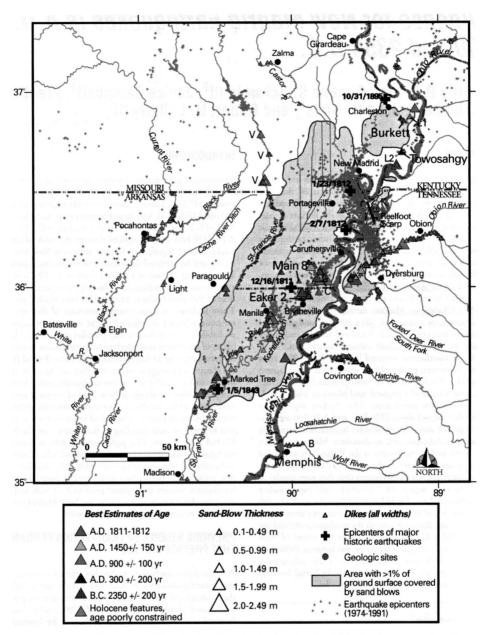

Figure 7.7 Map of earthquake liquefaction deposits in the New Madrid seismic zone region (from Tuttle *et al.*, 2005). For color version, see Plates section.

(2005) (Figure 7.7). Large aftershocks diminished after 1812 and the seismic threat was largely forgotten until the 1970s when the Mississippi River valley was evaluated for the construction of nuclear power plants. Installation of seismometers since the 1970s has clearly defined the NMSZ (Figure 7.1) (Chiu *et al.*, 1992; Mueller and Pujol, 2001).

Magnitudes of the 1811–1812 earthquakes have been estimated from intensity data with maximum magnitude estimates having ranged from M 8.1 (Johnston, 1996; Johnston and Schweig, 1996) to low 7 magnitude (Gomberg, 1993; Newman *et al.*, 1999; Hough *et al.*, 2000).

7.3.2 *Geological structure of the NMSZ*

NMSZ earthquakes are generally believed to occur along reactivated, and commonly inverted, Cambrian Reelfoot Rift faults (Van Arsdale, 2009). In some cases faults are illuminated by seismic reflection data while in other cases earthquakes occur along planar surfaces that we interpret as fault planes. Specifically, the southern NMSZ seismicity arm is occurring along the Axial (Cottonwood Grove) fault (oriented 46°, 90°); earthquakes occurring along the central northwest-trending arm appear to define a southwest-dipping Reelfoot reverse fault that is divided into the Reelfoot South fault (150°, 44° SW) and Reelfoot North fault (167°, 30° SW), the Risco fault seismicity (92°, 82° N), and the New Madrid North fault seismicity (29°, 72° SE) (Csontos and Van Arsdale, 2008). Fault plane solutions indicate that the Axial and New Madrid North faults are right-lateral strike-slip faults, the Risco fault is a left-lateral strike-slip fault, and the Reelfoot fault is a thrust fault at depths of 4 to 14 km (Csontos and Van Arsdale, 2008). Seismic reflection data and fault modeling (Purser and Van Arsdale, 1998) reveal that the Reelfoot thrust fault steepens into a reverse fault with a southwest dip of 72° above 4 km (Figure 7.8) and Champion *et al.* (2001) model the Reelfoot fault above ~500 m depth as a trishear fault propagation fold. The Reelfoot North fault has a hanging wall horst that is manifest at the ground surface as the Lake County Uplift with the Tiptonville dome culmination.

Although seismically quiet in recent times there are two other faults with Quaternary displacement within the immediate NMSZ area. The 135 km long Bootheel fault is a transpressional right-lateral strike-slip fault with at least 13 m of strike-slip offset and ~3 m of up-to-the-east displacement (Figure 7.2) (Guccione *et al.*, 2005). Immediately northeast of the Lake County uplift is the New Markham fault. Odum *et al.* (1998) interpret the New Markham fault to be a near-surface continuation of the Reelfoot fault. However, the sense of displacement on this fault is up to the east, which is opposite to Reelfoot fault displacement. Van Arsdale *et al.* (1998) propose that the New Markham fault is a transpressive strike-slip fault with up-to-the-east reverse displacement. The Bootheel and New Markham faults may be linking faults (Wesnousky, 1988) that bypass the Reelfoot North fault stepover zone and may eventually "straighten out" the New Madrid shear zone (Schweig and Ellis, 1994; Van Arsdale *et al.*, 1998).

Considering the regional tectonic setting it appears that the NMSZ is a right-lateral strike-slip fault system with two compressional left stepovers (Figure 7.2) (Csontos *et al.*,

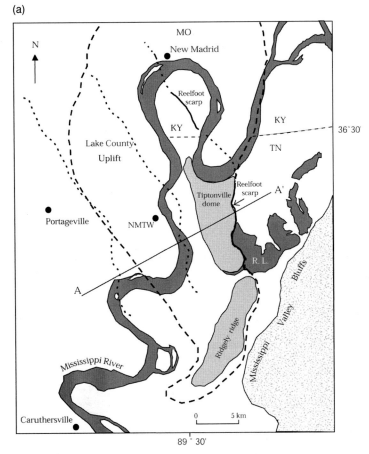

(a)

MO
New Madrid
Reelfoot scarp
KY
KY
36°30'
Lake County Uplift
TN
Reelfoot scarp
A'
Tiptonville dome
Portageville
NMTW
R.L.
Ridgely ridge
Mississippi River
Caruthersville
0   5 km
89° 30'

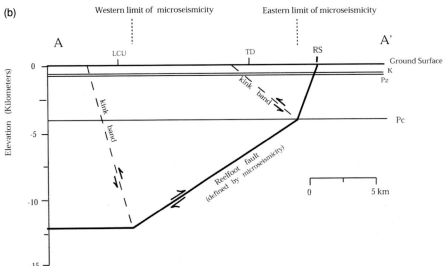

(b)

Western limit of microseismicity   Eastern limit of microseismicity

A   A'
LCU   TD   RS
                                                            Ground Surface
0   K
                                                                    Pz
kink band
                                                                    Pc
kink band
-5

Reelfoot fault
(defined by microseismicity)
0   5 km
-10

-15

Elevation (Kilometers)

Figure 7.8 (a) The Lake County uplift and vicinity (from Purser and Van Arsdale, 1998). The dashed line marks the boundary of the Lake County uplift as defined by Russ (1982), and the dotted lines are kink bands (back thrusts) on the west side of the Tiptonville dome and Lake County uplift. Line A–A' is the line of cross-section in B. RL, Reelfoot Lake. (b) Cross-section A–A' with its kink bands (back thrusts). K, top of Cretaceous; Pz, top of Paleozoic; Pc, top of Precambrian; LCU, Lake County uplift western margin; TD, Tiptonville dome western margin; RS, Reelfoot scarp. No vertical exaggeration. Reverse fault displacement not visible at this scale.

2008). The major faults in the NMSZ are the southeastern Reelfoot Rift margin faults (Chiu *et al.*, 1997), the Axial fault, the New Madrid North fault (northwest Reelfoot Rift margin), and the Reelfoot fault. The Reelfoot fault consists of two distinct segments: the Reelfoot South fault is a left stepover linking the southeastern Reelfoot Rift margin and the Axial fault, whereas the Reelfoot North fault is a left stepover linking the Axial fault with the New Madrid North fault (northwestern Reelfoot Rift margin). Northeast of New Madrid, Missouri, the northwestern Reelfoot Rift margin has 36 m of Quaternary displacement on the Charleston uplift (Figure 7.2) (Pryne *et al.*, 2013).

The Commerce Geophysical Lineament is a very long basement feature (Figure 7.2) that may be an outlier fault of the Reelfoot Rift (Langenheim and Hildenbrand, 1997; Clendenin and Diehl, 1999; Van Arsdale and Cupples, 2013). Faulting in the Thebes Gap area of Missouri (Harrison *et al.*, 1999) and Illinois (Nelson *et al.*, 1997) reveals that at least this portion of the Commerce Geophysical Lineament (Commerce fault) is predominantly a right-lateral strike-slip fault that has been active throughout the Phanerozoic and has Quaternary movement. In the Western Lowlands (west of Crowley's Ridge) southwest of Thebes Gap, Vaughn (1994) has identified four paleoliquefaction deposits (23,000–17,000 yr BP, 13,430–9,000 yr BP, AD 240–1020, and AD 1440–1540) that may be due to prehistoric earthquakes on a north-northwest-trending fault underlying the St. Francis River (Figure 7.2).

Research also indicates Quaternary faulting east of the NMSZ (Parrish and Van Arsdale, 2004). Cox *et al.* (2001a) present evidence for Quaternary ground tilting based on analysis of drainage-basin asymmetry that they attribute to basement block tilting in southwestern Kentucky, western Tennessee, and northwestern Mississippi. Subsurface mapping of the Upland Complex in western Kentucky and Tennessee reveals Quaternary east–west normal faulting (Van Arsdale and Cupples, 2013). Faulting has also been mapped south of Memphis along the Big Creek fault zone (Spitz and Schumm, 1997; Harris and Sorrells, 2006; Harris, 2009), immediately west of Memphis along the Meeman–Shelby fault (Hao *et al.*, 2013), and beneath Memphis that is Quaternary in age (Velasco *et al.*, 2005; Martin, 2008; Van Arsdale *et al.*, 2012).

Immediately north of New Madrid, Missouri, is Sikeston Ridge (Figure 7.1), which has been interpreted to be an erosional remnant of Pleistocene ancestral Mississippi River incision. Although there is no evidence of surface faulting along the margins of Sikeston Ridge, it is possible that Sikeston Ridge is an erosionally modified horst (Sexton, 1992; Csontos, 2007).

Regionally, we see other examples of compressional stepover zones within the Reelfoot Rift (Csontos *et al.*, 2008). The southern portion of Crowley's Ridge is a fault-bounded block that has been interpreted to be a compressional stepover between the southeastern and northwestern Reelfoot Rift margins (Figure 7.2). A stepover origin has also been proposed for the sub-alluvial Joiner Ridge and its northwestern surface continuation of the Manila High (Odum *et al.*, 2010).

The Reelfoot fault hanging wall horst, Joiner Ridge, the southern portion of Crowley's Ridge, and possibly Sikeston Ridge are nearly parallel uplifted blocks (Figures 7.1 and 7.2).

An obvious question is: what is controlling this structural pattern? It appears that these stepover zones are related to basement faults of the Reelfoot Rift. The Reelfoot fault extends between the rift margins, is divided into two segments by the Axial fault, and appears to overlie the Grand River tectonic zone (Csontos, 2007). Sikeston Ridge originates at the intersection of Reelfoot fault (Grand River tectonic zone) and the northwestern Reelfoot Rift margin; Joiner Ridge appears to be related to the intersection of the southeastern Reelfoot Rift margin and the Bolivar Mansfield tectonic zone; and the southern portion of Crowley's Ridge extends across the entire Reelfoot Rift, wherein the northern end originates at the Bolivar Mansfield tectonic zone/northwest margin of the Reelfoot Rift intersection and its southern end originates at the White River fault zone/southeastern margin of the Reelfoot Rift intersection. The relatively uniform spacing, parallelism, and the fact that the stepovers terminate at basement fault intersections suggest that the basement fault intersections are controlling the positions and orientations of the compressional stepovers (Talwani, 1999; Hildenbrand *et al.*, 2001; Gangopadhyay and Talwani, 2005; Van Arsdale, 2009).

7.3.3 Reelfoot fault segments

The Reelfoot fault consists of two fault segments with different strikes and dips based on earthquake foci locations. Between the depths of 4 and 14 km earthquake foci illuminate the Reelfoot North fault (167°, 30° SW) and the Reelfoot South fault (150°, 44° SW) (Csontos and Van Arsdale, 2008). In the upper 1 km, both faults have been imaged with seismic reflection data (Purser and Van Arsdale, 1998; Van Arsdale *et al.*, 1998) and dip 73° SW. It is believed that both faults retain the 73° SW dip to a depth of approximately 4 km (Figure 7.8) (Purser and Van Arsdale, 1998; Csontos and Van Arsdale, 2008). These two fault segments also differ at the ground surface. Whereas the Reelfoot North fault has a 10 m high monoclinal scarp, there is no surface scarp along the Reelfoot South fault (Van Arsdale *et al.*, 1999). Although there are subtle geological indicators that the Reelfoot South fault moved in 1812 (Van Arsdale *et al.*, 1999), the Pleistocene Hatchie River terraces of the Obion River (a west-flowing tributary of the Mississippi River) are essentially flat where they overlie the subsurface Reelfoot South fault. It should also be noted that the Reelfoot North fault monocline has its maximum height of 10 m at its mid-point near its intersection with the Mississippi River and the scarp height diminishes to zero north at New Madrid and south at the southeastern margin of Reelfoot Lake (Csontos and Van Arsdale, 2008). The uplift that has been occurring on the Reelfoot North fault over the past 2,600 years appears to be truncated by the New Madrid North fault (Western Reelfoot Rift margin) at its northern end and the Axial fault at its southern end, with no significant surface displacement on the Reelfoot South fault. A possible explanation for the absence of a scarp along the Reelfoot South fault may be that the blind reverse fault underlies a 50 m higher landscape and the fault has simply not propagated high enough to warp the ground surface (Van Arsdale *et al.*, 1999). It is also possible that the two faults are contiguous and are acting semi-independently or they are not continuous.

Absence of a scarp along the Reelfoot South fault and absence of displacement at the southern end of the Reelfoot North fault may indicate that the Reelfoot South fault did not rupture during the New Madrid earthquakes of 1811–1812 (Hough and Page, 2011). Alternatively, the February 1812 earthquake may have been the only rupture that has occurred along the full length of the Reelfoot fault during the Holocene.

7.3.4 Coseismic regional deformations

A number of landforms were formed or enhanced during the New Madrid earthquakes of 1811–1812 (Fuller, 1912; Mihills and Van Arsdale, 1999; Guccione, 2005; Csontos, 2007). Most obvious is the subsidence of Reelfoot Lake and uplift of the adjacent Lake County uplift (Fuller, 1912; Russ, 1982; Stahle *et al.*, 1992). Uplift along the Blytheville arch in northeastern Arkansas and subsidence of the area immediately northwest of the arch occurred in 1811 and during prehistoric faulting events (Figure 7.2). To be more specific, uplift has been documented along the northwestern flank of the subcropping Blytheville arch at the Manila High (Guccione *et al.*, 2000; Odum *et al.*, 2001) and the Marked Tree High (Guccione, 2005). Uplift of the Manila High partially impounded the Little River to form the Big Lake sunkland, and uplift of the Marked Tree High partially impounded the St. Francis River to form the Lake St. Francis sunkland.

7.3.5 Seismicity and Reelfoot Rift faults

Only particular segments of the Reelfoot Rift faults are currently seismically active (Chiu *et al.*, 1992, 1997) and/or have Holocene fault displacement. These include the Axial fault (Guccione, 2005), Reelfoot fault (Mueller and Pujol, 2001), New Madrid North fault (Baldwin *et al.*, 2005), Commerce fault (Harrison *et al.*, 1999), and the central segment of the southeastern Reelfoot Rift margin (Figure 7.2) (Cox *et al.*, 2001b, 2006; Williams *et al.*, 2001). Each of these seismically active segments is truncated by basement faults. The Reelfoot fault seismicity is truncated by the Reelfoot Rift margin faults, the Axial fault seismicity is truncated on its north by the Grand River tectonic zone and on its south by the Boliver Mansfield tectonic zone, the New Madrid North seismicity is truncated on its southern end by the Grand River tectonic zone with its northern end possibly truncated by the Charleston uplift, and there is Holocene displacement along the southeastern Reelfoot Rift margin between the Grand River tectonic zone and the Boliver Mansfield tectonic zone (Figure 7.2 and Table 7.1). Thus, it appears that (1) the Precambrian basement faults divide the basement into structural blocks (Figure 7.2), (2) basement fault intersections control near-surface faulting and geomorphology (e.g., southern Crowley's Ridge, Lake County uplift, and Joiner Ridge), and (3) active fault segment lengths are controlled by basement fault intersections. This suggests earthquake size may be controlled by the length of fault segments that bound the basement blocks of Figure 7.2.

Reelfoot Rift fault segments appear to have turned on and off through geological time. The Commerce fault segment of the Commerce Geophysical Lineament at Thebes Gap

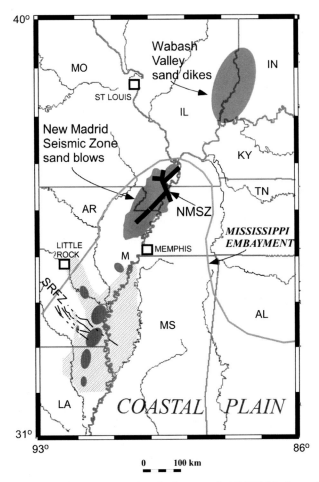

Figure 7.9 Seismically induced Quaternary liquefaction areas in NMSZ, Marianna (M) (Tuttle *et al.*, 2006), and southeastern Arkansas and northeastern Louisiana (from Cox *et al.*, 2010). SRFZ, Saline River fault zone.

(where the Mississippi River crosses Crowley's Ridge at the Missouri–Illinois border) is relatively seismically quiet now, but has experienced extensive Quaternary faulting (Harrison *et al.*, 1999; Harrison and Schultz, 2002). Similarly, the southeast Reelfoot Rift margin north of Dyersburg is seismically quiet, but at Union City, Tennessee, there is a west-facing fault scarp that reveals Pleistocene faulting of this portion of the rift margin (Figure 7.2) (Cox *et al.*, 2001b, 2006). Sikeston Ridge, the southern portion of Crowley's Ridge (Spitz and Schumm, 1997; Van Arsdale *et al.*, 1995), and Joiner Ridge are seismically quiet but their bounding faults may have been active during the Pleistocene. Seventy-five kilometers southwest of Memphis, Tennessee, near the town of Marianna, Arkansas, is an area of low seismicity that has major earthquake liquefaction that occurred between 5,000 and 7,000 years ago (Figure 7.9) (Al-Shukri *et al.*, 2005; Tuttle *et al.*, 2006).

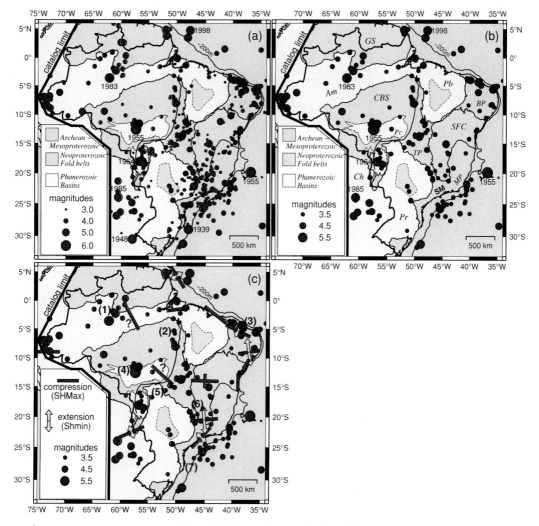

Figure 3.2 For caption, see text, p. 54.

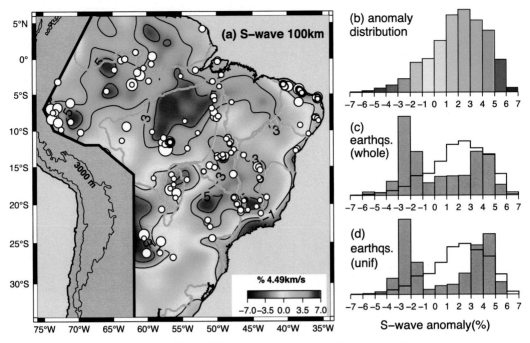

Figure 3.3 For caption, see text, p. 58.

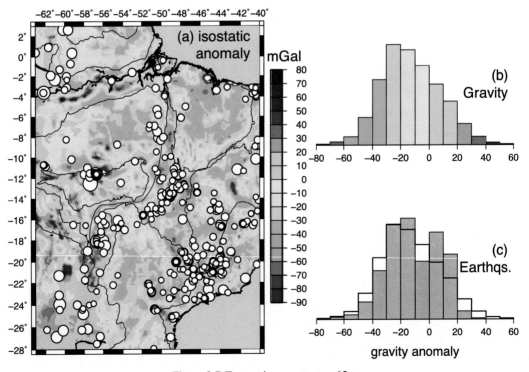

Figure 3.7 For caption, see text, p. 65.

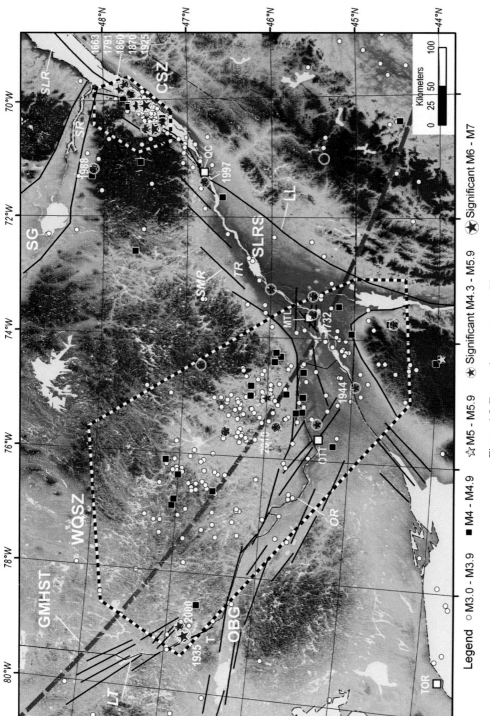

Figure 4.2 For caption, see text, p. 78.

Legend   ○ M3.0 - M3.9   ■ M4 - M4.9   ☆ M5 - M5.9   ★ Significant M4.3 - M5.9   ✪ Significant M6 - M7

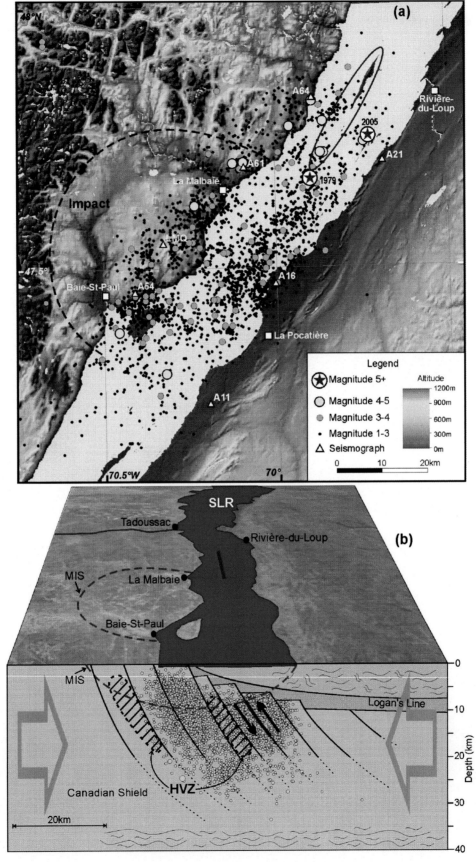

Figure 4.3 For caption, see text, p. 80.

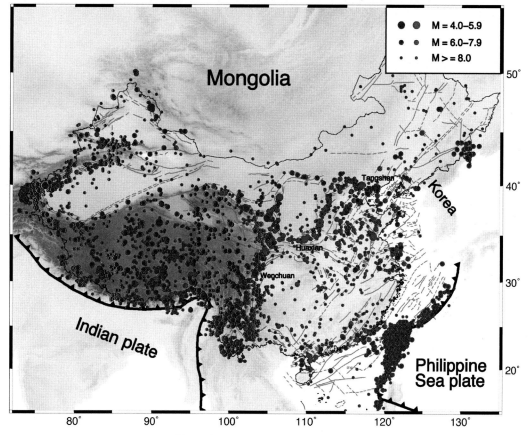

Figure 5.1 For caption, see text, p. 98.

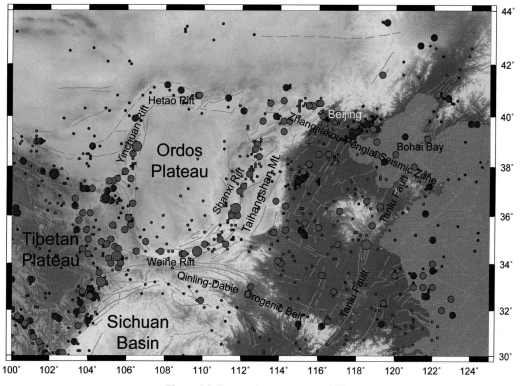

Figure 5.2 For caption, see text, p. 100.

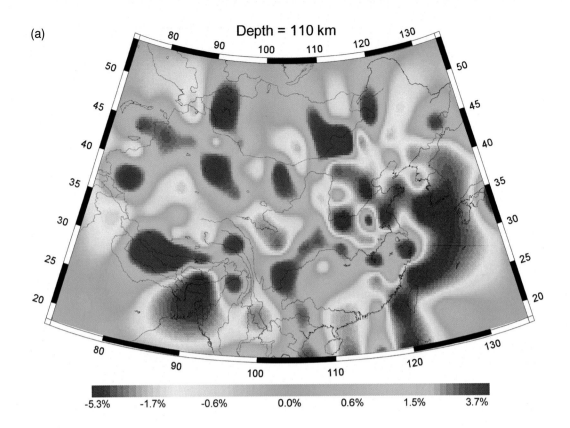

(a)

Depth = 110 km

-5.3%   -1.7%   -0.6%   0.0%   0.6%   1.5%   3.7%

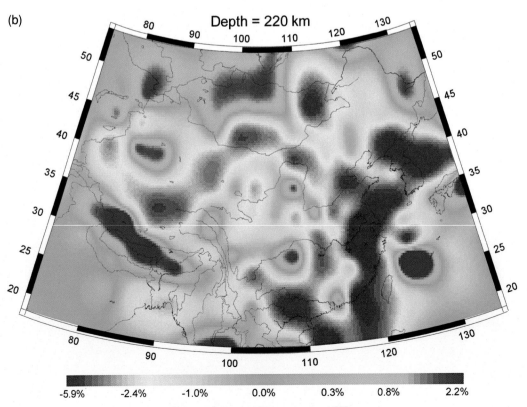

(b)

Depth = 220 km

-5.9%   -2.4%   -1.0%   0.0%   0.3%   0.8%   2.2%

Figure 5.3 For caption, see text, p. 102.

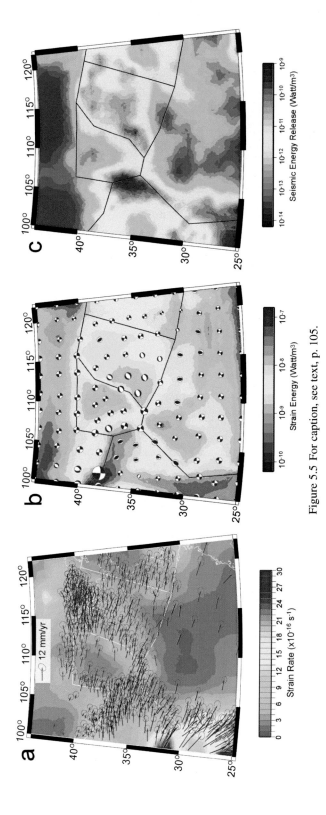

Figure 5.5 For caption, see text, p. 105.

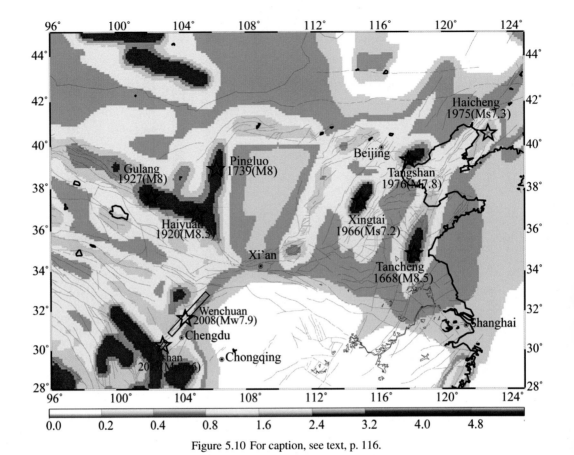

Figure 5.10 For caption, see text, p. 116.

Figure 6.5 For caption, see text, p. 135.

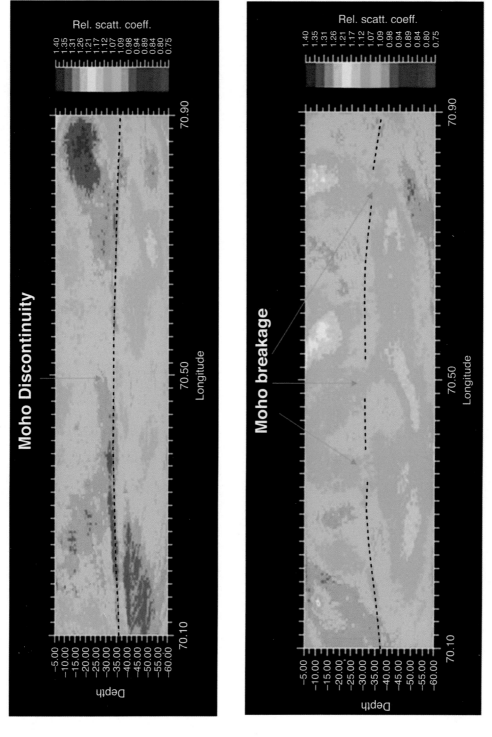

Figure 6.6 For caption, see text, p. 136.

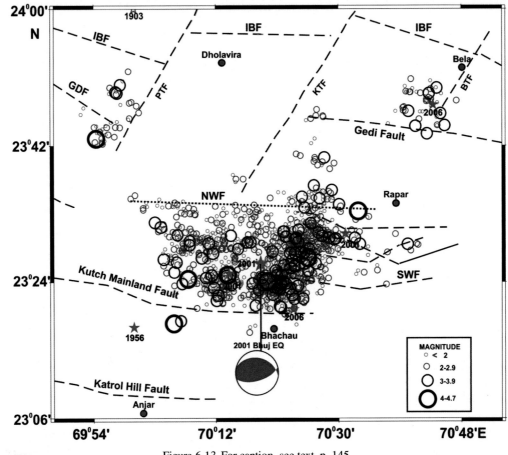

Figure 6.13 For caption, see text, p. 145.

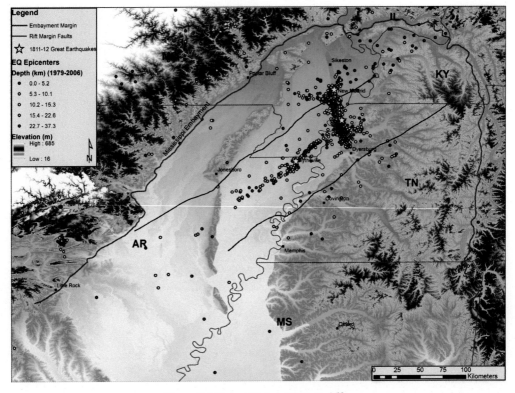

Figure 7.1 For caption, see text, p. 163.

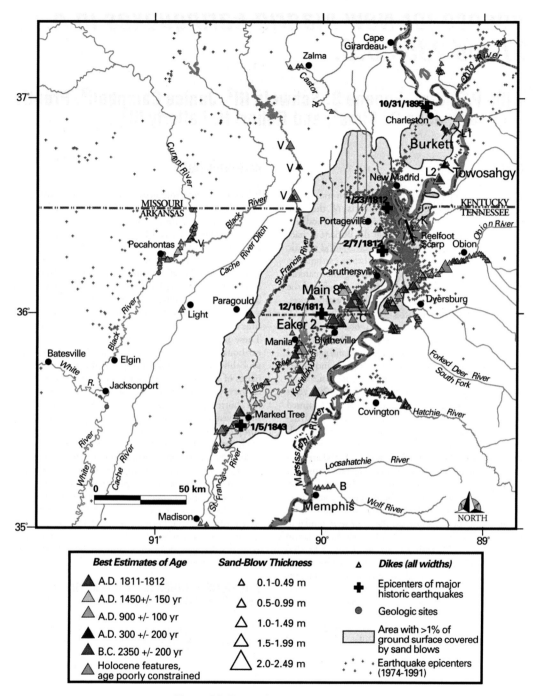

Figure 7.7 For caption, see text, p. 174.

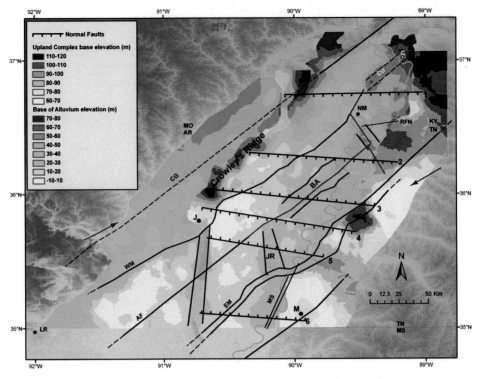

Figure 7.11 (a) For caption, see text, p. 183.

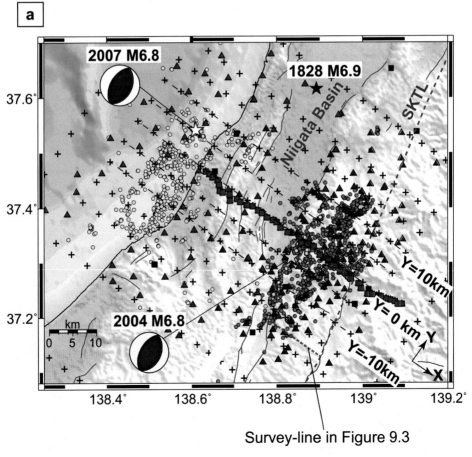

Survey-line in Figure 9.3

Figure 9.2 (a) For caption, see text, p. 236.

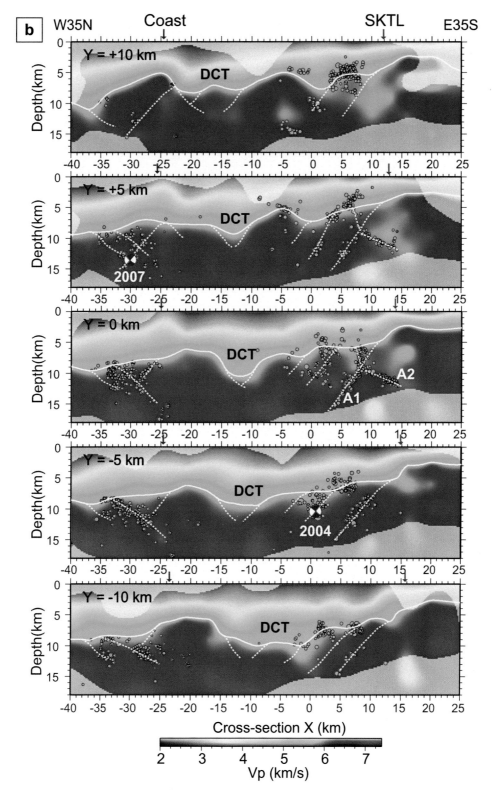

Figure 9.2 (b) For caption, see text, p. 237.

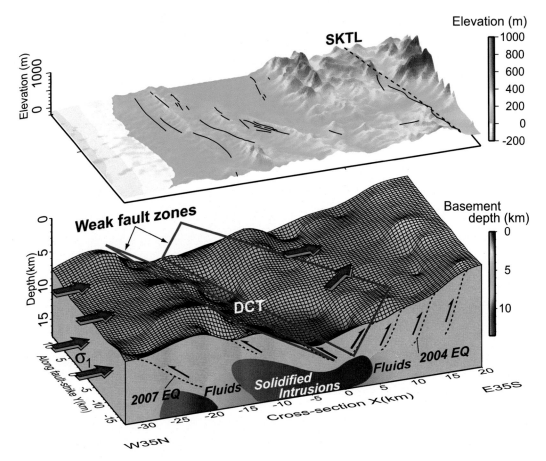

Figure 9.5 For caption, see text, p. 243.

(a)

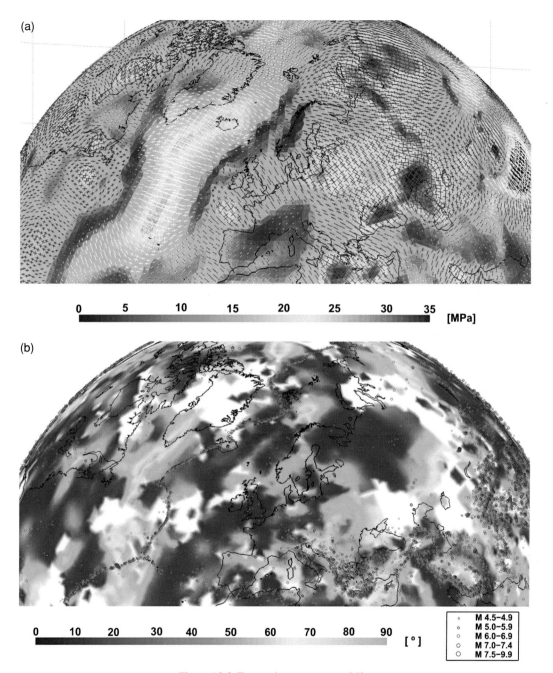

0   5   10   15   20   25   30   35   [MPa]

(b)

0   10   20   30   40   50   60   70   80   90   [°]

○ M 4.5–4.9
○ M 5.0–5.9
○ M 6.0–6.9
○ M 7.0–7.4
○ M 7.5–9.9

Figure 10.2 For caption, see text, p. 262.

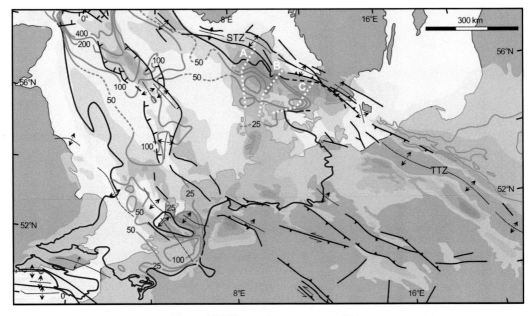

Figure 10.3 For caption, see text, p. 264.

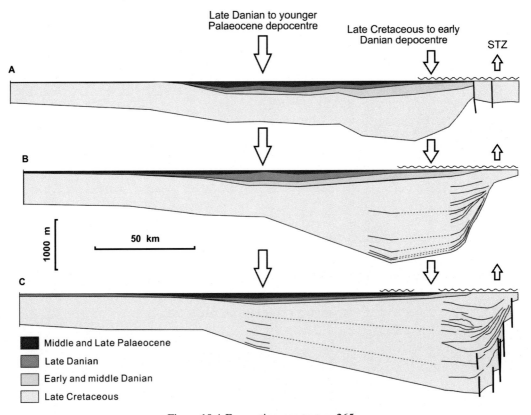

Late Danian to younger
Palaeocene depocentre

Late Cretaceous to early
Danian depocentre

STZ

A

B

1000 m

50 km

C

■ Middle and Late Palaeocene

■ Late Danian

□ Early and middle Danian

□ Late Cretaceous

Figure 10.4 For caption, see text, p. 265.

Movement on one of the southeastern Reelfoot Rift margin faults is believed responsible for the Marianna liquefaction. Further southwest near Monticello, Arkansas, but still within the projection of the Reelfoot Rift, Cox *et al.* (2000) have identified Holocene faulting and earthquake liquefaction (Cox *et al.*, 2004) along the northwest-trending Saline River fault zone (Figure 7.9). Although this area is also underlain by the Ouachita–Appalachian thrust belt, Cox *et al.* (2000) and Cox (2010) believe that the northwest-trending basement fault responsible for this Quaternary faulting is the Alabama–Oklahoma transform fault (Figure 7.4) – the probable southern terminus of the Reelfoot Rift.

7.3.6 New Madrid seismic zone fault activation models

Various models have been proposed to explain the NMSZ. Gomberg and Ellis (1994) numerically model far-field (plate tectonic scale) and locally derived driving strains for the NMSZ faults. Applying elastic dislocation theory Tavakoli *et al.* (2010) model the NMSZ as a 240 km long transpressional flower structure with the principal fault being a right-lateral shear zone rooted in the lower crust that is located along the axis of the Reelfoot Rift (Axial fault). The driving mechanism for the Tavkoli *et al.* model is a drag force at the base of the North American plate. Pratt (2012) combines an analog sandbox model and computer models of a restraining stepover within a N28°E-trending right-lateral shear zone to analyze the NMSZ (Figure 7.10). Deformation displayed in Figure 7.10 is the predicted map view of surface displacements caused by three N47°E upper-crustal faults above a N28°E lower crustal shear zone under a uniform N70°E regional compression. The driving mechanism for the Pratt model is the horizontal N60°E to N80°E maximum compressive stress due to ridge push. Van Arsdale and Cupples (2013) believe that ridge push and/or basal drag is the driving force, but using hundreds of well logs they show that Quaternary right-lateral shear extends across the entire Reelfoot Rift to include the outboard Commerce and Big Creek faults (Figure 7.11). This shear has produced east–west-trending normal faults and north–south-trending compressional stepovers.

Within the Reelfoot Rift region of the Eastern United States are the seismically quiet Rough Creek graben, Rome Trough, Mid Continent Rift, and Southern Oklahoma aulacogen (Figure 7.4). This raises the fundamental questions: why is the Reelfoot Rift seismically active when its neighbors are not, and why does activity within the Reelfoot Rift migrate among different faults during the Quaternary? A number of contributing factors have been cited, some of which are unique to the Reelfoot Rift. Two of these factors are (1) the Reelfoot Rift faults have broken completely through the crust (Nelson and Zhang, 1991; Bartholomew and Van Arsdale, 2012), and (2) these faults are favorably oriented to experience shear in the regional stress field (Zoback and Zoback, 1989; Ellis, 1994; Heidbach *et al.*, 2008). Regionally, a N60°E to N80°E horizontal maximum compressive stress is due to the plate driving force of ridge push (Zoback, 1992; Richardson, 1992) and/or basal drag from lower mantle flow (Liu and Bird, 2002).

A third factor argued by a number of authors is that the rift pillow in the base of the crust beneath the Reelfoot Rift causes a local stress concentration. Grana and Richardson

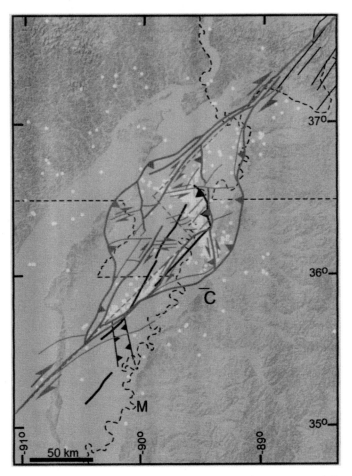

Figure 7.10 Model of the NMSZ combining sandbox analog and computer modeling (from Pratt, 2012). Map view of predicted surface displacements caused by three N47°E upper-crustal faults above a N28°E lower-crustal shear zone under a uniform N70°E regional compression. Seismicity shown as white dots. M, Memphis, Tennessee.

(1996) argue that body forces acting on the rift pillow contribute to the present-day stress field, whereas Stuart *et al.* (1997) suggest that slip along a subhorizontal detachment fault near the domed top of the rift pillow creates a stress concentration in the upper crust above the rift pillow dome. Pollitz *et al.* (2001) propose that sudden weakening caused by Wisconsin deglaciation caused the rift pillow to start sinking, which in turn initiated a downward pull, leading to Reelfoot thrust faulting. Although the rift pillow has been cited as a factor in causing the NMSZ, as discussed above, the rift pillow's age is in question. The apex of the rift pillow underlies the "bull's-eye" of the Cambrian–Upper Cretaceous unconformity beneath the NMSZ. (See the top of the Paleozoic animation at www.geosociety.org/pubs/ft2009.htm [Csontos *et al.*, 2008; Van Arsdale, 2009]). It seems unlikely that this superposition is a coincidence, and this Cretaceous unconformity pattern

(a)

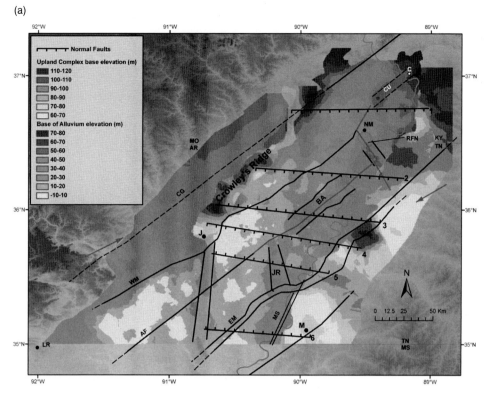

Figure 7.11 (a) Reelfoot Rift and its outboard faults (from Van Arsdale and Cupples, 2013). Right-lateral simple shear is occurring along the N45°E basement faults driven by the regional horizontal maximum compressive stress (red arrows). The resulting strain field has caused north-striking compressional uplifts bound by black fault lines (e.g., Joiner ridge) and the late Pliocene to perhaps Holocene west-striking normal faults with barbs on the down-dropped side. CG, Commerce geophysical lineament/fault; WM, northwestern Reelfoot Rift margin; AF, Axial fault (Cottonwood Grove fault); EM, southeastern Reelfoot Rift margin; BC, Big Creek/Ellendale fault; CU, Charleston uplift; RFN, Reelfoot North fault/Lake County uplift; NMN, New Madrid North fault; BA, Blytheville arch; JR, Joiner ridge; MS, Meeman-Shelby fault zone; LR, Little Rock, Arkansas; C, Cairo, Illinois; NM, New Madrid, Missouri; M, Memphis, Tennessee. For color version, see Plates section.

strongly suggests that the Reelfoot Rift pillow was either intruded or reactivated during mid Cretaceous (Grana and Richardson, 1996; Csontos *et al.*, 2008).

A fourth possible contributing factor for the contemporary faulting may be that the Reelfoot Rift faults have been intermittently active since the mid Cretaceous. When the central Mississippi River valley passed over the Bermuda hotspot, the Reelfoot Rift was heated and uplifted 3 km as part of the Mississippi Embayment arch (Cox and Van Arsdale, 1997, 2002) and the rift faults were probably reactivated to accommodate the uplift. Subsequent Late Cretaceous cooling subsidence of the Mississippi Embayment that occurred after passing off the hotspot continued at least through the Paleocene, which may have kept the Reelfoot Rift faults active during the Paleogene (e.g., Van Arsdale *et al.*, 1995; Luzietti

(b)

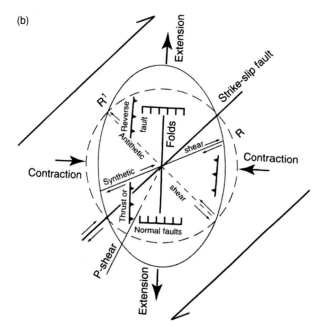

Figure 7.11 (b) The strain ellipse illustrates the nature and orientation of structures that form within an area undergoing right-lateral simple shear (map view). Local east–west compression results in north-striking folds and reverse faults and local north–south extension results in east-striking normal faults as seen in (a) (from Keller and Pinter, 2002).

et al., 1995). Neogene movement has been documented on the Crittenden County fault (Luzetti *et al.*, 1995), the Reelfoot fault (Kelson *et al.*, 1996; Van Arsdale, 2000), Axial fault (Guccione, 2005), Bootheel fault (Guccione *et al.*, 2005), New Madrid North fault (Baldwin *et al.*, 2005; Pryne *et al.*, 2013), bounding faults of Crowley's Ridge (Van Arsdale *et al.*, 1995), the Commerce fault (Harrison *et al.*, 1999), and the southeastern Reelfoot Rift margin faults (Cox *et al.*, 2006) (Figure 7.2 and Table 7.1). Thus, the Reelfoot Rift is unique among its neighboring rifts in that although the total amount of displacement on individual Reelfoot Rift faults has been small (e.g., 70 m on the Reelfoot fault since the Cretaceous) many of the rift faults (perhaps all) have experienced Cenozoic movement.

The fifth reason, which is also unique to the Reelfoot Rift, is that the very large Mississippi River passes directly over the NMSZ and most of the rift. Van Arsdale *et al.* (2007) have discussed the late Pliocene through Holocene erosional history of the central Mississippi River valley that they believe may be tied to the late Holocene onset of the NMSZ, which is discussed below.

7.3.7 Earthquake recurrence

Paleoseismologic investigations have revealed a recurrence interval of approximately 500 years for large NMSZ earthquakes over the past ~1,700 years (Kelson *et al.*, 1996;

Tuttle *et al.*, 2002). Trenches excavated across the Reelfoot scarp (monocline) identify three earthquakes (Russ, 1982; Kelson *et al.*, 1996) and regional paleoliquefaction studies identify five earthquake sequences, including 1811–1812, that have occurred within the last ~4,300 years (AD 1811–1812, 1450 ± 150, 900 ± 100, 300 ± 200, and 2350 BC ± 200 years) (Tuttle *et al.*, 2002, 2005). Straightening of the Mississippi River upstream from the Lake County Uplift at approximately AD 900 and between 2244 and 1620 BC has been interpreted by Holbrook *et al.* (2006) to be due to uplift of the Reelfoot fault across the course of the river, which further supports the paleoliquefaction dates of AD 900 and 2350 BC. Paleoliquefaction data also indicate that the AD 1450, AD 900, and 2350 BC earthquake periods had multiple earthquakes within a short period of time like the 1811–1812 sequence (Tuttle *et al.*, 2002, 2005).

The maximum structural amplitude of the Reelfoot scarp is ~11 m, which has occurred within the last 2,600 years (Champion *et al.,* 2001; Carlson and Guccione, 2010). At depth beneath the Reelfoot scarp, the unconformity at the base of the Holocene alluvium is displaced 16 m, top of the Eocene Wilcox Group 31 m, top of the Paleocene Midway Group 42 m, top of Cretaceous 63 m, and top of Paleozoic 73 m. The Reelfoot fault has moved 16 m in the past 10,000 years with 11 m occurring in the past 2,600 years, implying a slip rate of 1.8 mm/yr throughout the entire Holocene and 6.2 mm/yr over the past 2,600 years (Van Arsdale, 2000). In subsequent studies of the Reelfoot scarp, a late Holocene slip rate of 3.9 ± 1 mm/yr was interpreted by Champion *et al.* (2001), while Carlson and Guccione (2010) estimate a late Holocene slip rate of 13 mm/yr calculated from the AD 1450 and 1812 faulting events. The subsurface fault displacement history (Sexton and Jones, 1986; Van Arsdale, 2000; Champion *et al.*, 2001) further indicates that the Reelfoot fault turns on and off through time. This raises the question: what caused the fault to become active during the Holocene?

7.3.8 Proposed Holocene triggering mechanisms of the NMSZ

Kenner and Segall (2000) model the NMSZ as having an elastic lithosphere with a weak lower-crustal zone and show that a prolonged sequence of large earthquakes can result from a local perturbation of the stress field by changes in thermal state, pore pressure, or most likely recession of the Laurentian ice sheet 14,000 years ago. Grollimund and Zoback (2001) proposed that glacial unloading north of the NMSZ increased seismic strain rates in the NMSZ at the end of the Wisconsin; however, Wu and Johnston (2000) previously argued that glacial unloading is unlikely to have triggered the large earthquakes in New Madrid. A problem with having the Wisconsin ice sheet retreat cause the onset of Holocene seismicity in the NMSZ is that there have been perhaps as many as 20 ice sheet advances and retreats during the Pleistocene (Easterbrook, 1999), yet there has been only 31 m of displacement on the Reelfoot fault since the Eocene. There appears to have been something unique about the late Wisconsin or early Holocene that initiated the Holocene reactivation of the NMSZ.

Schweig and Ellis (1994) speculate that the NMSZ may have initiated between 11 and 3 Ma due to a change in North American plate motion and consequent rotation of the stress field that resulted in higher resolved shear stress on the Reelfoot Rift faults. However, this model raises the question of why other Eastern United States rifts have not also been reactivated. The model also does not account for the Holocene initiation of seismicity within the NMSZ and surrounding Reelfoot Rift. More recently Forte *et al.* (2007) argue that contemporary descent of the Farallon slab in the lower mantle beneath the Central United States is responsible for New Madrid seismicity. In the Forte *et al.* model, downwelling mantle flow, viscously coupled to the ancient Farallon slab, is the driving mechanism. However, their calculated S_{Hmax} in the NMSZ appears to be of the same magnitude and direction as much of the Midcontinent Rift System where there is no seismicity or evidence of Quaternary faulting.

Liu and Zoback (1997) relate New Madrid faulting to local high heat flow and local differences in lithospheric strength. McKenna *et al.* (2007) refute the existence of the proposed heat flow anomaly and conclude that there is no fundamental difference in lithospheric strength between the NMSZ and the surrounding area. Even if subsequent research should support local high heat flow in the NMSZ, there is no apparent reason why heat flow should have increased during the Quaternary.

Excess fluid pressure in the NMSZ due to regional groundwater flow has been proposed to explain its seismicity (McKeown and Diehl, 1994). High pore pressure would reduce the normal stress across the faults and promote faulting. However, the Holocene groundwater flow pattern in the NMSZ appears to have been in existence since the formation of the Mississippi Embayment in the Late Cretaceous and thus this model does not provide an explanation as to why faulting turned on during the Holocene.

Van Arsdale *et al.* (2007) have discussed the late Pliocene through Pleistocene erosional history of the central Mississippi River valley, which they suggest may be tied to the late Holocene onset of NMSZ activity. The ancestral Mississippi/Ohio River system used to flow south across this region at an elevation that was 100 m higher than today's Mississippi River (Figure 7.6). Entrenchment of the ancestral Mississippi/Ohio River system started at ~4 Ma with the most recent 6 m of entrenchment occurring between 12,000 and 10,000 years ago (Calais *et al.*, 2010). Calais *et al.* (2010) argue that this most recent entrenchment reduced the vertical stress, which reduced the horizontal stress that kept the NMSZ faults locked. In this model, the erosion between 12–10 ka reactivated the stressed NMSZ faults.

The Calais *et al.* (2010) model provides both an explanation for the onset of the NMSZ and for the onset of the formation of the Lake County Uplift. Denudation of the Mississippi River valley at the end of the Pleistocene caused right-lateral slip on the Axial and New Madrid North faults thereby causing uplift of the Lake County Uplift compressional stepover. As discussed above, Sikeston Ridge, Joiner Ridge, and the southern portion of Crowley's Ridge may also be compressional stepovers. If indeed these are all compressional stepovers then they could all be a consequence of right-lateral shear across the Reelfoot Rift that occurred as the valley was denuded over the past ~4 Ma. The denudation-driven model provides a mechanism to distribute minor displacement among Reelfoot Rift faults at different times within the Quaternary since denudation would have occurred at different times and different locations in the Mississippi valley. This model is supported by the fact that Crowley's Ridge is older than 18,000 years; Sikeston Ridge is younger than 18,000 years; the Lake County Uplift has come up within the last 2,600 years; and both Joiner Ridge and the Charleston Uplift are apparent at the base of the Mississippi River alluvium (Csontos *et al.*, 2008; Odum *et al.*, 2010; Pryne *et al.*, 2013) but have not yet emerged as topographic ridges.

7.3.9 Is elastic strain energy accumulating in the Reelfoot Rift faults

Paleoseismologic trench studies indicate that the Reelfoot North fault has ruptured three times (Russ, 1979; Kelson *et al.*, 1996; Champion *et al.*, 2001) and paleoliquefaction identifies four large earthquake sequences over the past 1,700 years (Tuttle *et al.*, 2002) implying an earthquake recurrence interval of ~500 years. A ~500-year recurrence interval for $M > 7.5$ earthquakes is inconsistent with the rate of M ~6 earthquakes since 1812. Two earthquakes estimated to be in the low M 6 range have occurred in the NMSZ, one in 1843 near Marked Tree, Arkansas, and the second in 1895 near Charleston, Missouri (Hopper and Algermissen, 1980; Guccione, 2005). If even one of the 1811–1812 earthquakes was M 7.5 then, based on the Gutenberg–Richter frequency–magnitude relationship, there should have been four M ~6.5 earthquakes in the NMSZ since 1812 (Atkinson *et al.*, 2000). A more significant problem with a continuing short recurrence interval is that GPS measurements reveal very low strain rates (0.2 ± 2.4 mm/yr) in the NMSZ (Newman *et al.*, 1999; Calais *et al.*, 2005; Stein, 2010). At these strain rates it would take ~10,000 years to accumulate sufficient strain energy to generate a M 7 earthquake (Stein, 2010). These arguments suggest that the 1811–1812 earthquakes were in the low M 7 range (Hough *et al.*, 2000; Mueller and Pujol, 2001; Hough and Page, 2011). An alternative explanation is that the NMSZ turns on for a brief period of geological time and turns off for very long periods of geological time and that we are at or near the end of a cluster of faulting/seismic activity (Coppersmith, 1988; Van Arsdale, 2000).

The denudation model by Van Arsdale *et al.* (2007) and Calais *et al.* (2010) appears to provide an explanation that fits both the geological and historical records. In the denudation model, normal stress on the Reelfoot Rift faults is reduced due to erosion of sediment from above the NMSZ faults, thereby increasing the ratio of shear stress to normal stress and allowing favorably oriented faults to slip. The reduction in normal stress is controlled by the erosional history of the Mississippi River. Once a certain threshold of erosion is passed then the critical shear stress/normal stress ratio is exceeded and faulting occurs. If this is indeed the driving mechanism, it would occur as a burst of faulting (seismic) activity separated by perhaps millions of years of quiescence until another erosional period occurs. Additionally, we expect that in the case of a basement rift and large river system there would be minor fault displacement over a large area within short geological time. It is possible that this mechanism has been active in the deep geological past of the Reelfoot Rift. I now consider the history of the Reelfoot North fault.

There has been 16 m of Quaternary displacement, 15 m of post-Wilcox/pre-Quaternary displacement, 11 m of Wilcox displacement, 21 m of Midway displacement, and 10 m of Late Cretaceous displacement on the Reelfoot North fault (Van Arsdale, 2000). Based on these displacements and estimates of their respective time frames one can calculate a Late Cretaceous slip rate of 0.0007 mm/yr, Paleocene slip rate of 0.002 mm/yr, Wilcox Group slip rate of 0.001 mm/yr, post-Wilcox/pre-Holocene slip rate of 0.0003 mm/yr, and a Holocene slip rate of 1.8 mm/yr. If we spread the 16 m of Holocene displacement over the entire time frame of the Paleocene, then we would have a slip rate of 0.001 mm/yr. Secondly, the magnitudes of the slip for these time periods separated by unconformities are similar. This similarity implies that perhaps any single cluster of activity may not exceed ~20 m of fault displacement on the Reelfoot North fault. From either perspective it appears that since the plate tectonic setting of the NMSZ has not changed substantially since essentially the Cretaceous, we should expect uniform behavior over time. However, this does not get the NMSZ off the hook. If the Reelfoot North fault can slip as much as 20 m during the current cycle then there may be 9 more meters of slip before this fault shuts down perhaps for millions of years.

7.4 Conclusions

Earthquakes on the NMSZ occur along faults within the Cambrian Reelfoot Rift that appear to have cut across northwest-striking Proterozoic faults (Figures 7.1 and 7.2). At a more regional scale the Commerce Geophysical lineament and the Big Creek fault zone (Figure 7.11) are interpreted to be outboard faults of the rift. Right-lateral shear across the Reelfoot Rift is causing right-lateral displacement along the Southeastern Reelfoot Rift margin faults, Axial fault, and New Madrid North fault, and reverse displacement on the Reelfoot North and Reelfoot South faults. Thus, there are three strike-slip faults and two left stepovers that are currently seismically active. There is also evidence of Quaternary faulting along portions of the Southeastern Reelfoot Rift margin faults, Big Creek fault, the Commerce Geophysical lineament, the Bootheel fault, the New Markham fault, Crowley's Ridge, the Saline River fault zone, Joiner Ridge, and the Meeman–Shelby fault (Figure 7.2 and Table 7.1). In effect, the Reelfoot Rift is a Cambrian structure that has been active over much of its area within the Quaternary. Proposed explanations for the driving force for the faulting and resultant earthquakes include a sinking rift pillow, ridge push, a subducting Farallon slab, local weakening of the crust, excess pore-fluid pressure, glacial isostatic adjustment, and erosional unloading. The absence of significant Neogene deformation in the region requires Quaternary onset of the most recent episode of deformation and the paleoseismic record indicates that faulting began in the NMSZ during the Holocene. Holocene reactivation of the NMSZ faults has been attributed to local stress perturbation due to melting of the Wisconsin ice sheet, a change in North American plate motion, or a reduction in normal stress across strained faults due to denudation by the Mississippi River. The denudation model explains the Holocene onset of the NMSZ and, I believe, also provides the most satisfactory explanation as to the cause, distribution, displacement magnitude, and timing of Quaternary faulting of the Reelfoot Rift faults throughout the lower Mississippi River valley. Quaternary incision of the Mississippi River system locally diminishes the stress on the faults produced by ridge push and releases the stored elastic strain energy at different times and different locations, thereby turning Reelfoot Rift faults on and off with relatively minor total Quaternary displacement occurring on individual faults.

References

Al-Shukri, H. J., Lemmer, R. E., Mahdi, H. H., and Connelly, J. B. (2005). Spatial and temporal characteristics of paleoseismic features in the southern terminus of the New Madrid seismic zone in eastern Arkansas. *Seismological Research Letters*, 76, 502–511.

Anderson, K. H. (1979). *Geological Map of Missouri: 1:500,000*. Missouri Geological Survey.

Atekwana, E. A. (1996). Precambrian basement beneath the central Midcontinent United States as interpreted from potential field imagery. In *Basement and Basins of Eastern North America*, ed. B. A. van der Pluijm and P. A. Catacosinos. Geological Society of America Special Paper 308, pp. 33–44.

Atkinson, G., and 24 others (2000). Reassessing the New Madrid seismic zone. *Eos, Transactions, American Geophysical Union*, 81, 402–403.

Autin, W. J., Burns, S. F., Miller, B. J., Saucier, R. T., and Snead, J. I. (1991). Quaternary geology of the lower Mississippi Valley. In *Quaternary Nonglacial Geology: Conterminous, U.S., K-2*, ed. R. B. Morrison. Geological Society of America, pp. 547–582.

Baldwin, J. N., Harris, J. B., Van Arsdale, R. B., *et al.* (2005). Constraints on the location of the Late Quaternary Reelfoot and New Madrid North faults in the northern New Madrid seismic zone, central United States. *Seismological Research Letters*, 76, 772–789.

Bartholomew, M. J., and Hatcher, R. D. (2010). The Grenville orogenic cycle of southern Laurentia: unraveling sutures, rifts, and shear zones as potential piercing points for Amazonia. *Journal of South American Earth Sciences*, 29, 4–20.

Bartholomew, M. J., and Van Arsdale, R. B. (2012). Structural controls on intraplate earthquakes in the eastern United States. In *Recent Advances in North American Paleoseismology and Neotectonics East of the Rockies*, ed. R. T. Cox, M. P. Tuttle, O. S. Boyd, and J. Locat. Geological Society of America Special Paper 493, pp. 165–190, doi:10.1130/2012.2493(08).

Calais, E., Mattioli, G. S., DeMets, C., *et al.* (2005). Tectonic strain in plate interiors? Comment on "Space geodetic evidence for rapid strain rates in the New Madrid seismic zone of the central USA by Smalley, *et al.*, 2005, Nature." *Nature*, 438, E9–10, doi: 10.1038/nature04428.

Calais, E., Freed, A .M., Van Arsdale, R., and Stein, S. (2010). Triggering of New Madrid seismicity by late-Pleistocene erosion. *Nature*, 466, 608–611.

Campbell, C. E., Oboh-Ikuenobe, F. E., and Eifert, T. L. (2008). Megatsunami deposit in Cretaceous-Paleogene boundary interval of southeastern Missouri. In *The Sedimentary Record of Meteorite Impacts*, ed. K. R. Evans, J. W. Horton Jr., D. T. King Jr., and J. R. Morrow. Geological Society of America Special Paper 437, pp. 189–198.

Carlson, S. D., and Guccione, M. J. (2010). Short-term uplift rates and surface deformation along the Reelfoot fault, New Madrid seismic zone. *Bulletin of the Seismological Society of America*, 100, 1659–1677.

Champion, J., Mueller, K., Tate, A., and Guccione, M. (2001). Geometry, numerical models and revised slip rate for the Reelfoot fault and trishear fault-propagation fold, New Madrid seismic zone. *Engineering Geology*, 6, 31–49.

Chiu, J. M., Johnston, A. C., and Yang, Y. T. (1992). Imaging the active faults of the central New Madrid seismic zone using PANDA array data. *Seismological Research Letters*, 63, 375–393.

Chiu, S. C., Chiu, J. M., and Johnston, A. C. (1997). Seismicity of the southeastern margin of Reelfoot rift, central United States. *Seismological Research Letters*, 68, 785–794.

Clark, P. U., Nelson, A. R., McCoy, W. D., Miller, B. B., and Barnes, D. K. (1989). Quaternary aminostratigraphy of Mississippi Valley loess. *Geological Society of America Bulletin*, 101, 918–926.

Clendenin, C. W., and Diehl, S. F. (1999). Structural styles of Paleozoic intracratonic fault reactivation: a case study of the Grays Point fault zone in southeastern Missouri, USA. *Tectonophysics*, 305, 235–248.

Coppersmith, K. J. (1988). Temporal and spatial clustering of earthquake activity in the central and eastern United States. *Seismological Research Letters*, 59, 299–304.

Cox, R. T. (1988). Evidence for Late Cenozoic activity along the Boliver-Mansfield tectonic zone, midcontinent, USA. *The Compass*, 65, 207–213.

Cox, R. T. (2010). *Holocene Faulting and Liquefaction Along the Southern Margin of the North American Craton (Alabama-Oklahoma Transform)*. Final Technical Report, U.S. Geological Survey.

Cox, R. T., and Van Arsdale, R. B. (1997). Hotspot origin of the Mississippi embayment and its possible impact on contemporary seismicity. *Engineering Geology*, 46, 5–12.

Cox, R. T., and Van Arsdale, R. B. (2002). The Mississippi Embayment, North America: a first order continental structure generated by the Cretaceous superplume mantle event. *Journal of Geodynamics*, 34, 163–176.

Cox, R. T., Van Arsdale, R. B., Harris, J. B., *et al.* (2000). Late Quaternary faulting in the southern Mississippi embayment and implications for regional neotectonics. *Geological Society of America Bulletin*, 112, 1724–1735.

Cox, R. T., Van Arsdale, R. B., and Harris, J. B. (2001a). Identification of possible Quaternary deformation in the northeastern Mississippi Embayment using quantitative geomorphic analysis of drainage-basin asymmetry. *Geological Society of America Bulletin*, 113, 615–624.

Cox, R. T., Van Arsdale, R. B., Harris, J. B., and Larsen, D. (2001b). Neotectonics of the southeastern Reelfoot rift zone margin, central United States, and implications for regional strain accommodation. *Geology*, 29, 419–422.

Cox, R. T., Larsen, D., Forman, S. L., *et al.* (2004). Preliminary assessment of sand blows in the southern Mississippi embayment. *Bulletin of the Seismological Society of America*, 94, 1,125–1,142.

Cox, R. T., Cherryhomes, J., Harris, J. B., *et al.* (2006). Paleoseismology of the southeastern Reelfoot Rift in western Tennessee and implications for intraplate fault zone evolution. *Tectonics*, 23, 1–17.

Cox, R. T., Gordon, J., Forman, S., *et al.* (2010). Paleoseismic sand blows in north Louisiana and south Arkansas. *Seismological Research Letters*, 81, 1032–1047.

Cox, R. T., Van Arsdale, R., Clark, D., Hill, A., and Lumsden, D. (2013). A revised paleo-earthquake chronology on the southeast Reelfoot rift margin near Memphis, Tennessee. *Seismological Research Letters*, 84, 402–408.

Csontos, R. M. (2007). Three dimensional modeling of the Reelfoot rift and New Madrid seismic zone. Unpublished Ph.D. thesis, The University of Memphis, Memphis, Tennessee.

Csontos, R., and Van Arsdale, R. (2008). New Madrid seismic zone fault geometry. *Geosphere*, 4, 802–813.

Csontos, R., Van Arsdale, R., Cox, R., and Waldron, B. (2008). The Reelfoot Rift and its impact on Quaternary deformation in the central Mississippi River Valley. *Geosphere*, 4, 145–158.

Culotta, R. C., Pratt, T., and Oliver, J. (1990). A tale of two sutures: COCORP's deep seismic surveys of the Grenville province in the eastern U.S. mid-continent. *Geology*, 18, 646–649.

Dart, R., and Swolfs, H. S. (1998). Contour mapping of relic structures in the Precambrian basement of the Reelfoot Rift, North American Midcontinent. *Tectonics*, 17, 235–249.

Dickinson, R., and Gehrels, G. E. (2009). U-Pb ages of detrital zircons in Jurassic eolian and associated sandstones of the Colorado Plateau: Evidence for transcontinental dispersal and intraregional recycling of sediment. *Geological Society of America Bulletin*, 121, 408–433.

Easterbrook, D. (1999). *Surface Processes and Landforms*. Upper Saddle River: Prentice Hall.

Ellis, W. L. (1994). *Summary and Discussion of Crustal Stress Data in the Region of the New Madrid Seismic Zone*. U.S. Geological Survey Professional Paper 1538-B, B1–B13.

Erwin, C. P., and McGinnis, L. D. (1975). Reelfoot Rift: reactivated precursor to the Mississippi Embayment. *Geological Society of America Bulletin*, 86, 1287–1295.

Forte, A. M., Mitrovica, J. X., Moucha, R., Simmons, N. A., and Grand, S. P. (2007). Descent of the ancient Farallon slab drives localized mantle flow below the New Madrid seismic zone. *Geophysical Research Letters*, 34, L04308.

Fuller, M. L. (1912). The New Madrid earthquake. *Bulletin U.S. Geological Survey*, 494.

Gangopadhyay, A., and Talwani, P. (2005). Fault intersections and intraplate seismicity in Charleston, South Carolina: insights from a 2-D numerical model. *Current Science*, 88, 1609–1616.

Ginzburg, A., Mooney, W. D., Walter, A. W., Lutter, W. J., and Healy, J. H. (1983). Deep structure of northern Mississippi embayment. *American Association of Petroleum Geologists Bulletin*, 67, 2031–2046.

Gomberg, J. (1993). Tectonic deformation in the New Madrid seismic zone: inferences from map view and cross-sectional boundary element models. *Journal of Geophysical Research*, 98, 6639–6664.

Gomberg, J., and Ellis, M. (1994). Topography and tectonics of the central New Madrid seismic zone: results of numerical experiments using a three-dimensional boundary element program. *Journal of Geophysical Research*, 89, 20,299–20,310.

Grana, J. P., and Richardson, R. M. (1996). Tectonic stress within the New Madrid seismic zone. *Journal of Geophysical Research*, 101, 5445–5458.

Grollimund, B., and Zoback, M. D. (2001). Did glaciation trigger intraplate seismicity in the New Madrid seismic zone? *Geology*, 29, 175–178.

Guccione, M. J. (2005). Late Pleistocene and Holocene paleoseismology of an intraplate seismic zone in a large alluvial valley, the New Madrid seismic zone, central USA. *Tectonophysics*, 408, 237–264.

Guccione, M. J., Van Arsdale, R. B., and Hehr, L. H. (2000). Origin and age of the Manila high and associated Big Lake "Sunklands", New Madrid seismic zone, northeastern Arkansas. *Geological Society of America Bulletin*, 112, 579–590.

Guccione, M. J., Marple, R., and Autin, W. J. (2005). Evidence for Holocene displacement on the Bootheel Fault (lineament) in southeastern Missouri: seismotectonic implications for the New Madrid region. *Geological Society of America Bulletin*, 117, 319–333.

Haley, B. R., Glick, E. E., and Bush, W. V. (1993). *Geological Map of Arkansas: 1: 500,000.* U.S. Geological Survey.

Hamilton, R. M., and Johnston, A. C. (1990). Tecumseh's prophecy: preparing for the next New Madrid earthquake. *U.S. Geological Survey Circular*, 1066.

Hao, Y., Magnani, M. B., McIntosh, K., Waldron, B., and Guo, L. (2013). Quaternary deformation along the Meeman–Shelby fault, near Memphis, Tennessee, imaged by high-resolution marine and land seismic reflection profiles. *Tectonics*, 32, 1–15, doi:10.1002/tect.20042.

Harris, J. B. (2009). Hammer-impact SH-wave seismic reflection methods in neotectonic investigations: general observations and case histories from the Mississippi Embayment, U.S.A. *Journal of Earth Science*, 20, 513–525.

Harris, J. B., and Sorrells, J. L. (2006). Shear-wave seismic reflection images of the Big Creek fault zone near Helena, Arkansas. *Society of Exploration Geophysics New Orleans Annual Meeting*, 1500–1503.

Harrison, R. W., and Schultz, A. (2002). Tectonic framework of the southwestern margin of the Illinois Basin and its influence on neotectonism and seismicity. *Seismological Research Letters*, 73, 698–731.

Harrison, R. W., Hoffman, D., Vaughn, J. D., *et al.* (1999). An example of neotectonism in a continental interior: Thebes Gap, Midcontinent, United States. *Tectonophysics*, 305, 399–417.

Heidbach, O., Tingay, M., Barth, A., *et al.* (2008). The release 2008 of the World Stress Map (available online at www.world-stress-map.org).

Hildenbrand, T. G. (1985). Rift structure of the northern Mississippi from the analysis of gravity and magnetic data. *Journal of Geophysical Research*, 90, 12607–12622.

Hildenbrand, T. G., and Hendricks, J. D. (1995). Geophysical setting of the Reelfoot Rift and relation between rift structures and the New Madrid seismic zone. In *Investigations of the New Madrid Seismic Zone*, ed. K. M. Shedlock and A.C. Johnson. U.S. Geological Survey Professional Paper, 1538-E.

Hildenbrand, T. G., Stuart, W. D., and Talwani, P. (2001). Geologic structures related to New Madrid earthquakes near Memphis, Tennessee, based on gravity and magnetic interpretations. *Engineering Geology*, 62, 105–121.

Holbrook, J., Autin, W. J., Rittenour, T. M., Marshak, S., and Goble, R. J. (2006). Stratigraphic evidence for millennial-scale temporal clustering of earthquakes on a continental-interior fault: Holocene Mississippi river floodplain deposits, New Madrid seismic zone, USA. *Tectonophysics*, 420, 431–454.

Holm, D. K., Anderson, R., Boerboom, T. J., *et al.* (2007). Reinterpretation of Paleoproterozoic accretionary boundaries of the north-central United States based on a new aeromagnetic-geologic compilation. *Precambrian Research*, 157, 71–79.

Hopper, M., and Algermissen, S. (1980). *An Evaluation of the Effects of the October 31, 1895, Charleston, Missouri, Earthquake*. U.S. Geological Survey, Open-File Report 80–778.

Hough, S. E., and Page, M. (2011). Toward a consistent model for strain accrual and release for the New Madrid seismic zone, central United States. *Journal of Geophysical Research*, 116, BO3311, 17 pp.

Hough, S. E., Armbruster, J. G., Seeber, L., and Hough, J. F. (2000). On the modified Mercalli intensities and magnitudes of the 1811–1812 New Madrid, central U.S. earthquakes. *Journal of Geophysical Research*, 105, 23,839–23,864.

Howe, J. R. (1985). Tectonics, sedimentation, and hydrocarbon potential of the Reelfoot aulocogen. Unpublished MS thesis, University of Oklahoma, Norman.

Jibson, R. W., and Keefer, D. K. (1989). Statistical analysis of factors affecting landslide distribution in the New Madrid seismic zone, Tennessee and Kentucky. *Engineering Geology*, 27, 509–542.

Johnson, P. R., Zietz, I., and Thomas, W. A. (1994). Possible Neoproterozoic – early Paleozoic grabens in Mississippi, Alabama, and Tennessee. *Geology*, 22, 11–14.

Johnston, A. C. (1996). Seismic moment assessment of earthquakes in stable continental regions-III. New Madrid 1811–1812, Charleston 1886 and Lisbon 1755. *Geophysical Journal International*, 126, 314–344.

Johnston, A. C., and Schweig, E. S. (1996). The enigma of the New Madrid earthquakes of 1811–1812. *Annual Review of Earth and Planetary Sciences*, 24, 339–384.

Keller, E. A., and Pinter, N. (2002). *Active Tectonics: Earthquakes, Uplift, and Landscape*. Upper Saddle River, New Jersey: Prentice-Hall.

Kelson, K. I., Simpson, G. D., Van Arsdale, R. B., *et al.* (1996). Multiple Late Holocene earthquakes along the Reelfoot fault, central New Madrid seismic zone. *Journal of Geophysical Research*, 101, 6151–6170.

Kenner, S. J., and Segall, P. (2000). A mechanical model for intraplate earthquakes: application to the New Madrid seismic zone. *Science*, 289, 2329–2332.

Langenheim, V. D., and Hildenbrand, T. G. (1997). Commerce geophysical lineament: its source, geometry, and relation to the Reelfoot rift and New Madrid seismic zone. *Geological Society of America Bulletin*, 109, 580–595.

Liu, L., and Zoback, M. D. (1997). Lithospheric strength and intraplate seismicity in the New Madrid seismic zone. *Tectonics*, 16, 585–595.

Liu, P., and Bird, P. (2002). North America plate is driven westward by lower mantle flow. *Geophysical Research Letters*, 29, 17-1–17-4.

Luzietti, E. A., Kanter, L. R., Schweig, E. S., Shedlock K. M., and Van Arsdale, R. B. (1995). *Shallow Deformation Along the Crittenden County Fault Zone Near the Southeastern Boundary of the Reelfoot Rift, Northeast Arkansas*. U.S. Geological Survey Professional Paper 1538-J.

Markewich, H. H., Wysocki, D. A., Pavich, M. J., *et al.* (1998). Paleopedology plus TL, 10Be, and 14C dating as tools in stratigraphic and paleoclimatic investigations, Mississippi River valley, U.S.A. *Quaternary International*, 51/52, 143–167.

Martin, R. V. (2008). Shallow faulting of the southeast Reelfoot rift margin. Unpublished Ph.D. thesis, Department of Earth Sciences, University of Memphis, Memphis, Tennessee.

McCracken, M. H. (1971). *Structural Features of Missouri*. Missouri Geological Survey and Water Resources, Report Investigation 49.

McKenna, J., Stein, S., and Stein, C. A. (2007). Is the New Madrid seismic zone hotter and weaker than its surroundings? In *Continental Intraplate Earthquakes: Science,*

Hazard, and Policy Issues, ed. S. Stein, and S. Mazzotti. Geological Society of America Special Paper 425, 167–175.

McKeown, F. A., and Diehl, S. F. (1994). *Evidence of Contemporary and Ancient Excess Fluid Pressure in the New Madrid Seismic Zone of the Reelfoot Rift, Central United States*. U.S. Geological Survey Professional Paper 1538-N.

Mihills, R. K., and Van Arsdale, R. B. (1999). Late Wisconsin to Holocene New Madrid seismic zone deformation. *Bulletin of the Seismological Society of America*, 89, 1019–1024.

Mooney, W. D., Andrews, M. C., Ginzburg, A., Peters, D. A., Hamilton, R. M. (1983). Crustal structure of the Northern Mississippi Embayment and comparison with other continental rift zones. *Tectonophysics*, 94, 327–338.

Mueller, K., and Pujol, J. (2001). Three-dimensional geometry of the Reelfoot blind thrust: implications for moment release and earthquake magnitude in the New Madrid seismic zone. *Bulletin of the Seismological Society of America*, 91, 1563–1573.

Nelson, K. D., and Zhang, J. (1991). A COCORP deep reflection profile across the buried Reelfoot rift, south-central United States. *Tectonophysics*, 197, 271–293.

Nelson, W. J., Denny, F. B., Devera, J. A., Follmer, L. R., and Masters, J. M. (1997). Tertiary and Quaternary tectonic faulting in southernmost Illinois. *Engineering Geology*, 46, 235–258.

Newman, A., Stein, S., Weber, J., *et al.* (1999). Slow deformation and lower seismic hazard at the New Madrid seismic zone. *Science*, 284, 619–621.

Nuttli, O. W. (1973). The Mississippi Valley earthquakes of 1811–1812, intensities, ground motion and magnitudes. *Bulletin of the Seismological Society of America*, 63, 227–248.

Obermeier, S. F. (1989). *The New Madrid Earthquakes: An Engineering-Geologic Interpretation of Relict Liquefaction Features*. U.S. Geological Survey Professional Paper 1336-B.

Odum, J. K., Stephenson, W. J., Shedlock, K. M., and Pratt, T. L. (1998). Near-surface structural model for deformation associated with the February 7, 1812, New Madrid, Missouri, earthquake. *Geological Society of America Bulletin*, 110, 149–162.

Odum, J., Stephenson, W. J., Williams, R. A., *et al.* (2001). High resolution seismic-reflection imaging of shallow deformation beneath the northeast margin of the Manila high at Big Lake, Arkansas, New Madrid seismic zone, central USA. *Engineering Geology*, 62, 91–103.

Odum, J. K., Stephenson, W. J., and Williams, R. A. (2010). Multisource, high-resolution seismic-reflection imaging of Meeman–Shelby fault and a possible tectonic model for a Joiner Ridge–Manila High stepover structure in the upper Mississippi embayment region. *Seismological Research Letters*, 81, 647–660.

Parrish, S., and Van Arsdale, R. (2004). Faulting along the southeastern margin of the Reelfoot Rift in northwestern Tennessee revealed in deep seismic-reflection profiles. *Seismological Research Letters*, 75, 784–793.

Penick, J. L. (1981). *The New Madrid Earthquakes*. Columbia University: University of Missouri Press.

Pollitz, F. F., Kellogg, L., and Burgmann, R. (2001). Sinking mafic body in a reactivated lower crust: a mechanism for stress concentration at the new Madrid seismic zone. *Bulletin of the Seismological Society of America*, 91, 1882–1887.

Pratt, T. L. (2012). Kinematics of the New Madrid seismic zone, central U.S., based on stepover models. *Geology*, 40, 371–374.

Pratt, T. L., Williams, R. A., Odum, J. K., and Stephenson, W. J. (2012). Origin of the Blytheville arch, and long-term displacement on the New Madrid seismic zone, central U.S. In *Recent Advances in North American Paleoseismology and Neotectonics East of the Rockies*, ed. R. T. Cox, M. P. Tuttle, O.S. Boyd, and J. Locat. Geological Society of America Special Paper 493, pp. 1–16, doi:10.1130/2012.2493(01).

Pryne, D., Van Arsdale, R., Csontos, R., and Woolery, E. (2013). Northeastern extension of the New Madrid North fault – New Madrid seismic zone, central United States. *Bulletin of the Seismological Society of America*, 103, 2277–2294.

Purser, J. L., and Van Arsdale, R. B. (1998). Structure of the Lake County Uplift: New Madrid seismic zone. *Bulletin of the Seismological Society of America*, 88, 1204–1211.

Rankin, D. W., and 12 others (1993). Proterozoic rocks east and southeast of the Grenville front. In *Precambrian: Conterminous U.S., Geology of North America, v. C-2*, ed. J. C. Reed Jr., *et al.* Boulder, Colorado: Geological Society of America, pp. 335–462.

Richardson, R. M. (1992). Ridge forces, absolute plate motions, and the intraplate stress field. *Journal of Geophysical Research*, 97, 11,739–11,748.

Russ, D. P. (1979). Late Holocene faulting and earthquake recurrence in the Reelfoot Lake area, northwestern Tennessee. *Geological Society of America Bulletin*, 90, 1013–1018.

Russ, D. P. (1982). Style and significance of surface deformation in the vicinity of New Madrid, Missouri. In *Investigations of the New Madrid Earthquake Region*, ed. F. A. McKeown and L. C. Pakiser, U.S. Geological Survey Professional Paper 1236, pp. 95–114.

Saucier, R. T. (1994). *Geomorphology and Quaternary Geologic History of the Lower Mississippi Valley*. U.S. Army Engineer Waterways Experiment Station, Vicksburg, Mississippi, 1.

Schweig, III, E. S., and Ellis, M. A. (1994). Reconciling short recurrence intervals with minor deformation in the New Madrid seismic zone. *Science*, 264, 1308–1311.

Sexton, J. L. (1992). *Collaborative Research SIUC-USGS, High Resolution Seismic Reflection Surveying of the Bootheel Lineament in the New Madrid Seismic Zone*. NEHRP Final Report.

Sexton, J. L., and Jones, P. B. (1986). Evidence for recurrent faulting in the New Madrid seismic zone from Mini-Sosie high-resolution reflection data. *Geophysics*, 51, 1760–1788.

Shoemaker, M., Vaughn, J. D., Anderson, N. L., *et al.* (1997). A shallow high-resolution seismic reflection study of Dudley ridge, south-east Missouri. *Computers & Geosciences*, 23, 1113–1120, doi:10.1016/S0098-3004(97)00096-4.

Spitz, W. J., and Schumm, S. A. (1997). Tectonic geomorphology of the Mississippi Valley between Osceola, Arkansas, and Friars Point, Mississippi. *Engineering Geology*, 46, 259–280.

Stahle, D. W., Van Arsdale, R. B., and Cleaveland, M. K. (1992). Tectonic signal in baldcypress trees at Reelfoot Lake, Tennessee. *Seismological Research Letters*, 63, 439–448.

Stark, J. T. (1997). The East Continent rift complex: evidence and conclusions. In *Middle Proterozoic to Cambrian Rifting, Central North America*, ed. R. W. Ojakangas, A. B. Dickas, and J. C. Green. Geological Society of America Special Paper 312, pp. 253–266.

Stein, S. (2010). *Disaster Deferred: How New Science Is Changing Our View of Earthquake Hazards in the Midwest*. New York: Columbia University Press.

Stuart, W. D., Hildenbrand, T. G., and Simpson, R. W. (1997). Stressing of the New Madrid seismic zone by a lower crust detachment fault. *Journal of Geophysical Research*, 102, 27,623–27,633.

Talwani, P. (1999). Fault geometry and earthquakes in continental interiors. *Tectonophysics*, 305, 371–379.

Tavakoli, B., Pezeshk, S., and Cox, R. T. (2010). Seismicity of the New Madrid seismic zone derived from a deep-seated strike-slip fault. *Bulletin of the Seismological Society of America*, 100, 1646–1658.

Thomas, W. A. (1988). The Black Warrior Basin. In *The Geology of North America, v. D-2: Sedimentary Cover – North American Craton, U.S.*, ed. L. L. Sloss. Geological Society of America, pp. 471–492.

Thomas, W. A. (1991). The Appalachian-Ouachita rifted margin of southeastern North America. *Geological Society of America Bulletin*, 103, 415–431.

Thomas, W. A. (1993). Low-angle detachment geometry of the late Precambrian-Cambrian Appalachian-Ouachita rifted margin of southeastern North America. *Geology*, 21, 921–924.

Thomas, W. A. (2006). Tectonic inheritance at a continental margin. *GSA Today*, 16, 4–11.

Tollo, R. P., Corriveau, L., McLelland, J., and Bartholomew, M. J. (2004). Introduction. In *Proterozoic Tectonic Evolution of the Grenville Orogen in North America*, ed. R. P. Tollo, L. Corriveau, J. McLelland, and M. J. Bartholomew. The Geological Society of America Memoir 197, pp. 1–18.

Tuttle, M. P., Schweig, E. S., Sims, J. D., *et al.* (2002). The earthquake potential of the New Madrid seismic zone. *Bulletin of the Seismological Society of America*, 92, 2080–2089.

Tuttle, M. P., Schweig III, E. S., Campbell, J., *et al.* (2005). Evidence for New Madrid earthquakes in A.D. 300 and 2350 B.C. *Seismological Research Letters*, 76, 489–501.

Tuttle, M. P., Al-Shukri, H., and Mahdi, H. (2006). Very large earthquakes centered southwest of the New Madrid seismic zone 5,000–7,000 years ago. *Seismological Research Letters*, 77, 755–770.

Van Arsdale, R. B. (2000). Displacement history and slip rate on the Reelfoot fault of the New Madrid seismic zone. *Engineering Geology*, 55, 219–226.

Van Arsdale, R. B. (2009). *Adventures Through Deep Time: The Central Mississippi River Valley and Its Earthquakes*. Geological Society of America Special Paper 455.

Van Arsdale, R., and Cox, R. (2007). The Mississippi's curious origins. *Scientific American*, 296, 76–82.

Van Arsdale, R. B., Cupples, W. B. (2013). Late Pliocene and Quaternary deformation of the Reelfoot Rift. *Geosphere*, 9(6), doi:10.1130/GES00906, 1819–1831.

Van Arsdale, R. B., Williams, R. A., Schweig, E. S., *et al.* (1995). The origin of Crowley's Ridge, northeastern Arkansas: erosional remnant or tectonic uplift? *Bulletin of the Seismological Society of America*, 85, 963–986.

Van Arsdale, R. B., Purser, J. L., Stephenson, W., and Odum, J. (1998). Faulting along the southern margin of Reelfoot Lake, Tennessee. *Bulletin of the Seismological Society of America*, 88, 131–139.

Van Arsdale, R. B., Cox, R. T., Johnston, A. C., Stephenson, W. J., and Odum, J. K. (1999). Southeastern extension of the Reelfoot fault. *Seismological Research Letters*, 70, 348–359.

Van Arsdale, R. B., Bresnahan, R. P., McCallister, N. S., and Waldron, B. (2007). The Upland Complex of the central Mississippi River valley: its origin, denudation, and possible role in reactivation of the New Madrid seismic zone. In *Continental Intraplate*

Earthquakes: Science, Hazard, and Policy Issues, ed. S. Stein, and S. Mazzotti. Geological Society of America Special Paper 425, pp. 177–192.

Van Arsdale, R. B., Arellano, D., Stevens, K. C., *et al.* (2012). Geology, geotechnical engineering, and natural hazards of Memphis, Tennessee, USA. *Environmental & Engineering Geoscience*, 18, 113–158, doi:10.2113/gseegeosci.18.2.113.

Van Schmus, W. R., and 24 others (1993). Transcontinental Proterozoic provinces. In *Precambrian: Conterminous U.S., Geology of North America, v. C-2*, ed. J. C. Reed, *et al.* Boulder, Colorado: Geological Society of America, pp. 171–334.

Van Schmus, W. R., Bickford, M. E., and Turek, A. (1996). Proterozoic geology of the east-central Midcontinent basement. In *Basement and Basins of Eastern North America*, ed. B. A. van der Pluijm and P. A. Catacosinos. Geological Society of America Special Paper 308, pp. 7–32.

Van Schmus, W. R., Schneider, D. A., Holm, D. K., Dodson, S., and Nelson, B. K. (2007). New insights into the southern margin of the Archean-Proterozoic boundary in the north-central United States based on U-Pb, Sm-Nd, and Ar-Ar geochronology. *Precambrian Research*, 157, 80–105.

Vaughn, J. D. (1994). *Paleoseismological Studies in the Western Lowlands of Southeastern Missouri*. Final Technical Report, U.S. Geological Survey.

Velasco, M., Van Arsdale, R., Waldron, B., Harris, J., and Cox, R. (2005). Quaternary faulting beneath Memphis, Tennessee. *Seismological Research Letters*, 76, 598–614.

Wesnousky, S. G. (1988). Seismological and structural evolution of strike-slip faults. *Nature*, 335, 340–343.

Williams, R. A., Stephenson, W. J., Odum, J. K., and Worley, D. M. (2001). Seismic-reflection imaging of Tertiary faulting and post-Eocene deformation 20 km north of Memphis, Tennessee. *Engineering Geology*, 62, 79–90.

Wu, P., and Johnston, P. (2000). Can deglaciation trigger earthquakes in North America? *Geophysical Research Letters*, 27, 1323–1326.

Zoback, M. L. (1992). Stress field constraints on intraplate seismicity in eastern North America. *Journal of Geophysical Research*, 92, 11,761–11,782.

Zoback, M. L., and Zoback, M. D. (1989). Tectonic stress field of the continental United States. In *Geophysical Framework of the Continental United States*, ed. L. Pakiser and W. Mooney. Geological Society of America Memoir 172, pp. 523–540.

8

The impact of the earthquake activity in Western Europe from the historical and architectural heritage records

THIERRY CAMELBEECK, PIERRE ALEXANDRE, ALAIN SABBE, ELISABETH
KNUTS, DAVID GARCIA MORENO, AND THOMAS LECOCQ

Abstract

This chapter presents and discusses the impact of the earthquake activity in the plate interior region of Western Europe extending from the Lower Rhine Embayment to southern England. The present study is based on methodologies combining the historical and architectural heritage records to better quantify moderate and extensive damage from past earthquakes. These methodologies have been applied to seven destructive earthquakes with magnitudes ranging from 4.5 to 6.0, characteristic of the seismic activity of this area.

The extremely high seismic vulnerability of this region is illustrated by the destruction resulting from small shallow earthquakes such as the 1983 $M = 4.6$ Liège (Belgium) and 1884 $M = 4\frac{3}{4}$ Colchester (England) earthquakes and the elevated financial losses produced by the 1992 $M = 5.3$ Roermond (the Netherlands) earthquake, despite the low observed intensities. This vulnerability is directly related to the very high density of population and to the substantial fraction of poorly constructed masonry dwellings in the building inventory in most of the cities of Western Europe. Indeed, the consequences of a rare $M \sim 6.0$ seismic event, such as the Verviers (Belgium) 1692 earthquake, could certainly be catastrophic in terms of victims and destruction.

Comparing the damage caused by past earthquakes in large structures such as churches or castles and classical dwellings provides information on their source and regional site effects. On the one hand, the earthquakes that caused moderate to heavy damage to large structures located far away from the epicentre had magnitudes greater than 5.0. On the other hand, the observation that churches were damaged in some localities of the Lower Rhine Embayment or the Brabant Massif while typical houses located in the same localities suffered less or had no damage appears to be related to the presence of a sedimentary cover with a

Intraplate Earthquakes, ed. Pradeep Talwani. Published by Cambridge University Press. © Cambridge University Press 2014.

low Q-factor that suppresses the high-frequency content of the seismic source, and (or) a thickness corresponding to a soil fundamental period that enhances ground motion in the frequency range of the large building natural resonance frequency.

8.1 Introduction

The damage caused by past earthquakes is an important source of information about the seismic vulnerability of a region and its study may be very helpful to validate seismic risk studies. Unfortunately, the available data on the impact of destructive earthquakes in plate interior regions, such as the tectonically stable Europe, are often imprecise, because strong earthquakes are rare and only a few of them have occurred since the developments of modern seismology.

The area of Western Europe extending from the Lower Rhine Embayment to the southern North Sea (Figure 8.1) is characterised by a moderate seismic activity, with the occurrence of 14 earthquakes with estimated magnitude greater than or equal to 5.0 since the fourteenth century (Camelbeeck *et al.*, 2007). The most important of these earthquakes occurred on 18 September 1692, in the northern part of the Belgian Ardenne. Its magnitude

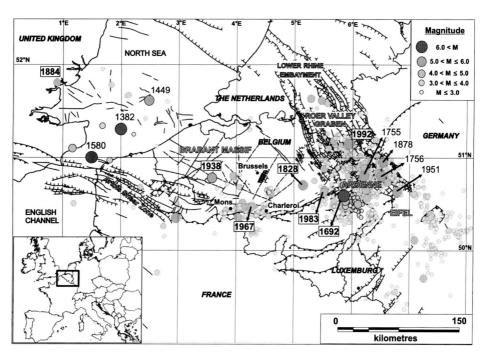

Figure 8.1 Seismic activity in the region extending from the Lower Rhine Embayment to southern North Sea from 1350 to 2012. Rectangles enclose dates of earthquakes studied in this chapter.

has been estimated as 6¼ (Alexandre *et al.*, 2008). The heavy damage caused by the two recent M 4.6 and 5.3 earthquakes in Liège (Belgium) on 8 November 1983 and in Roermond (the Netherlands) on 13 April 1992 (De Becker, 1985; Plumier *et al.*, 2005; Berz, 1994) enhances the need to quantify the possible consequences of the seismic activity in this part of Europe as the basis for earthquake mitigation strategies and policies of prevention.

An optimum dataset to evaluate the impact of an earthquake would be that for which specific damage is reported and described for each of the damaged buildings. In Western Europe, such detailed information has only been collected by the Belgian Federal Calamity Centre for the localities affected by the destructive earthquake that occurred in Liège (Belgium) on 8 November 1983. A sufficient dataset should include quantitative damage information that would allow accurate intensity values to be assigned. Even for recent earthquakes this kind of data is not necessarily available, resulting sometimes in bad intensity evaluation.

During the past ten years we have significantly improved our knowledge of the impact of some damaging past earthquakes in Western Europe by enlarging the historical data archive and developing the methodologies to identify earthquake traces in historical buildings. In this chapter, we present and discuss this methodological framework and the impact of seismic activity on the region from the Lower Rhine area to southern England based on damage information gathered for seven typical past destructive earthquakes. We also examine the influence of the sedimentary cover on the intensity of the damage caused by these earthquakes in the northern part of the study area.

8.2 Seismic activity between the Lower Rhine Embayment and the North Sea

The study area is the most seismically active region of Western and Central Europe to the north of the Alps (Figure 8.1). An important part of this activity is concentrated in the Roer Valley graben, part of the Lower Rhine Embayment that crosses the border region between Belgium, Germany, and the Netherlands (Hinzen and Oemisch, 2001; Camelbeeck *et al.*, 2007). Seven earthquakes with M ≥ 5.0 have occurred there since 1350, the most significant had a magnitude M ~ 5¾ and affected the region of Düren in Germany on 18 February 1756. More recent strong earthquakes occurred on 14 March 1951, near Euskirchen in Germany (M = 5.3) and on 13 April 1992 in Roermond, the Netherlands (M = 5.3). Ahorner (1975) provided a comprehensive seismotectonic study establishing the relationship between this seismic activity and normal faults in the Roer Valley graben offsetting Quaternary deposits up to 175 m. Paleoseismic investigations (Camelbeeck and Meghraoui, 1996, 1998; Vanneste *et al.*, 1999, 2001; Meghraoui *et al.*, 2000; Vanneste and Verbeeck, 2001) suggest that coseismic surface ruptures have occurred during earthquakes with magnitude in the range 6.0–7.0 during the Holocene and late Pleistocene. A synthesis of these results is given in Camelbeeck *et al.* (2007).

To the west of the graben, seismic activity is also well established in the north of the Belgian Ardenne and in the Eifel Mountains in Germany. The most significant event of the whole study region is the M = 6¼ earthquake that occurred in this zone on 18 September 1692, near the city of Verviers (Alexandre *et al.*, 2008). At the northern limit of the Ardenne, two recent damaging earthquakes also occurred in the city of Liège on 21 December 1965 (M = 4.3) (Van Gils, 1966) and 8 November 1983 (M = 4.6).

Another zone of concentrated seismic activity is the region of the Hainaut province of Belgium, between the cities of Mons and Charleroi. Despite the fact that none of these earthquakes exceeded magnitude 4.5, some of them caused some damage locally due to their shallow depths (≤3–5 km). Reported damage during these earthquakes concerns mainly the collapse or partial collapse of chimneys, broken windows, cracked ceilings and walls. The tectonic origin of these earthquakes is debated, as it is possible that part of the reported activity in this area could be related to the extensive mining works that took place in this area from the end of the nineteenth century to the beginning of the 1970s (Descamps, 2009).

Another particularity of the study area is that part of the seismic activity appears to be spatially diffuse and sporadic, as is common in many plate interior regions worldwide. For instance, in the Strait of Dover and southern North Sea, seismic activity has been weak since the beginning of the seventeenth century; however, three historical earthquakes that occurred in this area on 21 May 1382, 23 April 1449 and 6 April 1580 produced significant damage in southern England, northern France, and Belgium (Melville *et al.*, 1996). In the same way, the Brabant Massif region has not shown any significant earthquake activity since the installation of the modern seismic network in Belgium in 1985, but this region was violently shaken on 23 February 1828 and on 11 June 1938 by two M ~5.0 earthquakes that caused widespread damage.

8.3 The background and methodologies of historical seismicity in Western Europe

The retrieval and analysis of information on earthquake effects, whatever the epoch of the earthquake, are part of the research field called "historical seismicity". Most of the information on historical earthquakes in northwestern Europe, spanning a period from the earliest available sources (*c*. AD 700) to the present, can be found in the database that we have developed at the Royal Observatory of Belgium, which is partly accessible on the Internet (Camelbeeck *et al.*, 2009).

In this chapter, we will focus specifically on the information concerning reported damage with the purpose of evaluating the local and regional impacts of destructive earthquakes. This objective requires more detailed information on the damage characteristics and their spatial distribution than that of classical investigations evaluating earthquake location and magnitude from intensity datasets. According to the differences in the type of documentation that can be collected on damaging earthquakes, in this study we differentiate between the information that was collected before and since the development of regional and national seismic observatories, which corresponds roughly to the year 1900.

8.3.1 The period from 1900 to the present

In the study area of Western Europe, many permanent seismological observatories were established at the end of the nineteenth century or at the beginning of the twentieth century. For example, in France, the first seismological measurements were conducted in 1892 in

Strasbourg and in 1908 in the Paris Parc Saint-Maur Observatory. In Belgium, the seismic station of Uccle has been operational since 1899, while the seismic station located in the Netherlands Meteorological Institute in De Bilt was installed in 1907. These observatories published their seismic phase measurements and earthquake observations in the form of seismic bulletins that are still available. For the earthquakes that have occurred since this epoch, part of the available information describing their effects and the damage they caused comes from scientific reports published by these national or regional seismic observatories.

Nevertheless, there are few earthquakes that occurred in Western Europe for which data of sufficient quality exist to evaluate quantitatively the importance of the damage. The 1983 Liège earthquake is probably the best documented one, with detailed damage reports existing for each of the 16,000 buildings for which the owners asked for a contribution from the Belgian State for the repair costs of the damage. This dataset is stored as paper files at the Belgian Federal Calamity Centre and has been used in different investigations on the earthquake (Jongmans and Campillo, 1984; Jongmans, 1989; Jongmans and Plumier, 2000; Garcia Moreno and Camelbeeck, 2013).

The dataset established by Pappin *et al.* (1994) after a three-day damage survey in the 1992 Roermond earthquake epicentral area is less detailed, but it provides the percentage of the buildings of selected types that experienced slight and moderate damage from the earthquake. Residential masonry was the only building type sufficiently common to allow a statistical analysis.

Since 1932, every time an earthquake occurs the Royal Observatory of Belgium sends a standard questionnaire to the authorities of the Belgian localities affected by the seismic event. This document provides quantitative information on the damage to chimneys in each locality. This data is extremely useful to quantify the intensity of the damage as it can be assumed that the proportion of damaged chimneys is equivalent to the proportion of buildings presenting at least moderate damage (see Section 8.4). Therefore, these official inquiries provide relevant information allowing a comparison of the percentage of moderate and extensive damage for the best documented 1983 Liège and 1992 Roermond earthquakes with older ones. In Section 8.5, we will discuss such an evaluation carried out for the $M = 4.3$, 28 March 1967 earthquake that occurred in the Hainaut region and the $M = 5.0$, 11 June 1938 earthquake that strongly shook the western part of Belgium and northern France.

These data are complemented with information provided by scientific studies, such as for the British Colchester 1884 earthquake, for which the number of repaired buildings in the more affected localities is available in a scientific report performed at the time of the earthquake (Musson *et al.*, 1990).

8.3.2 Pre-1900 period: historical seismicity and the architectural heritage

There are two main differences in the data available to study earthquakes before 1900 compared to those for the twentieth century. First, the earlier earthquakes were not recorded by seismic instruments and, second, there are far fewer scientific studies and less documentation describing their effects and damage. Hence, their investigation requires collecting accounts and eyewitness reports in different types of historical sources. Often, the most interesting sources are the individual narratives found in diaries, letters, and brief notes. Local chronicles and annotations in parish registers generally mention the seismic events, but often without detailed information. In fact, there is more chance of finding quantitative data on the damage in administrative sources, including reports from local authorities, account books indicating repair costs, or official civil or ecclesiastic reports asking for funds to repair damaged buildings or churches. Reports written some decades later can also include relevant information on the damage produced by an earthquake as indirect first-hand sources transmitted to us as second-hand sources. Newspapers and other periodical contemporaneous publications are another common source of information, after their origins during the seventeenth century in Western Europe (Alexandre *et al.*, 2007).

In essence, all the historical reports are incomplete in the sense that they provide very little quantitative information on earthquake effects and, for a given earthquake, the geographical distribution of the localities for which they provide information often is not homogeneous. To improve our knowledge of the impact of historical earthquakes, it is thus necessary to collect more eyewitness accounts in record offices, in order to better characterize local damage and geographically extend our knowledge of earthquake effects.

Another way to better characterize the damage caused by past earthquakes is to look for repairs or weaknesses in present-day buildings that already existed at the epoch of the earthquake, in order to evaluate whether these disturbances can be explained by an earthquake. Such evidence is complementary to the historical reports and is an invaluable source of information on the destructiveness of past earthquakes, as will be shown in this chapter. Large buildings such as churches or castles appear to be the most appropriate structures for which to investigate such relationships, even if they are not the most adequate for evaluating intensity (Grünthal *et al.*, 1998). Indeed, these buildings more easily survive the effects of aging than individual houses. Moreover, the construction and maintenance data for these buildings are generally noted in the archives of the local parish or local administration. Hence, there is more chance of retrieving information on their different phases of construction or reconstruction than for particular houses.

Establishing a link between an earthquake and possible damage to a specific building from observed pathologies or (and) repair traces needs a strict methodological approach. The first aspect to consider concerns the estimation of the age of the buildings and the dating of repairs and existing pathologies. Second, pathologies and repairs can result from numerous different causes other than the earthquake hypothesis. Therefore, it is necessary to confront the observations with the different possible hypotheses and to accept the earthquake origin only if sufficient scientific arguments can be given. A beautiful example in archeoseismology illustrating the importance of such detailed investigation, including numerical modelling, is given by the study by Hinzen *et al.* (2010) on the Lycian sarcophagus of Arttumpara in Turkey, which deciphered human and earthquake actions on the sarcophagus.

Part of this chapter is devoted to an analysis of the impact of damaging earthquakes on architectural heritage buildings in Belgium. In this regard, we conducted two different investigations to better evaluate the importance of the damage caused by two past earthquakes that demonstrate the real benefit of such an approach. We first investigated the repairs and pathologies of the churches in the epicentral area of the 23 February 1828 earthquake (Philipront, 2007), which are presented and discussed in Section 8.5.6 of this chapter. Second, we studied the houses in the centre of the village of Soiron, located in the epicentral area of the $M = 6\frac{1}{4}$, 1692 Verviers earthquake, to estimate their age and identify pathologies and repairs that could be associated with this earthquake (Dewattines, 2010). This information is used in Section 8.5.7 to formulate hypotheses on the intensity of the earthquake in the village, which are discussed in the light of the information collected in historical sources.

8.4 Damage quantification

Most of the buildings damaged during the destructive twentieth- and nineteenth-century earthquakes that occurred in the study area can be classified as low-rise unreinforced brick-masonry buildings. That is because, even if the building types have evolved to either steel or reinforced concrete construction, low-rise brick masonry (or mixed concrete–brick) continues to be the type of construction typically chosen by the population. Thus, the building inventory continues to include masonry structures that are sometimes very old. Hence, the main information at our disposal concerns this kind of building. Nevertheless, knowing the range of vulnerabilities of such unreinforced masonry buildings and other types of buildings, it should always be possible to evaluate the potential damage for the other types of buildings by comparison with the damage observed to masonry buildings.

Because comparison of the impact of historical earthquakes and more recent ones can only be done using seismic intensity values, it seems advisable to define the seriousness of damage to the buildings by a damage scale associated with the intensity scale used. For this reason, we quantified the degree of damage according to the European Macroseismic Scale EMS-98 (Grünthal *et al.*, 1998). This scale distinguishes five different degrees of damage, where degree 1 (D1) corresponds to negligible to slight damage, degree 2 (D2) to moderate damage (cracks in numerous walls, partial falling of chimneys, etc.), degree 3 (D3) to heavy damage (significant cracks in most of the walls, chimneys rupture at the junction with the roof, etc.), degree 4 (D4) to very heavy damage (serious weakness of the walls, partial structural failure of roofs and floors), and degree 5 (D5) to complete or nearly complete building collapse.

Furthermore, by using the HAZUS fragility curves (FEMA, 1999) on the 1983 Liège earthquake dataset, Garcia Moreno and Camelbeeck (2013) were able to associate effective losses for this earthquake with the damage scale associated with HAZUS, providing a way to evaluate seismic risks (Figure 8.2). Therefore, it is also logical to use the damage scale defined by HAZUS for the present study. Fortunately, there is a good correlation between the damage scales associated with EMS-98 and the HAZUS fragility curves (FEMA, 1999)

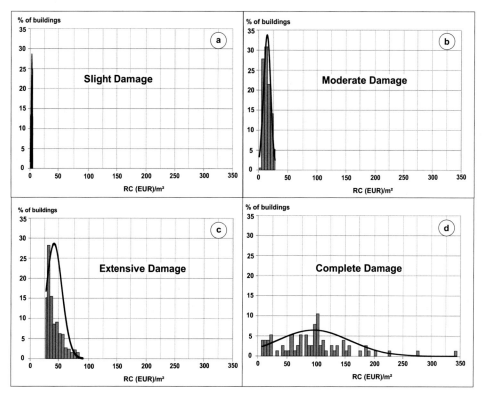

Figure 8.2 Statistics of the repair costs per unit surface for the different damage grades for the 1983 Liège earthquake (based on Garcia Moreno and Camelbeeck, 2013): (a) slight damage (SL), equivalent to damage grade 1 of the EMS-98 macroseismic scale (Grünthal *et al.*, 1998); (b) moderate damage (MD), equivalent to damage grade 2 of the EMS-98 macroseismic scale; (c) extensive damage (ED), equivalent to damage grade 3 of the EMS-98 macroseismic scale; (d) complete damage (CD), including damage grades 4 and 5 of the EMS-98 macroseismic scale. The cost information has been converted from Belgian Francs (BEF) to Euros (EUR). At the epoch of the earthquake, the cost of a two-floor unreinforced brick-masonry house with a ground surface area of 100 m² was around 1 million BEF or 25,000 EUR.

for unreinforced masonry (Hill and Rossetto, 2008). There is equivalence between EMS-98 degree 1 and slight damage in HAZUS, degree 2 and moderate damage, degree 3 and extensive damage, and degrees 4 and 5 with complete damage.

In the case of the 1983 Liège earthquake, Garcia Moreno and Camelbeeck (2013) showed that the repair costs begin to be important at the degree of damage D2 on the EMS-98 macroseismic scale, which also corresponds to moderate damage on the HAZUS damage scale (Figure 8.2). Therefore, this information should allow evaluation of the global repair costs of an earthquake. Hence, in Figure 8.3 we present maps reporting the percentage of damage greater than or equal to moderate in each affected locality for the five earthquakes for which we retrieved this detailed information, rather than the intensity.

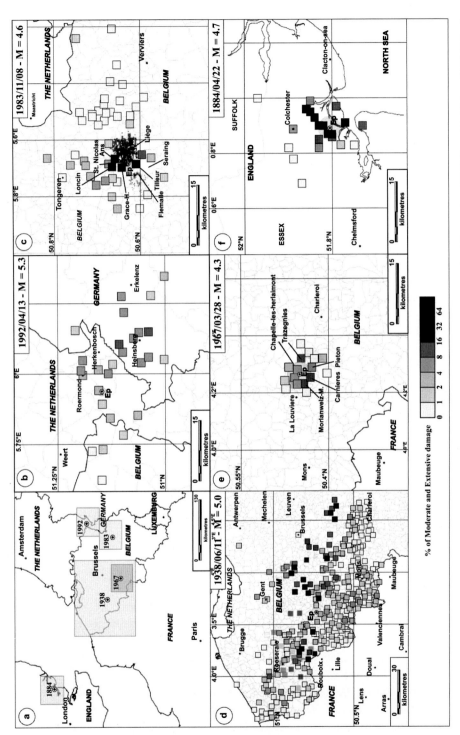

Figure 8.3 Statistics of damage greater than or equal to moderate for the 13 April 1992 Roermond (b), 8 November 1983 Liège (c), 11 June 1938 Nukerke (d), 28 March 1967 Carnières (e) and 22 April 1884 Colchester (f) earthquakes.

Considering that most of the low-rise unreinforced masonry buildings commonly found in these regions of Western Europe should be classified as vulnerability A and B (for the less vulnerable part of the stock), the quantitative information of moderate or extensive damage allows us to fix the intensity on the EMS-98 macroseismic scale: Intensity VI thus corresponds to grade 2 or moderate damage in a few buildings (up to 10%) of vulnerability class A and B. Intensity VII is associated with damage of grade 3 or extensive damage in many buildings (more than 10%) of vulnerability class A. Few of these buildings suffered grade 4 or complete damage. To reach intensity VIII, many of the vulnerability class A buildings would have to suffer degrees of damage 4, and a few of them should collapse.

For historical earthquakes, there is no or little quantitative information on moderate or more extensive damage. Therefore, in order to assign intensities, it is important to be aware of the often poor information and the necessity to use intensity ranges, indicating the uncertainty of the interpretation. Another significant problem is the lack of information on the vulnerability of the buildings for which descriptions of damage are given in the historical sources. Despite the great diversity of the materials used and the construction methods during historical times, it is often supposed that the vulnerability class of traditional houses ranges between classes A and B of the EMS-98 intensity scale, similar to that for the low-rise unreinforced masonry buildings typical of the twentieth century. This may not be totally correct, and is clearly an aspect that should be investigated more deeply when studying past earthquakes. In the next section, we will present our evaluation of the intensity based on the aforementioned hypothesis in the localities of the epicentral area of the $M = 5$, 23 February 1828 central Belgium and $M = 6\frac{1}{4}$, 18 September 1692 Verviers earthquakes.

For each of the studied earthquakes, we have compiled an inventory of the damage caused to large buildings of the architectural heritage, mainly churches. For these churches, we have considered the damage scale proposed by Meidow and Ahorner (1994) for the occasion of the 1992 Roermond earthquake. They define slight damage as "fine cracks in plaster, fall of small pieces of plaster and the loosening of pinnacles or comparable building parts", moderate damage as "small cracks in walls and vaults, cracks between church tower and nave and the falling of pinnacles", and heavy as "large and deep cracks in walls and vaults, and damage to load-bearing parts". We added a fourth grade, D4, when parts of the church collapsed.

8.5 Typical damaging earthquakes in Western Europe

In this section, we present the results of an investigation on the impact of some destructive earthquakes that occurred in the study area. The list of seismic events studied is not exhaustive. It includes seven earthquakes presenting typical characteristics of the seismic activity of this part of Europe. This allows us to discuss the main characteristics of the damage caused by moderate and large earthquakes in the area.

8.5.1 The 13 April 1992 Roermond earthquake

The $M = 5.3$, 1992 Roermond (the Netherlands) earthquake is one of the largest earthquakes observed during historical times in the Lower Rhine Embayment (Figure 8.1). Its focal mechanism is pure normal faulting at a depth of 17 km on the Peel fault, which is the eastern border fault of the Roer Valley graben (Camelbeeck and van Eck, 1994; Ahorner, 1994).

This earthquake was felt over a large area of western and central Europe with a mean radius of perceptibility of about 440 km (Meidow and Ahorner, 1994). Bouwkamp (1994) noticed that the main damage was restricted to failure and cracking of brick masonry chimneys and parapets. There were very few cases in which the upper portion of chimneys had toppled over and damaged the tiles on the roof. A partial collapse of gable ends, inadequately tied to the wooden roof structures, was also noticed in a few cases, as well as diagonal shear cracking of exterior masonry walls. This damage was mainly observed in old masonry structures, some of them dating back to the eighteenth century. More modern steel and reinforced concrete constructions were far less affected by the earthquake.

Meidow and Ahorner (1994) also mentioned damage on 150 churches in Germany, predominantly in the epicentral area, but also at larger distances in Cologne, Bonn, and Koblenz (Figure 8.4a). Damage was also observed on different churches located in the Netherlands, with the most affected one being situated in Herkenbosch (Bouwkamp, 1994).

The official macroseismic inquiry (Haak *et al.*, 1994) reported a maximum intensity of VII (on the MSK scale), but there is no description of the way this intensity value was calculated. Another analysis of the seismic intensities caused by this earthquake was performed by Meidow and Ahorner (1994), who established a macroseismic map for the German territory. They also assessed the maximum intensity as VII (in the MSK scale) and justified this value based on the number of buildings that were significantly damaged in the localities to which this intensity has been attributed. They also mentioned that eight buildings were uninhabitable in Germany and had to be evacuated.

The only study that evaluated the amount of damage experienced by buildings in the localities affected by the Roermond earthquake is that of Pappin *et al.* (1994). The analysis of their observations is available at the Cambridge Earthquake Impact Database: www. ceqid.org/. These authors provided a statistical analysis of the damage caused to residential masonry buildings located in the Netherlands and western Germany. In the most affected localities, the percentage of moderate damage (EMS-98 degree 2) did not exceed 17% of the buildings and there are only two localities where a little (2 and 3%) extensive damage (EMS-98 degree 3) was observed (Figure 8.3b). From these numbers, they assigned a maximum intensity value of VI (on the MSK scale) to the most affected localities. We agree with this interpretation because in order to reach intensity VII (on the EMS-98 scale) many buildings (more than 10%) of vulnerability class A should have experienced extensive damage (EMS-98 grade 3), and this is clearly not the case. Note that Pappin *et al.* (1994) did not visit the locality of Herkenbosch where significant damage was notified.

The Roermond earthquake presents two particularities that are important to take into consideration for regional seismic risk studies. The first one, underlined by Meidow and Ahorner (1994), is that the observed maximal intensity is low for its magnitude by comparison with other earthquakes in the Lower Rhine Embayment. Its focal depth, which is deeper than usual in the Lower Rhine Embayment, is the first factor that explains the low epicentral intensity. A second factor is the strong seismic energy absorption by the 1500 m

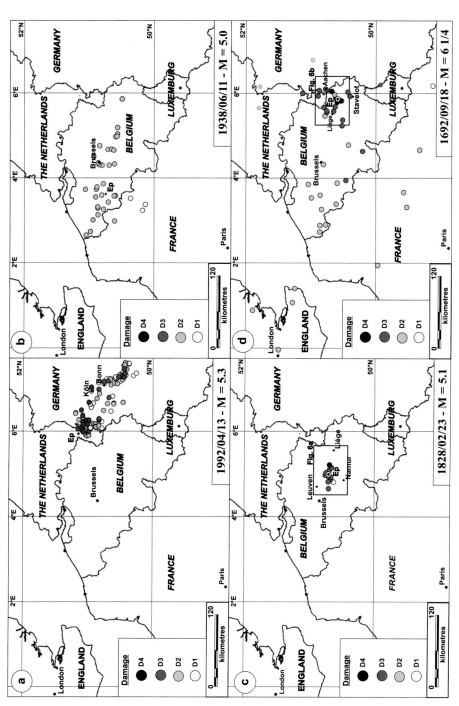

Figure 8.4 Observed damage to churches during M ≥ 5.0 earthquakes: the 13 April 1992 Roermond (a), the 11 June 1938 Nukerke (b), the 23 February 1828 Hamnut (c) and the 18 September 1692 Verviers (d) earthquakes. The damage scale is that defined by Ahorner and Meidow (1994) for the Roermond 1992 earthquake. The rectangles in maps (c) and (d) indicate the epicentral area of the 1828 and 1692 earthquakes shown in Figure 8.5.

thick layer of soft Tertiary and Quaternary sediments in the Roer Valley graben in the region of Roermond.

The second particularity is that, despite the small value of the maximum observed intensity, the losses caused by this earthquake were evaluated at the time of the earthquake as around 125 million EUR (Berz, 1994). This may be due to the extent of the moderately damaged area, which is directly related to the earthquake magnitude and focal depth, in a region of dense population, industry, and important natural hazards insurance coverage.

A significant part of the estimated losses, corresponding to 24%, is actually due to the damage caused to churches. It is generally assumed (Grünthal *et al.*, 1998) that, even though they may be better built, monumental buildings such as churches may be more likely to sustain damage than ordinary buildings. This is probably true for slight damage, but in the present case, moderate and extensive damage have been observed in some churches, even in locations where houses suffered very slight or no damage.

This apparent difference in the behaviour of low-rise buildings and churches is essentially linked to the fact that seismic wave energy in the high-frequency range, which corresponds to the first mode natural frequency of houses, is largely attenuated before the waves reach the Earth surface by a significant thickness of sediments covering the bedrock. On the contrary, at lower frequency the energy content of the seismic signal is less modified by propagation through poorly consolidated sediments. As the Roermond earthquake source is large enough to generate seismic energy at frequencies around 1 Hz, in the frequency band of the first mode of vibration of large buildings such as churches, the seismic energy is sufficient to cause damage to churches, even at large distance.

Reports and eyewitness accounts from the newspapers about the Roermond earthquake can be found via the link: http://seismologie.be/cup2014.html.

8.5.2 The 8 November 1983 Liège earthquake

The earthquake that shook the region of Liège (Belgium) on 8 November 1983 at 0 h 49 m (UT) had unusual consequences considering its low magnitude, 4.6. Significant damage left more than 1,000 people homeless, causing serious logistical problems.

Within an area of 10 km^2 around the locality of Saint-Nicolas (Figure 8.3c), at the earthquake epicentre, more than 16,000 buildings were damaged by this earthquake (De Becker, 1985; Phillips, 1985; Plumier *et al.*, 2006). The most apparent damage was the fall of chimneys, bricks or ornamental features. The fall of these features caused damage to roofs and many cars in the street, suggesting that if the earthquake had occurred during the day the human consequences (two deaths and some dozen injuries) would have been more tragic. Many walls were also intensively cracked and it was necessary to shore up many of them. The worst structural damage was observed in older and poorly built structures. In the two most affected localities, Saint-Nicolas and Liège, 129 houses were declared uninhabitable and 37 were demolished. Well-constructed masonry buildings fared relatively well, with damage usually restricted to chimneys. The damage in two churches in the epicentral area was sufficient for the authorities to declare a temporary decree of uninhabitability. There was little damage to industrial installations, as expected, because most of them are located outside the most affected area.

Garcia Moreno and Camelbeeck (2013) collected and provided an overview of the available data on the damage caused by the 1983 Liège earthquake. The damage is especially well documented for each of the 16,000 houses for which the owners asked for a contribution from the Belgian State to help with the repair costs. These data are stored at the Belgian Federal Calamity Centre. Unfortunately, the informatics structure of this database has not been preserved to the present day, and the paper copies of these files are the only documents currently available. This dataset allowed Garcia Moreno and Camelbeeck (2013) to estimate the statistical distribution of damage in the localities of Saint-Nicolas, Liège, and Flémalle. The results from that study have been synthesized in Figure 8.3c, where the percentage of houses presenting moderate to extensive damage is shown in squares of 200 by 200 metres in these localities. For the other localities, the damage quantification is based on the official inquiry of the Royal Observatory of Belgium and is given by locality.

In the most affected squares in Saint-Nicolas, Liège, and Flémalle, the percentage of buildings with moderate or stronger than moderate (extensive and complete) damage is respectively 78%, 100%, and 75%, while the median values of the observed percentages of damage for the complete set of these squares are respectively 32%, 25%, and 4%.

In the localities close to Liège and Saint-Nicolas, the percentage of damaged chimneys is reported in the official inquiry done by the Royal Observatory of Belgium. It is 25% in Ans, 20% in Grâce-Hollogne, 21% in Loncin, 50% in Tilleur, and 14% in Ougrée. The number of damaged buildings in those different localities corresponds to half of the total damage reported to the Belgian Federal Calamity Centre.

This significant damage, which reached EMS-98 intensity of VII, has been attributed to the shallow depth of the earthquake, the amplification of ground shaking due to the local geology, and the consequences of former mining activity on the ground surface, which had already affected some of the buildings located in this area (Jongmans and Campillo, 1984; Monjoie, 1985; Jongmans, 1989; Jongmans and Plumier, 2000).

The reports and eyewitness accounts from the newspapers about the Liège earthquake can be found via the link: http://seismologie.be/cup2014.html. The statistics of damage averaged over the 200 m by 200 m squares are included as an electronic supplement to the paper by Garcia Moreno and Camelbeeck (2013). The whole dataset can be obtained by request to the first author of this chapter [TC].

8.5.3 *The 28 March 1967 Carnières earthquake*

This earthquake is typical of the damaging earthquakes that occurred in the Hainaut seismic zone during the twentieth century, which typically present magnitudes between 4.0 and 4.5 and are at a shallow focal depth.

This earthquake was strongly felt in the region between the cities of La Louvière and Charleroi (Figures 8.1 and 8.3e). The shaking frightened a large part of the population; some ran outdoors, while others took refuge in their cellars. In some localities, the power supply suddenly stopped. Reported damage included the collapse or partial collapse of chimneys, broken windows, cracked ceilings and walls (source: *La Nouvelle Gazette*, 29 March 1967).

From the official inquiry of the Royal Observatory of Belgium, the most affected localities were Carnières (20% of damage equal to or greater than moderate), Trazegnies (10%), Morlanwez-Mariemont, Piéton, and Chapelle-lez-Herlaimont (7%) (Figure 8.3e). Hence, this type of earthquake causes moderate and extensive damage at a local scale.

The reports and eyewitness accounts from the newspapers about this earthquake can be found via the link: http://seismologie.be/cup2014.html.

8.5.4 The 11 June 1938 Nukerke earthquake

On 11 June 1938 at 11 h 57 m (local time) an earthquake strongly shook western and central Belgium, and northern France. It was also felt in the Netherlands, southeastern England, the extreme west of Germany and the Grand Duchy of Luxemburg. The epicentre was located more or less 50 km to the west of Brussels (Figure 8.1). With a magnitude $M = 5.0$ and a focal depth around 20 km (Camelbeeck, 1993), this earthquake caused damage over a large area (Figure 8.3d). The investigation performed by Somville (1939) at the time of the earthquake describes the main effects produced by it. The main damage to buildings was numerous falling chimneys; roughly 17,500 chimneys were damaged in Belgium alone, which caused serious destruction of roofs and verandas. Many walls were cracked; some of them collapsed. Inside houses, ceilings were cracked, some fell down; windows were broken; numerous objects such as chimneypieces, large mirrors attached to walls, frames or plates were dislodged and broken when falling to the ground. Large pieces of furniture or beds were displaced.

In many localities, people inside buildings were frightened and ran outdoors, thinking that their houses were near to collapse. In the fields, farmers felt the soil oscillating under their feet and found it difficult to stand. Three people died and several dozen were injured.

Based on the inquiry done by the Royal Observatory of Belgium, the quantity of moderate to extensive damage in all the localities of western Belgium is shown in Figure 8.3d. The most affected localities were Rekkem (48% of the houses), Kerksen (44%), Kortrijk (40%), Bousval (37.5%), Outrijve (32%), and Court-Saint-Etienne (32%). The damage caused to houses is unequally distributed on the two sides of an axis elongated in a more or less northwest–southeast alignment with a length of 150 km centred on the earthquake epicentre. To the south of this axis, moderate damage was observed in Belgium and southwards to the French border. Unfortunately, there is no quantitative information from France to extend the observations into northern France. The most significant damage occurred more or less along the axis, sometimes at distances as large as 70 km (localities of Bousval and Court-Saint-Etienne). To the north of this line, damage appeared along an axis oriented northeast–southwest, which corresponds to river valleys that incise the Meso-Cenozoic sediments of the Brabant Massif. This damage distribution suggests that it is linked to the thickness of the soft sediment cover and corresponds to the region where the soft sediment cover ranges from a few metres to about 50 m (Nguyen *et al.*, 2004).

Heavy material fell down from high buildings, mainly from churches. An inventory of the churches damaged by the earthquake has been compiled by consulting the newspapers and testimonies sent to the Royal Observatory of Belgium at the time of the earthquake. Thirty-two churches damaged by the 1938 earthquake have been identified in Belgium, with damage states ranging from slight to moderate (Figure 8.4b). This list is of course not exhaustive. Indeed, when comparing the location of these churches with the areas presenting the majority of damaged houses (Figure 8.3d), they appear to be not exactly identical. The damaged church area is slightly shifted to the north by comparison with the damage region defined by the damaged classical low-rise buildings. The damaged church zone actually corresponds to the quasi east–west region where the natural periods of soil resonance are between 1 and 1.5 seconds, which are in agreement with the range of the fundamental mode of oscillation of this type of large structure (Nguyen *et al.*, 2004). To the west of the epicentre, there appears to be a better overlap between the damage distributions of churches and houses. This area corresponds to the valley of two large rivers, the Lys and the Schelde-Escaut, and therefore to places assumed to present soil conditions less resistant to seismic action.

The reports and eyewitness accounts from the newspapers about this earthquake can be found via the link: http://seismologie.be/cup2014.html.

8.5.5 The 22 April 1884 Colchester earthquake

Despite its relatively low estimated magnitude of around $M = 4\frac{3}{4}$, the Colchester earthquake is considered to be the most damaging British earthquake in the last 400 years (Figure 8.1). This event is very well documented due to the large number of local newspapers available and also the existence of local amateur scientific societies at the end of the nineteenth century. Musson *et al.* (1990) provided a complete study of the earthquake, including most of the reports written at the epoch of the earthquake and their associated sources. For the purpose of our study, we have considered the Musson *et al.* (1990) interpretation of the damage provided by the scientific work of Meldola and White (1885), the most prominent study at the time of the earthquake. This information is reported in Figure 8.3f.

Musson *et al.* (1990) concluded that most of the damage in the epicentral area of this earthquake ranged between EMS-98 damage grades 2 and 3. There is also evidence of grade 4 damage associated with gaps in walls or the collapse of parts of buildings, but they are rare and confined to buildings of the poorest districts. In fact, the importance and the large geographical extent of the damage generated by the 1884 Colchester earthquake appears relatively similar to those observed during the 1983 Liège earthquake in Belgium (Figure 3c). On the other hand, unlike the Liège earthquake, the Colchester seismic event caused moderate and extensive damage to churches in the epicentral area: the fall of parts of towers, damage to roofs, and cracks in masonry and plaster.

8.5.6 The 23 February 1828 Hannut earthquake

We have revisited the 23 February 1828 earthquake that occurred in the central part of Belgium (Figure 8.1), which was the first earthquake worldwide for which a scientist,

Egen, drew up a macroseismic map applying an intensity scale (Gisler *et al.*, 2008). For that purpose, Egen used a scale he invented himself (Egen, 1828). We have synthesized and critically assessed the information provided by this and other contemporaneous scientific studies and newspapers of the Low Countries. To improve our knowledge of the damage and earthquake effects in its epicentral area, we also undertook a systematic survey of the official, private, and religious historical sources of this region. All these original witness accounts, as well as a description of their historical sources, are available at http://seismologie.be/cup2014.html.

8.5.6.1 Analysis of historical data

Based on some of the reports about the earthquake, it is possible to classify the effects of the 1828 earthquake as well as to approximate the intensity values on the EMS-98 macroseismic scale for the different affected localities (Grünthal *et al.*, 1998).

The descriptions of damage indicate that in the epicentral area houses experienced moderate to extensive damage. For many localities situated in the epicentral area, the descriptions concentrate on the amount of damage to chimneys or cracks in the walls, and do not mention the collapse or complete destruction of houses, as was the case, for example, for the 1692 Verviers earthquake, which happened in the north of the Belgian Ardenne (Alexandre *et al.*, 2008), discussed in the next section. This suggests that unreinforced masonry buildings presenting grades of damage 2 and 3 (moderate to extensive) may have been widespread in these cities. As already mentioned, Belgian traditional houses of this epoch were mainly unreinforced masonry buildings, which can be classified as vulnerability classes ranging from A to B. Hence, if buildings were all classified as vulnerability A, the corresponding intensity should have reached values of VII. On the other hand, if they had vulnerability B, the intensity could have been one order of magnitude higher (VIII). In either scenario, few buildings should have presented damage states equal to or greater than 4, which agrees with the damage observations. Some reports suggest that damage grade 4 could have been attributed to some of the building stock of Héron, Lens-Saint-Rémy, and Petit-Hallet. In these cases, it would thus be realistic to consider that the most affected buildings were of vulnerability class A; therefore we assessed an intensity of VII in these localities.

In view of the local reports, we have also attributed intensity VII to the following localities: Berloz, Crehen, Gelinden, Héron, Jauche, Lens-Saint-Rémy, Marilles, Petit-Hallet, Tienen, Waremme, and the nearby villages of Longchamps, Froidebise, and Walquin (Figure 8.5a).

We also associated a possible intensity range of VI–VII with the localities for which at least one report mentioned that many houses may have lost their chimneys or were cracked or damaged. This is the case for Andenne and Bilzen. Intensity values of VI–VII have also been assessed for the village of Grand-Hallet. For this locality, there is no report on the damage caused by the 1828 earthquake to the houses, but the repair costs of the church are close to those paid for the church of Berloz, suggesting significant damage.

For other localities situated in the epicentral area of this earthquake or at its periphery, damage is mentioned, but it appears as quantitatively less important. We assessed, nevertheless, intensity VI for those where a certain number of chimneys collapsed. This

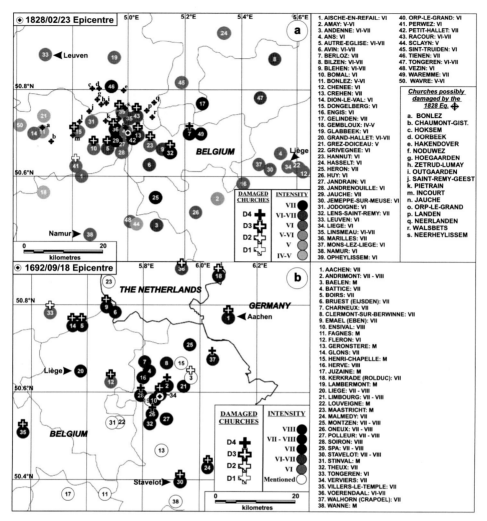

Figure 8.5 Macroseismic map of the epicentral area of the 23 February 1828 Hannut (a) and 18 September 1692 Verviers (b) earthquakes. The mapped areas correspond to the rectangles drawn on Figure 8.4c and d. Each locality for which a historical source mentioned the two earthquakes is identified by a number inside a circle. The estimated intensity (on the EMS-98 macroseismic scale) is easily identifiable by the tone of the circle. The churches for which damage is reported in historical accounts of the two earthquakes are indicated by a cross, showing the estimated damage grade on the damage scale defined by Ahorner and Meidow (1994). We also indicate the churches in which traces of repairs and pathologies could be associated to the 1828 seismic event.

was the case in the cities of Liège, Namur, Glabeek, Leuven, Hannut, Hasselt, Jemeppe-sur-Meuse, Opheylissem, Perwez, and Sint-Truiden.

In many reports from the 1828 earthquake epicentral area, it is also mentioned that frightened people ran away from the churches in the localities of Andenne, Hannut, Huy, Lens-Saint-Servais, Liège, Namur, Sclayn, Perwez, and Tongeren. In other localities, it is mentioned that all the people left their homes in a hurry: Andenne, Jodoigne, Namur, Sainte-Marguerite (suburb of Liège), Sint-Truiden, and Tongeren. These observations certainly correspond to intensity equal to or greater than VI on the EMS-98 macroseismic scale, confirming the range of intensities assessed from the damage discussed above.

8.5.6.2 Damaged churches during the 1828 earthquake

The earthquake that occurred 23 February 1828 caused severe damage to the architectural heritage, mainly the churches of Central Belgium. Several historical sources mentioned damage caused by the earthquake on such buildings: pieces of ceilings falling down, cracks on walls and vaults, collapse of vaults, the ruin of some parts of buildings, etc. This damage led to repairs that are sometimes mentioned in the historical sources too: cramp irons on facades, placing of ties on vaults. We have elaborated an interdisciplinary methodology to recognize the traces of damage produced by this earthquake on these buildings with the objective of improving our knowledge of other specific historical earthquakes, and also evaluating the vulnerability of the architectural heritage in the perspective of its preservation.

Seismic damage in churches attested by historical sources The first step in this analysis consisted of a survey of the 18 churches for which historical sources mentioned some damage (Figure 8.6). Visits to the churches that still exist allowed pathologies corresponding to those described in the reports on the earthquakes to be identified in some cases. Most of the damage has been repaired, but it can still be identified in the buildings and associated with repairs mentioned in historical documents. This is the case for the church of **Autre-Eglise** (Figure 8.6A): "The building took damage from the 1828 earthquake. The front was repaired in 1830; the cramp irons visible in the choir to the east up to a window also date from this epoch" (Tarlier and Wauters, *Géographie et histoire des communes belges*, 1872). At present, it can be observed that this church is built in stone except for a large part of the front, which is built in brick and was rebuilt after the earthquake.

In the village of **Marilles**, the tower of the church was also damaged by the 1828 earthquake, as attested by the many cramp irons presently holding the different walls of the tower (Figure 8.6B). Some of them at the top bear the date 1831, which is the date of the end of the repair work that followed the earthquake.

The church of **Petit-Hallet** (Figure 8.6C) was repaired in 1829, although it still presents recognizable damage, as substantial as cracks above and under the windows of the nave and the choir, many cramp irons in the face of the tower, and cracks in the middle of the arches of the vaults. This last observation can be directly linked to the descriptions in historical sources that mentioned that parts of the vault collapsed. Of course, not all the present pathologies are necessarily linked to the earthquake, and a specific investigation on the Petit-Hallet church was performed to show which of the existing pathologies and repair traces could be due to the 1828 earthquake (Philipront, 2007).

The historical sources mention major damage in the church of **Berloz** as well: cracks on walls, collapse of the vaults, and ruin of the tower. The two aisles had to be rebuilt in

Figure 8.6 The impact of the M = 5 23 February 1828 earthquake on churches. (A) Photograph of Autre-Eglise church showing the gable end formed on a sub-foundation of stone before the earthquake with its upper part rebuilt in bricks after the earthquake, as noted in historical sources (see the inscription 1830 on the small white stone above the door, which indicates the date of the end of the repair work). The cramp irons visible on the tower are also repairs following the earthquake. (B) Photograph of Marilles church tower, damaged by the earthquake. Cramp irons indicate the date of the end of the repair work (1831). (C) Photograph of an opening in the nave wall of Petit-Hallet church, repaired in 1829, with cracks in the masonry above the opening. All the openings in the nave show the same kind of disturbance. One can also observe the introduction of a small stone to prevent the keystone from falling. (D) Photograph of an opening in the nave wall of Jauche church with double cracking of the upper-level masonry, as well as cross-shaped cracks. One can also observe small spiral cramp irons to prevent rocking of the keystone.

1829. The date is written on a stone of one of the windows of the choir. The church is presently in a good state, which makes the search for earthquake traces difficult. We could, however, tell that the side aisles were built more recently by their materials and better maintenance than the choir and the tower on which several cramp irons are visible.

In **Racour** the historical sources mention that the church was also damaged by the 1828 earthquake, especially its tower, which was cracked and of which the spire had to be pulled down. The examination of the present tower shows a significant number of cramp irons on one of its walls, which are also bulging.

In some other cases, the historical reports mention damage in churches without giving any detail, but a visit to these places revealed pathologies and repairs that correspond to damage typically observed with this earthquake. This is the case for the church of **Melin**.

In the case of the church of **Grand-Hallet**, there is no mention of specific damage, but the church council reported that the building repair costs were around 1500 florins. This cost is relatively similar to that necessary to repair the badly damaged church of Berloz, suggesting that the damage in the church of Grand-Hallet caused by the earthquake was also significant. Indeed, several pathologies and repairs are presently visible: cracks under and above windows, a displaced keystone, damaged reflex angles, cracks in the vaults, placing of ties in the vaults, and a strange Y-shaped crack at the level of the choir. Some of these must be consequences of the 1828 earthquake.

Several churches are so well restored that damage or repairs mentioned by the historical sources are not visible today. This is the case for the church of **Opheylissem**. Historical sources mentioned damage to such an extent that the church had to be closed for some time after the earthquake.

It is interesting to notice that, among the churches known to have been affected by the 1828 earthquake, there are a few that present important pathologies whereas historical sources do not mention serious damage caused by this earthquake. Some of these pathologies look like pathologies of seismic origin. The church of **Perwez** is a good example of this. The historical sources only mention the fall of objects and pieces from the ceiling. However, the walls under and above all the windows of this church are cracked inside and outside of the building. The ceilings of the side aisles and the vaults are cracked too. The tower was restored at one point and some cramp irons were emplaced. The keystones of windows are lowered and one of them describes a rocking motion, which is characteristic of earthquake damage.

Some of the churches were demolished at approximately the time of the earthquake. We therefore suspect that some of these churches could have been so badly damaged by the 1828 earthquake that they needed to be demolished afterwards. This seems to be the case for the village of **Dion-le-Val**, where a new church was built in 1837: "... an earthquake in 1828 significantly cracked the bell tower. The building was in danger of collapsing and it was necessary to build a new one." (*Bulletin du Cercle historique de Chaumont-Gistoux*, 44, 2003, p. 16).

In other localities, new churches were built, but without any relationship to the earth-quake, even if damage and repairs related to this event were observed. This was the case in Dongelberg, Glabeek, and Lens-Saint-Remy.

Pathologies and repairs in churches not confirmed by historical sources The interpretation of all the damage observations discussed in the previous section provided a set of damage characteristics that earthquakes similar to the 1828 one might produce in churches. This damage dataset has been used to identify in the existing pathologies or visible repairs the consequences of the 1828 earthquake in other churches located in the earthquake epicentral area, and for which no report has been found describing the consequences of this seismic event. This damage has been classified in seven categories from CD1 to CD7. We list in Table 8.1 the churches where these different types of pathologies, damage or repairs have been observed.

The most frequently observed damage (CD1) is cracks above and under windows and especially at the level of the first window of the nave beyond the tower. One can observe such cracks in half of the inspected churches. Some of them even have all windows systematically cracked. These pathologies could arise from the earthquake but also from the problem of differential settling caused by the difference in height and weight of the tower in comparison with the nave, or from the problem of poor foundations or loose substrates. However, it is also possible that the settling of the ground may have been induced by the earthquake, which appears to be the case in the village of Jauche (Figure 8.6D).

Cracks on the tower or cracks at the intersection of two faces of the tower are also observed (CD2) in almost all churches. These types of damage are mentioned in historical reports for some of the church towers in the epicentral area of the 1828 earthquake. Their repair can be confirmed by the presence of cramp irons in the faces of some of these towers. They are used to hold a wall or two opposite faces together, preventing the splitting of walls and the detachment of the front from adjacent walls from bottom to top, which result from seismic action.

Face detachment (CD3) can also happen with the two perpendicular walls of the nave.

Cracks can also be observed in reflex angles (CD4) of the churches of Grand-Hallet, Perwez, Petit-Hallet, Hakendover, Incourt, and Jauche.

About 25% of the visited churches presented displacement of window keystones (CD5). In Jauche, one can see a similar principle of repair and consolidation with two small cramp irons, rounded spirals, at the two corners of the keystones (Figure 8.6d), which suggests repairs after the earthquake, as in Marilles and Racour.

Other pathologies and associated repair characteristics associated with earthquake shaking are cracked window sills (CD6) or cracks on the vaults (CD7).

From this investigation of churches in the central part of Belgium, we found evidence of pathologies and repair traces that could result from the 1828 earthquake because they are similar to those observed in the churches for which historical sources prove the damage. To confirm this hypothesis, we will first conduct a survey of churches outside the 1828

Table 8.1 *Types of damage in the churches of central Belgium related to the 23 February 1828 earthquake. The description of the different types of damage is given in the text.*

	Damage characteristics						
	CD1	CD2	CD3	CD4	CD5	CD6	CD7
Churches mentioned as damaged in historical sources							
Autre-Eglise		×	×				
Grand-Hallet	×			×	×		×
Marilles		×			×		
Melin	×	×				×	×
Perwez	×	×		×	×		×
Petit-Hallet	×	×		×	×	×	×
Racour		×			×		
Churches not mentioned in historical sources							
Bonlez	×						
Chaumont-Gistoux	×						
Hakendover	×	×	×	×			
Hoksem		×					
Hoegaarden							×
Incourt	×	×	×	×		×	
Jauche	×			×	×		
Laar						×	
Landen		×					
Lathuy	×						×
Neerheylissem		×					
Neerlanden	×	×			×		
Noduwez		×					
Oorbeeck	×	×					
Orp-le-Grand		×					
Outgarden					×		
Pietrain	×						
Saint-Remy-Geest		×					
Villers-le-peuplier	×	×					×
Walsbets		×	×				
Zetrud-Lumay		×					

earthquake epicentral area and analyze whether they present fewer pathologies and repairs than those in the area most affected by the seismic event. Second, we intend to use modelling to compare the observations on these buildings with the stress pattern expected from an earthquake with the same characteristics as the 1828 one.

8.5.7 The 18 September 1692 Verviers earthquake

The earthquake of 18 September 1692 is probably the largest known earthquake that has occurred in Western Europe north of the Alps (Figure 8.1). Recent investigations based on original reports of the effects produced by this earthquake suggest that its source was located in the region of Verviers in the northern part of the Belgian Ardenne and that its magnitude may have reached 6¼ (Alexandre and Kupper, 1997; Camelbeeck *et al.*, 2000; Alexandre *et al.*, 2005, 2008).

In the epicentral area, substantial to heavy damage and sometimes complete destruction are described for large buildings (castles and churches) and for houses. The ground shaking also appears to have been violently felt by the people (Alexandre *et al.*, 2008). For the most affected villages (Figure 8.5B) (e.g., Herve, Ensival, Soiron, Walhorn), the reports mention that some houses were ruined, with the consequence that inhabitants were injured or killed. This suggests that some buildings presented complete damage (EMS-98 grade 4 to 5) in the earthquake epicentral area. Considering that the most affected houses had vulnerability class A, "few" buildings presenting damage of grade 4 would correspond to intensity VII, whereas "few" buildings with damage of grade 5 would suggest intensity VIII. The latter is in agreement with Alexandre *et al.* (2008), who propose that the intensity reached VIII in these localities. Outside the epicentral area, one characteristic of the 1692 earthquake is the significant spatial extension of the damage to large buildings of the architectural heritage in Belgium, France, Germany, and southeastern England (Figure 8.4d).

The original reports of the witnesses and the description of the historical sources are available at http://seismologie.be/cup2014.html.

As usual for historical earthquakes, the accounts of the damage caused by the 1692 earthquake are poor from a quantitative point of view, which renders intensity evaluation uncertain. Soiron, one of the more damaged localities, is a village presenting a structure that has stayed unchanged since the eighteenth century. We took this as an opportunity to study the existing building stock in the centre of the village that dated from the seventeenth and the beginning of the eighteenth century to find complementary arguments to the historical data that may help to evaluate the earthquake intensity in this village (Dewattines, 2010).

Two historical sources describe the effects of the earthquake in Soiron. The priest wrote in his notes: "A horrible earthquake that brought down houses, chimneys..." (Servais Ronval, *Notes*). A second historical source written by the lord of Soiron depicted the very heavy structural damage caused to the castle. The structure of the church also was heavily damaged and, with the exception of its tower, a new church was built from 1723 to 1727 due to the consequences of this earthquake.

Figure 8.7A shows a map of the centre of Soiron with the houses that retain a complete or partial structure dating from the seventeenth century. There are about 36 such buildings. Most of these houses were partly or totally reconstructed after the 1692 earthquake (11 buildings, 30.5%) or show repairs or pathologies that could be attributed to this earthquake (14 buildings, 39%). During the eighteenth century, some other houses were rebuilt (11 buildings, 30.5%) (Figure 8.7B). These buildings are included in the basic structure

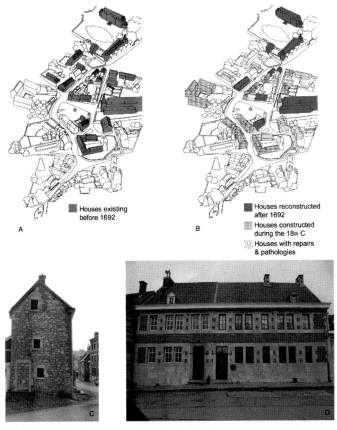

Figure 8.7 The impact of the M = 6¼ 18 September 1692 earthquake in the village of Soiron. (A) Isometric view of the historical centre of the village of Soiron, modified from Peters (1976), indicating the houses that existed during the seventeenth century before the 1692 earthquake occurred. (B) As for (A) but indicating the houses that our study evaluated as partially or totally rebuilt after the earthquake, the houses partially or totally built or transformed during the eighteenth century, and the houses showing disturbances or repair traces that could be attributed to the earthquake. (C) Photograph of the gable end of a house dating from the seventeenth century and showing disturbances that are characteristic of earthquake consequences: fall of a chimney taking away part of the upper wall and detachment of the facade from the gable end, with partial collapse of the facade, rebuilt with bricks after the earthquake. (D) Photograph of the main facade of two houses located on the church square, rebuilt after the earthquake in a style than we can attribute to the Late Mosan renaissance style (end of the seventeenth century).

of the village, suggesting that at least some of them may correspond to houses strongly damaged by the earthquake that were demolished and reconstructed or modified afterwards.

Regarding the repairs and pathologies still visible today, it seems that most of them concern fall of chimneys with parts of the pine walls (Figure 8.7C), stripping of the lateral walls and partial failure of roofs, damage that corresponds to extensive damage (grade 3 in the EMS-98 macroseismic scale).

For the houses presenting complete damage, equivalent to damage grade 4 or 5 on the EMS-98 macroseismic scale, it can be supposed that new houses were reconstructed. Some of them were reconstructed using the building style progressively adopted by richer families in this region since the second part of the seventeenth century (Figure 8.7D). The castle and the church were rebuilt in the years following the earthquake, but even if intensive works were required to stabilize the two edifices after the earthquake, they did not collapse (Nicolas Ignace de Woelmont, *Histoire de la maison de Woelmont*).

These observations and analysis suggest that most of the damage to the buildings in the centre of Soiron ranged from extensive to complete (grades 3 to 5 in the EMS-98 damage scale). Following our inventory, one-third of the damaged houses revealed extensive damage. These houses can be considered as the less vulnerable ones of the building stock of that epoch in Soiron, meaning very likely vulnerability class B at the best. Based on these hypotheses, the intensity can be evaluated as VIII on the EMS-98 macroseismic scale. This is in agreement with the intensity estimation from historical sources (Alexandre *et al.*, 2008) and not incompatible with the fact that the more vulnerable houses could have presented complete damage.

8.6 Discussion and conclusions

In this chapter, we have presented the methodological background associated with the investigation of the damage caused by past earthquakes in the region of Western Europe extending from the Lower Rhine Embayment to the southern North Sea. We have applied it to the study of seven earthquakes that represent typical seismic activity of this area and for which we have been able to retrieve relevant information on the characteristics, location, and extent of damage. Our study furnishes information on important seismological, geological, geographical, and architectural aspects influencing the consequences of earthquakes and helping to validate earthquake risk assessment.

8.6.1 Earthquake vulnerability of Western Europe

The precise inventory of the damage caused by the 1983 M = 4.6 Liège and 1992 M = 5.3 Roermond earthquakes has provided fundamental information on the present-day high seismic vulnerability of the major centres of population in this part of Europe. The vulnerability is especially high in regions where industrialization took place during the nineteenth and the early twentieth centuries. There are two main reasons for this high vulnerability. The first one is the very high population density, which explains the elevated estimated losses during the 1992 Roermond earthquake, despite the maximum macroseismic intensity barely reaching VI (EMS-98 scale). The second aspect to take into consideration is the extremely high vulnerability of the building stock composed of low-rise unreinforced brick-masonry constructions, which are sometimes very old. This has been dramatically demonstrated for the 1983 Liège earthquake. Of course, the more modern buildings constructed in steel or reinforced concrete were far less affected by the Liège and Roermond earthquakes and the application of the paraseismic Eurocode-8 building code should improve the resistance of new buildings in the future. Nevertheless, the risks will be present for a long time because poorly constructed masonry buildings will remain a significant part of the building stock in the old cities of Western Europe.

8.6.2 Damage and intensity

Using the Royal Observatory of Belgium macroseismic inquiry for earthquakes felt in Belgium since 1932, which includes a report on the percentage of damaged chimneys in each affected locality, we were also able to evaluate the quantity of damaged buildings presenting greater than or equal to moderate damage states caused in Belgium by the 11 June 1938 and 28 March 1967 earthquakes. We also added information provided by Musson *et al.* (1990) on the epicentral area of the British Colchester 1884 seismic event in our dataset. The complete dataset, presented in Figure 8.3, is more precise than the intensity for the destructiveness of the earthquakes and can be more directly related to the associated losses (Figure 8.2). This dataset is also, in essence, a basis on which to evaluate intensity in localities where damage was observed. It is very useful because, in the case of the most affected localities in the Netherlands and Germany during the 1992 Roermond earthquake, this allows us to evaluate the intensity as equal to VI on the EMS-98 macroseismic scale based on the damage statistics, which is less than the value of VII proposed by the official inquiry published after the earthquake (Haak *et al.*, 1994).

8.6.3 The complementarity of studying damage in classical houses and churches

The fundamental mode of vibration of traditional masonry houses ranges from a few to 10 Hz, which means that these structures are sensitive to the high-frequency range of seismic energy. On the other hand, the major buildings of the architectural heritage such as churches and castles are of larger dimensions and often of very complex structure. Their fundamental mode is thus at lower frequency (~1 Hz). As an example, the fundamental mode of the Boussu church tower, located near the city of Mons, has been evaluated at a frequency of 1.41 Hz (Defaut and Deneyer, 1999). Therefore, analyzing the damage caused to these large structures and comparing it to that of traditional buildings can shed a different light on the factors influencing the damage produced by an earthquake (Figures 8.3 and 8.4).

   We observe that the spatial extent and the magnitude of the damage caused to churches increase significantly with earthquake magnitude. Two of the studied earthquakes caused moderate to heavy damage at large distances from their epicentre, the 1992 Roermond and 1692 Verviers earthquakes (Figure 8.4). This is a clear indication of the lower frequency content of the seismic energy generated by these earthquakes, which is representative of magnitude 5.5–6.0 earthquakes. The 1938 Nukerke and 1828 Hannut earthquakes also caused damage to churches, but their extent is more limited to the region of intensity greater than or equal to VI, while the three smaller earthquakes with magnitudes around 4.5 caused damage to churches only in the limited regions of significant damage affecting masonry buildings.

Note that the damage caused to churches during the 1938 earthquake was enhanced by the thickness of the sedimentary cover, corresponding to a soil fundamental mode period between 1.0 and 1.5 seconds, as discussed by Nguyen *et al.* (2004). Our analysis differentiates the damage caused to masonry buildings from that observed in large buildings of the architectural heritage during the 1938 earthquake, permitting us to show that they differ slightly by location. The damage to traditional houses is more important in the area where the soil fundamental period is less than 0.3 second (or frequency higher than 3 Hz) (Figures 8.3 and 8.4). This difference can be explained by the progressive absorption of the high-frequency seismic energy when the thickness of the sedimentary cover increases, while the low-frequency cover is preserved and even amplified at the soil natural frequency. This phenomenon also explains the low intensity in the epicentral area of the 1992 Roermond earthquake by comparison to its magnitude (Meidow and Ahorner, 1994). Hence, by studying separately the two types of buildings, we provide information concerning two different frequency ranges in the response spectra, which is of high scientific and engineering interest. These results suggest that it is fundamental to take into account the regional soil properties in the studied part of Western Europe and that classical spectra that differentiate the soil effects by their properties in the first 30 metres under the surface are not sufficient to model earthquake strong ground motions in parts of the study area where the soil thickness reaches several dozen to hundreds of metres.

8.6.4 Risks from small and moderate earthquakes

Thanks to the outstanding study of the 1884 Colchester earthquake by Musson *et al.* (1990), we were able to compare the consequences of this seismic event, considered as the most damaging earthquake during the last 400 years in United Kingdom, with six of the numerous damaging past earthquakes in the region between the Lower Rhine Embayment and the southern North Sea. The amount of moderate to extensive damage in the localities most affected by this Colchester earthquake ranges between around 10% and 70% in an area with a radius of less than 10 km. This is very similar to the observed damage during the 1983 Liège earthquake in Belgium. These two earthquakes warn us of the consequences of such shallow small magnitude earthquakes that could occur everywhere in Western Europe. The danger is particularly of concern if the event occurs in the vicinity of a large historical city, such as Liège.

This is also the case for the three studied earthquakes with magnitude between 5.0 and 5.3. Of course, for these moderate earthquakes, the focal depth plays a role in the significance of the damage and associated geographical extent. Typical effects from a deep earthquake are well illustrated by the Belgian 1938 earthquake. With an estimated focal depth of around

20 km (Camelbeeck, 1993), the seismic action on the buildings was less strong than for a shallower earthquake with a similar magnitude because the seismic energy reaching the surface is more attenuated, but of course the affected area is of larger dimension. Hence, the moderate and extensive damage caused to houses and churches during the 1828 earthquake appear as more important than for the 1938 seismic event, but they are located in a more restricted region. This is particularly evident when looking at Figure 8.4, by the comparison of damage to churches.

8.6.5 Risks from large earthquakes

Three earthquakes of magnitude around 6.0 have occurred in the studied area since 1350. These earthquakes have a destructive potential that is important to evaluate. This is why we focused part of our work on evaluating the destruction caused by the $M = 6\frac{1}{4}$, 1692 Verviers earthquake. The study of historical texts (Alexandre *et al.*, 2008) suggested that some buildings suffered damage of grade 4 and 5 in villages of the epicentral area. Based on a precise inventory of the houses and of their pathologies and repair traces in the centre of the village of Soiron, one of the most affected localities, we were able to formulate a hypothesis on the damage and destruction caused by this earthquake. This analysis is limited by a lack of knowledge on the vulnerability of the buildings. If the buildings are considered as a mixture of vulnerability classes A and B of the EMS-98 macroseismic scale, it is coherent to consider that the less vulnerable (of class B) were on average less damaged, probably corresponding to extensive damage states (one-third of the total number of houses). Together with the complete damage of the other part of the building stock, this is compatible with intensity equal to VIII.

Considering the high seismic vulnerability of part of the present-day building stock of the study area and its high population density, the consequences of an earthquake of this magnitude would certainly be catastrophic in terms of victims and destruction.

8.6.6 The importance of investigations of the architectural heritage

A few years ago, we found it hard to imagine that it was possible in Western Europe to retrieve traces of past earthquakes in heritage buildings. Our study on the village of Soiron in the epicentral area of the 18 September 1692 earthquake and the churches affected by the 23 February 1828 earthquake drastically changed our point of view on this problem. Our results suggest that most of the buildings of Belgium and its surrounding regions should present pathologies or repairs associated with earthquake activity if they were constructed before the end of the seventeenth century. Up to now, we have developed a naturalist methodology based on field observations and measurements. Future methodological advancements will require the evaluation of the seismic vulnerability of these buildings and numerical modelling of earthquake effects to compare them with the observations.

The results presented in this chapter are a strong motivation to investigate other destructive past earthquakes, and also to study in more detail the different aspects that influence earthquake risks in the area between the Lower Rhine Embayment and the southern North Sea that have been demonstrated in the study.

Acknowledgements

We thank Michel Cara and Michel Granet from the Bureau Central Séismologique Français (BCSF) for giving us access to the original files of the BCSF macroseismic inquiry of the 11 June 1938 earthquake. Anne-Marie Barszez, Amélie Philippront, and Jean-Sébastien Dewattines are acknowledged for their participation in the studies of the churches in Hesbaye (A-MB and AP) and in the village of Soiron (J-SD).

References

Ahorner, L. (1975). Present-day stress field and seismotectonic block movements along major fault zones in Central Europe. *Tectonophysics*, 29, 233–249.

Ahorner, L. (1994). Fault-plane solutions and source parameters of the 1992 Roermond, the Netherlands, mainshock and its stronger aftershocks from regional seismic data. *Geologie en Mijnbouw*, 73, 199–214.

Alexandre, P., and Kupper, J. L. (1997). Le tremblement de terre de 1692 et le miracle de Notre-Dame des Récollets à Verviers. *Feuillets de la cathédrale de Liège*, 28–32, Liège.

Alexandre, P., Kusman, D., and Camelbeeck, T. (2005). Le tremblement de terre du 18 septembre 1692 dans le nord de l'Ardenne (Belgique): impact sur le patrimoine architectural. In *Actes des VIèmes Rencontres du Groupe APS "Archéosismicité et Vulnérabilité. Environnement, bâti ancien et société" (Perpignan, 4–5/10/2002)*, ed. A. Levret, Perpignan: Groupe APS, pp. 1–10.

Alexandre, P., Kusman, D., and Camelbeeck, T. (2007). La presse périodique, une source pour l'histoire des tremblements de terre dans l'espace belge. *Archives et Bibliothèques de Belgique*, LXXVIII, 1–4, 257–278.

Alexandre, P., Kusman, D., Petermans, T., and Camelbeeck, T. (2008). The 18 September 1692 earthquake in the Belgian Ardenne and its aftershocks. In *Historical Seismology: Interdisciplinary Studies of Past and Recent Earthquakes*, ed. J. Fréchet, M. Meghraoui, and M. Stucchi, Springer, pp. 209–230.

Berz, G. (1994). Assessment of the losses caused by the 1992 Roermond earthquake, the Netherlands (extended abstract). *Geologie en Mijnbouw*, 73, 281.

Bouwkamp, J. G. (1994). The 1992 Roermond earthquake, the Netherlands: earthquake engineering. *Geologie en Mijnbouw*, 73, 291–298.

Camelbeeck, T. (1993). Mécanisme au foyer des tremblements de terre et contraintes tectoniques: le cas de la zone intraplaque belge. Ph.D. thesis, Université Catholique de Louvain.

Camelbeeck, T., and van Eck, T. (1994). The Roer Valley Graben earthquake of 13 April 1992 and its seismotectonic setting. *Terra Nova*, 6, 291–300.

Camelbeeck, T., and Meghraoui, M. (1996). Large earthquakes in northern Europe more likely than once thought. *Eos, Transactions, American Geophysical Union*, 77, 405–409.

Camelbeeck, T., and Meghraoui, M. (1998). Geological and geophysical evidence for large paleoearthquakes with surface faulting in the Roer Graben (northwest Europe). *Geophysical Journal International*, 132, 347–362.

Camelbeeck, T., Alexandre, P., Vanneste, K., and Meghraoui, M. (2000). Long-term seis-micity in regions of present day low seismic activity: the example of Western Europe. *Soil Dynamics and Earthquake Engineering*, 20, 405–414.

Camelbeeck, T., Vanneste, K., Alexandre, P., *et al.* (2007). Relevance of active faulting and seismicity studies to assessments of long-term earthquake activity and maximum magnitude in intraplate northwest Europe, between the Lower Rhine Embayment and the North Sea. In *Continental Intraplate Earthquakes: Science, Hazard, and Policy Issues*, ed. S. Stein, and S. Mazzotti. Geological Society of America Special Paper 425, pp. 193–224.

Camelbeeck, T., Knuts, E., De Vos, F., and Alexandre, P. (2009). The historical earth-quake database of the Royal Observatory of Belgium. *Cahiers du Centre Européen de Géodynamique et de Séismologie*, 28, 31–36.

De Becker, M. (1985). L'enquête macroséismique du séisme de Liège du 8 novembre 1983. In *Le séisme de Liège et ses implications pratiques*, ed. L. Breesch, T. Camelbeeck, M. De Becker, *et al.*, *Annales des travaux publics de Belgique*, 4.

Defaut, C., and Deneyer, A. (1999). Contribution à l'étude de la stabilité de l'église St-Gery de Boussu. M.Sc. thesis, University of Mons.

Descamps, L. (2009). Relations entre l'activité sismique dans le Hainaut et l'activité minière. M.Sc. thesis, University of Mons.

Dewattines, J.-S. (2010). Patrimoine et sismicité: étude des traces du tremblement de terre du 18 septembre 1692 dans le village de Soiron. M.Sc. thesis, University of Mons.

Egen, P. N. C. (1828). Ueber das Erdbeben in den Rhein und Niederlanden vom 23. February 1828. *Poggendorffs Annalen der Physik und Chimie*, 13, 153–163.

Federal Emergency Management Agency (FEMA) (1999). *Earthquake Loss Estimation Methodology*. HAZUS99 Service Release 2 (SR2) technical manual, Washington, D.C.

Garcia Moreno, D., and Camelbeeck, T. (2013). Comparison of ground motions estimated from prediction equations and from observed damage due to the M = 4.6 1983 Liege earthquake (Belgium). *Journal of Natural Hazards and Earth System Sciences*, 13, 1983–1997.

Gisler, M., Kozák, J., and Vaněk, J. (2008). The 1855 Visp (Switzerland) earthquake: a milestone in macroseismic methodology? In *Historical Seismology: Interdisciplinary Studies of Past and Recent Earthquakes*, ed. J. Fréchet, M. Meghraoui, and M. Stucchi. Springer, pp. 231–247.

Grünthal, G., Musson, R., Schwarz, J., and Stucchi, M. (1998). European Macroseis-mic Scale 1998 (EMS-98). *Cahiers du Centre Européen de Géodynamique et de Séismologie*, 15.

Haak, H., van Bodegraven, J., Sleeman, R., *et al.* (1994). The macroseismic map of the 1992 Roermond earthquake, the Netherlands. *Geologie en Mijnbouw*, 73, 265–270.

Hill, M., and Rossetto, T. (2008). Comparison of building damage scales and damage descriptions for use in earthquake loss modelling in Europe. *Bulletin of Earthquake Engineering*, 6, 335–365, doi:10.1007/s10518–007–9057-y.

Hinzen, K. G., and Oemisch, M. (2001). Location and magnitude from seismic intensity data of recent and historic earthquakes in the Northern Rhine area, central Europe. *Bulletin of the Seismological Society of America*, 91, 40–56.

Hinzen, K., Schreiber, S., and Yerli, B. (2010). The Lycian sarcophagus of Arttumpara, Pinara, Turkey: testing seismogenic and anthropogenic damage scenario. *Bulletin of the Seismological Society of America*, 100 (6), 3148–3164, doi:10.1785/0120100079.

Jongmans, D. (1989). Les phénomènes d'amplification d'ondes sismiques dus à des structures géologiques. *Annales de la Société géologique de Belgique*, 112, 369–379.

Jongmans, D., and Campillo, M. (1984). Répartition des dommages pendant le tremblement de terre de Liège du 8 Novembre 1983: effet de source et effet de site. *Colloque national de génie parasismique sur les mouvements sismiques pour l'ingénieur, Saint-Rémy les Chevreuses*, 16 March 1988, pp. 2.23–2.33.

Jongmans, D., and Plumier, A. (2000). *Etude pilote du risque sismique sur une partie de la ville de Liège (4 km²)*. Internal report, Faculté des Sciences Appliquées, Université de Liège.

Meghraoui, M., Camelbeeck, T., Vanneste, K., Brondeel, M., and Jongmans, D. (2000). Active faulting and paleoseismology along the Bree fault zone, Lower Rhine graben (Belgium). *Journal of Geophysical Research*, 105, 13.809–13.841.

Meidow, H., and Ahorner, L. (1994). Macroseismic effects in Germany of the 1992 earthquake and their interpretation. *Geologie en Mijnbouw*, 73, 271–279.

Meldola, R., and White, W. (1885). Report of the East Anglian earthquake of April 22nd 1884, *Essex Field Club Special Memoir*, 1.

Melville, C., Levret, A., Alexandre, P., Lambert, J., and Vogt, J. (1996). Historical seismicity of the Strait of Dover–Pas de Calais. *Terra Nova*, 8, 626–647.

Monjoie, A. (1985). La géologie de la région liégeoise et le tremblement de terre du 8.11.1983. *Annales des travaux publics de Belgique*, 4, 337–345.

Musson, R., Neilson, G., and Burton, P. (1990). *Macroseismic Reports On Historical British Earthquakes XIV: 22 April 1884 Colchester*. Seismology report WL/90/33, British Geological Survey, Edinburgh.

Nguyen, F., Van Rompaey, G., Teerlynck, H., *et al.* (2004). Use of microtremor measurement for assessing site effects in Northern Belgium: interpretation of the observed intensity during the MS = 5.0 June 11 1938 earthquake. *Journal of Seismology*, 8, 41–56.

Pappin, J. W., Coburn, A. R., and Pratt, C. R. (1994). Observations of damage ratios to buildings in the epicentral region of the 1992 Roermond earthquake, the Netherlands. *Geologie en Mijnbouw*, 73, 299–302.

Peters, F. (1976). *Soiron, un village du Pays de Herve*. Ministère de la Culture française.

Philipront, A. (2007). Quels sont les effets sur le patrimoine architectural des séismes importants de nos régions ? Applications aux églises de Hesbaye: inventaire, méthodologies et perspectives. Service d'Architecture et de Mines. M.Sc. thesis, Faculté Polytechnique de Mons.

Phillips, D. W. (1985). Macroseismic effects of the Liège earthquake with particular reference to industrial installations. In *Seismic Activity in Western Europe*, ed. P. Melchior, Dordrecht: Reidel, pp. 369–384.

Plumier, A., Doneux, C., Camelbeeck, T., *et al.* (2005). *Seismic Risk Assessment and Mitigation for Belgium in the Frame of EUROCODE 8: Final Report*. Brussels: Federal Science Policy, (SP1481).

Plumier, A., Camelbeeck, T., and Barszez, A.-M. (2006). Le risque sismique et sa prévention en région Wallonne. In *Les risques majeurs en Région Wallonne. Prévenir en aménageant: Aménagement et Urbanisme, 7*, Direction générale de l'Aménagement du territoire du logement et du patrimoine (DGATLP). pp. 240–273.

Somville, O. (1939). *Le tremblement de terre belge du 11 juin 1938*. Observatoire Royal de Belgique.

Van Gils, J.-M. (1966). Les séismes de 15 et 21 décembre 1965 en Belgique. *Bulletin de la classe des Sciences, Académie royale de Belgique*, 5ᵉ série, LII, 101–107.

Vanneste, K., and Verbeeck, K. (2001). Paleoseismological analysis of the Rurrand fault near Jülich, Roer Valley graben, Germany: coseismic or aseismic faulting history? *Netherlands Journal of Geosciences/Geologie en Mijnbouw*, 80, 155–169.

Vanneste, K., Meghraoui, M., and Camelbeeck, T. (1999). Late Quaternary earthquake-related soft-sediment deformation along the Belgian portion of the Feldbiss fault, Lower Rhine Graben system. *Tectonophysics*, 309, 57–79.

Vanneste, K., Verbeeck, K., Camelbeeck, T., *et al.* (2001). Surface rupturing history of the Bree fault escarpment, Roer Valley Graben: new trench evidence for at least six successive events during the last 150 to 185 kyr. *Journal of Seismology*, 5, 329–359.

9

Intraplate earthquakes induced by reactivation of buried ancient rift system along the eastern margin of the Japan Sea

AITARO KATO

Abstract

Utilizing a dense seismic network deployed immediately after recent large intraplate earthquakes along the eastern margin of the Japan Sea, we discovered that stepwise and tilted block structures of the basement, which are geophysical evidence of a Miocene rift system formed during the spreading of the Japan Sea, are widely distributed beneath the thick sedimentary basin in the Niigata region. A similar structure associated with the ancient rift system is imaged in the source area of the Noto-Hanto earthquake. Most aftershocks following the recent intraplate earthquakes align roughly along the tilted block boundaries of the basement and are controlled by weaknesses associated with buried rift systems. Furthermore, we discuss the stress loading mechanisms for source faults of intraplate earthquakes. The structural coincidence between the stress axis distribution and the velocity structure observed in the Niigata region raises the possibility that ductile deformation of the sediments can partially accumulate elastic strain in the brittle parts of the fault zone. In addition, low-velocity anomalies are localized beneath the seismogenic zones, indicating that fluids may have locally weakened the crust. This study therefore suggests that reactivation of pre-existing faults within ancient rift systems by stress loading through ductile flow in the upper crust and creeping of the locally weakened lower crust is a plausible mechanical explanation for intraplate earthquakes.

9.1 Introduction

Many intraplate earthquakes have occurred in the Japanese islands as a result of internal deformations of overlying plates. The Japanese islands are situated in a tectonically active zone, where two oceanic plates are subducting (Figure 9.1a). Beneath northeast Japan, the Pacific plate is subducting from the east through the Japan Trench at a convergence rate

Intraplate Earthquakes, ed. Pradeep Talwani. Published by Cambridge University Press. © Cambridge University Press 2014.

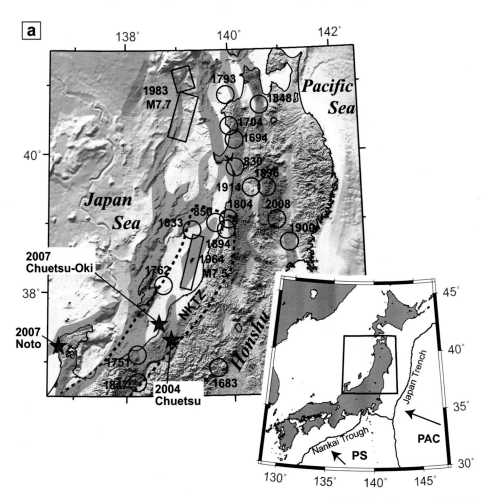

Figure 9.1 Seismotectonic setting in northern Honshu, Japan (modified after Kato *et al.*, 2009). (a) The Niigata–Kobe Tectonic Zone (NKTZ) is outlined by the broken line and contractional zones based on geological studies are drawn as gray-shade zones (Okamura *et al.* (2007)). Fault locations of recent major earthquakes and epicenters of historical large earthquakes with a magnitude greater than 7.0 are plotted as rectangles and circles, respectively (Usami, 2003). Three stars denote the epicenters of the 2004 Niigata-ken Chuetsu earthquake, the 2007 Chuetsu-Oki earthquake and the 2007 Noto-Hanto earthquake. The inset shows the location of the studied area (black framed rectangle). PAC, Pacific plate; PSP, Philippine Sea plate.

of ∼8.5 cm/yr, whereas beneath southwest Japan, the Philippine Sea plate is subducting from the southeast through the Nankai Trough at a convergence rate of ∼2–5 cm/yr (e.g., Loveless and Meade, 2010). Due to subduction of the two oceanic lithospheres, intraplate earthquakes in the Japanese islands are more frequent than in other regions in the world. Damage caused by the intraplate earthquakes is usually devastating due to their shallow depths (less than 15 km).

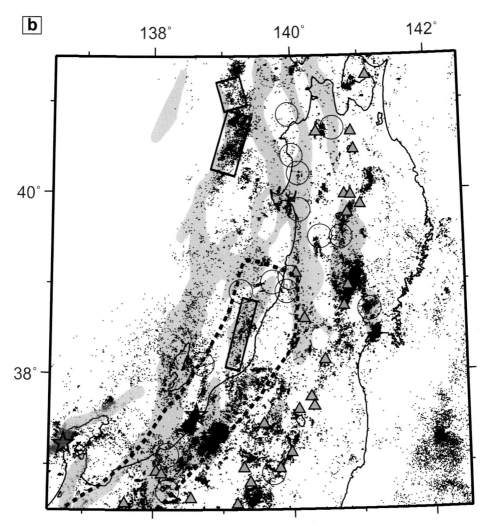

Figure 9.1 (b) Map of epicenters with magnitude greater than 1.0 reported by the JMA for the period from 2000 to 2010. Triangles denote locations of active volcanoes. Other symbols are the same as in (a).

Particularly in the eastern margin of the Japan Sea, historical and recent destructive intraplate earthquakes (e.g., the 1964 Niigata earthquake with Japan Meteorological Agency (JMA) magnitude (M_{JMA}) 7.5; the 1983 Japan Sea earthquake with M_{JMA} 7.7) have been concentrated along a zone of high east–west contractional strain rates detected by geodetic measurements (larger than 10^{-7} per year) and geological studies (Sagiya *et al.*, 2000; Okamura *et al.*, 2007) (Figure 9.1a). In addition, shallow microseismicity revealed by a state of the art nationwide high-sensitivity seismic network (e.g., Obara *et al.*, 2005) has been intensive along this contractional zone (shaded area in Figure 9.1b). Within the contractional zone, three destructive intraplate earthquakes showing reverse faulting with a strike of approximately N35 °E most recently occurred in the Niigata and Noto-Hanto regions. In addition, a shallow intraplate earthquake with M_{JMA} 6.7 was induced by the 2011 M_w 9.0 Tohoku-Oki earthquake at the south portion of this contractional zone.

These earthquakes are commonly located within Miocene–Pleistocene sedimentary basins. These sedimentary basins were formed as back-arc basins in a rift structure that developed during the opening stage of the Japan Sea (25–15 Ma) (Sato *et al.*, 1994). Several normal faults have subsequently been inverted as reverse faults owing to a change in the tectonic stress regime from extension to compression since 3.5 Ma. This stress inversion led to well-developed thrusts and related surface folding. The overlap between intraplate earthquakes and ancient rift systems beneath thick sediments suggests that ancient rift systems are important for nucleating present-day intraplate earthquakes in the compressional inverted basins. However, the details of the ancient buried rift structures and their potential effects on the seismogenesis of large intraplate earthquakes have not been fully understood. In addition, the driving force behind the generation of intraplate earthquakes (loading mechanism) remains a subject of controversy (e.g., Iio *et al.*, 2002, 2004). It has been argued that local heterogeneities in crustal structure play an important role in controlling the spatiotemporal evolution of seismicity and associated faulting behavior (e.g., Michael and Eberhart-Phillips, 1991; Chiarabba *et al.*, 2009; Kato *et al.*, 2009, 2010a; Zhao *et al.*, 2011). In order to illuminate these issues, it is thus critical to fully describe the crustal heterogeneity originating in the ancient rift system. Seismic tomography combined with a dense seismic network are powerful tools for imaging high-resolution crustal structures as well as precise hypocenters, and offer new insights into potential seismogenic structures. The dense and well-covered ray-paths from the many aftershocks triggered by each large earthquake provide us precious opportunities to (1) investigate the regional velocity structure and stress field in detail with a spatial resolution of ∼3–5 km, and (2) demonstrate that crustal heterogeneities associated with the Miocene rift structures significantly contribute to present-day seismogenesis along the eastern margin of the Japan Sea.

9.2 Data and method

Permanent seismic networks have been operated on the Japanese islands by the National Research Institute for Earth Science and Disaster Prevention (NIED), the Japan Meteorological Agency (JMA), and various universities. Nevertheless, the average spacing of these stations (from 20 to 30 km) is insufficient for resolving the detailed velocity structures in the source regions of inland earthquakes with magnitude less than 7.0. In order to obtain high-resolution three-dimensional seismic velocity tomographic images in source regions, as well as precise hypocenters, we have deployed a series of temporary dense seismic stations interpolating between the existing stations of the permanent seismic network. Recent technical advancement of portable seismometers and data acquisition systems has enabled us to carry out quick deployment of a dense seismic network for durations longer than 3 months. We usually deploy the temporary seismic stations with spatial intervals of several kilometers.

Each temporary seismic station continuously records three-component waveform signals at the sampling rate of 200 or 100 samples/second. Each station is equipped with a GPS receiver, which maintains the accuracy of the internal clock. After retrieving the waveforms, we then merge the huge earthquake datasets recorded by the temporary seismic network together with the corresponding data from permanent stations in each target region. We then manually pick P- and S-wave arrival times from earthquake waveforms detected based on the JMA catalog. We determine high-resolution three-dimensional velocity structures as well as precise hypocenters, applying the double-difference tomography method (Zhang and Thurber, 2003) to both the arrival time data and the differential arrival times obtained by the manually picked and waveform correlation method (e.g., Kato *et al.*, 2006a). In addition, we analyze high-resolution stress fields using focal mechanisms determined by the polarity data of P-wave first motion.

9.3 The 2004 Niigata-ken Chuetsu and 2007 Chuetsu-Oki earthquakes

Two neighboring destructive intraplate earthquakes (both with M_{JMA} 6.8) showing reverse faulting with a strike of approximately N35°E recently occurred in the Niigata region: the first, Niigata-ken Chuetsu earthquake on October 23, 2004, and the second, Niigata-ken Chuetsu-Oki earthquake on July 16, 2007 (Figure 9.1). The focal areas of the 2004 Chuetsu and 2007 Chuetsu-Oki earthquakes were located within a thick (locally >6 km deep) deformed Miocene–Pleistocene sedimentary basin (the Niigata Basin), which is characterized by NNE–SSW-trending faults and anticlinal fold hinges that form topographic hills (Figure 9.2a). This sedimentary basin was formed as a back-arc basin in a rift structure that developed during the opening stage of the Japan Sea (25–15 Ma). This basin is bounded to the east by the Shibata-Koide Tectonic Line (SKTL), where basement rocks dating back to more than 30 Ma are widely exposed. Geological studies (e.g., Sato, 1994) have inferred that parts of the normal faults within the rift system have subsequently been reactivated as a reverse fault system since the extensional tectonic stress regime changed to a compressional one in the late Pliocene (2–3 Ma), through a process of compressional inversion (Williams *et al.*, 1989). Although these shallow large earthquakes generated many fissures and landslides on the surface, only minor surface faulting was observed (Maruyama *et al.*, 2005).

We conducted a series of temporary seismic observations through a dense deployment of 145 portable stations after the 2004 earthquake (from October to November, 2004) (Kato *et al.*, 2006a, 2007) and 108 portable stations including a linear-seismic array on land and 20 ocean-bottom seismometers after the 2007 earthquake (from July to August, 2007) (Kato *et al.*, 2008a, 2009; Shinohara *et al.*, 2008) (Figure 9.2a).

9.3.1 Aftershock distribution and dynamic rupture process

Depth sections of P-wave velocity (V_p) models along W35°N–E35°S lines are shown in Figure 9.2b. Relocated aftershocks (gray circles) distributed within ±2.5 km of each line

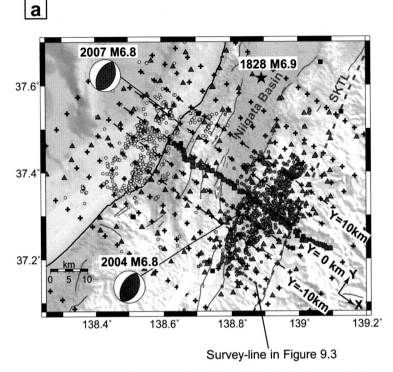

Survey-line in Figure 9.3

Figure 9.2 Dense seismic observations and regional V_p model in the Niigata region (modified after Kato *et al.*, 2009). (a) Map of relocated aftershocks of the 2004 Chuetsu and the 2007 Chuetsu-Oki earthquakes (gray circles observed in 2004, and yellow circles in 2007). Moment tensors for the two earthquakes are from the NIED. Blue squares denote the linear seismic array, and red and blue triangles represent temporary stations deployed after the earthquakes in 2004 and 2007, respectively. Black squares denote permanent stations. The grid used in the tomography analysis is plotted with crosses, and horizontal dashed lines correspond to the cross-sections in Figure 9.2b. Major active faults are drawn as red lines; a blue broken line indicates the SKTL. The gray broken line denotes a seismic survey line shown in Figure 9.3.

are superimposed. The striking aspect of the relocated aftershocks is a multi-segmented and complex distribution. The aftershock distributions associated with the 2004 mainshock reveal that the mainshock and the largest aftershock (A1) occurred on two 60° northwest-ward (NW) dipping planes, located approximately 5 km apart. Conversely, the October 27 event (A2) occurred on a southeast-ward (SE) dipping plane with a dip angle of 25° that was conjugate to the mainshock fault plane (Kato *et al.*, 2006a, 2009).

Similar conjugate sets of fault planes are also imaged in the aftershock sequences following the 2007 Chuetsu-Oki earthquake (e.g., Kato *et al.*, 2008a) (Figure 9.2b). From the center to the southern area, most of the aftershocks aligned along a SE-dipping plane with a low dip angle. Conversely, aftershocks in the north aligned along both NW- and SE-dipping planes. These planes were virtually normal to each other or conjugate. The

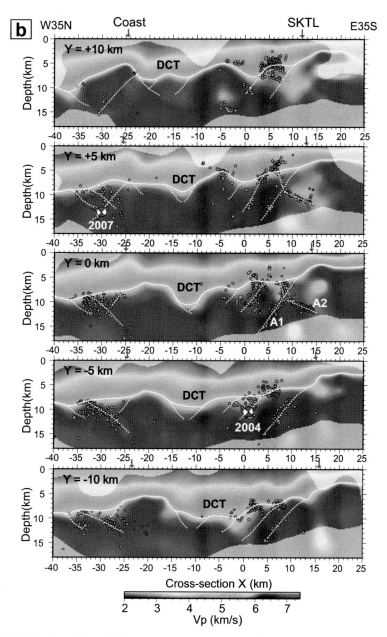

Figure 9.2 (b) Depth sections of the V_p model along W35N-E35S lines in Figure 9.2a. Relocated aftershocks (gray circles) distributed within ±2.5 km of each line are superimposed. Masked areas correspond to low model resolution. White curves denote iso-velocity contours of $V_p = 5.7$ km/s, and white broken lines show faults suggested from aftershock streaks, top surface geometries of the basement, or velocity changes within the basement. DCT, deep central trough. Blue and red arrows at the top of each section correspond to surface locations of the SKTL and the coastline. Moment tensor solutions for the 2004 and 2007 Niigata earthquakes (NIED) are shown using a lower hemisphere projection rotated into the plane of the section. For color version, see Plates section.

mainshock hypocenter of the 2007 Chuetsu-Oki earthquake appeared to be close to the bottom of a steeply NW-dipping plane. It has been proposed that the mainshock rupture initiated near the bottom of the NW-dipping fault plane and propagated to the southwest and then transferred to the SE-dipping plane across the central crosscutting area of the two faults (Kato *et al.*, 2008a; Takenaka *et al.*, 2009), inferring complex dynamic rupture.

Indeed, Aochi and Kato (2010) modeled the dynamic rupture propagation of the 2007 Chuetsu-Oki earthquake numerically along the inferred segmented fault system using a boundary integral equation method (BIEM). For angles between these fault planes from 80° to 95°, the possibility of rupture transfer from the NW-dipping fault plane to the SE-dipping is numerically demonstrated for any frictional level along fault planes, independently of the crosscutting distance in the center, suggesting two rupture modes. Simultaneous rupture transfer along the overlapping part is possible only under a high-stress load; however, this rupture mode yields an excessively large amount of coseismic slip. Otherwise, where regional stress is relatively low but pore pressure is high enough to cause the rupture (described as the low frictional coefficient case), the rupture transfer to the other fault segment does not occur until rupture terminates on the first fault segment regardless of the crosscutting distance between the two faults. Similarly, the 2009 Suruga Bay intraplate earthquake (M 6.4) in Japan, which occurred within the Philippine Sea plate and had a reverse fault component, showed a similar rupture transfer from one fault segment to the other conjugate segment (Aoi *et al.*, 2010). These recent studies of dynamic rupture process associated with large intraplate earthquakes suggest that rupture transfer or simultaneous ruptures are more common features than previously thought.

9.3.2 Ancient rift system buried beneath thick sedimentary basin

Depth sections of P-wave velocity (V_p) structures show strong lateral heterogeneity (Figure 9.2b), especially orthogonally to the fault strike (Kato *et al.*, 2006a, 2009). Seismic velocities in the hanging wall above the 2004 mainshock fault are lower than those in the footwall at depths shallower than 8 km. We consider that the low-velocity body in the hanging wall corresponds to soft sediments that have accumulated in half-grabens formed by crustal stretching during the opening of the Japan Sea. Conversely, the high-velocity body in the footwall is thought to correspond to the old basement rocks (30 Ma). Here, we define the basement as high-velocity bodies in which V_p is greater than 5.7 km/s (bounded below by the white curves in Figure 9.2b). Most aftershocks seem to be bounded by the basement. The centroid depth of the relocated aftershocks associated with the 2007 mainshock is slightly deeper than that of the 2004 mainshock. This lateral variation of aftershock depths well correlates with that of thickness of sedimentary layers, which on average increases with distance towards the west.

It is important to note that the top surface of the basement on the eastern side (-5 km < X < $+15$ km) shows clear stepwise structures that gradually deepen in the westward direction. In addition, westward-tilted block structures are interpreted by aftershock streaks, the top surface geometries of the basement, or velocity changes within the basement (broken white lines in Figure 9.2b). The boundaries between each stepwise and tilted block structure are mainly characterized by west-dipping faults with high dip angles. Near the center of the cross-sections (X ~ −10 km), the top surface of the basement reaches its deepest depth (~10 km thickness of sediments) and then rises sharply to the coastline (X = −25 km). This concave basement structure forms a deep central trough (DCT) filled with thick sediments between the source areas of the two Niigata earthquakes. In contrast, on the western side of the DCT (X < −20 km), the stepwise block structures are primarily tilted eastward with low dip angles (eastward-dipping low-velocity zones located just above the 2007 aftershock alignments along Y = −5 and −10 km), although westward-tilted structures also developed, especially in the north sections (Y > +5 km) showing complete inversion; in short, the west block appears to hang over the footwall. The stepwise and tilted block structures are observed along all of the cross-sections (Figure 9.2b), with block widths ranging from 5 to 10 km.

Furthermore, a very fine scale (~1 km grid) seismic tomography using a linear seismic array in the southwestern edge of the source region in the 2004 Chuetsu earthquake confirmed that the aftershocks appear to be aligned along pre-existing boundaries of the step-like array of tilted block structures (Kato *et al.*, 2010b; Figure 9.3). With deepening of the basement-cover contact to the west, the sedimentary succession in the hanging wall deepens to about 9 km. Indeed, the westward-tilt of the sedimentary strata was delineated by a seismic reflection survey conducted by the Japan National Oil Corporation (JNOC, 1988) in the vicinity of the studied area (Sato and Kato, 2005). The velocity model within the sedimentary basin correlates well with the seismic reflection profile. The surface elevation increases towards the southern part of the SKTL (Muikamachi fault), which illustrates that the cover sequence has been deformed by upward movements of the step-like tilted blocks in the basement. There seem to be three plateaus within the topography to the west of the Muikamachi fault trace (gray horizontal lines in the top graph in Figure 9.3). We therefore hypothesize that these plateaus might be created by both upward movements of the three tilted blocks and compressive deformations of the sedimentary strata in the hanging wall through compressional inversion.

The stepwise and tilted block structures as described above are clear evidence of the buried Miocene rift system formed during the spreading of the Japan Sea. It is worthwhile noting that the aftershock distributions associated with the two Niigata earthquakes correlated well with these complex and heterogeneous structures. Most aftershocks are aligned along the NW-dipping faults with high dip angles (~60°) or along the SE-dipping faults with low-dip angles (~35°) (Kato *et al.*, 2006a; Shinohara *et al.*, 2008). These fault planes are orthogonal to each other, and correspond roughly to the boundaries of the tilted blocks. These results suggest that the seismogenesis of the two Niigata earthquakes was due primarily to compressional inversion tectonics involving pre-existing structures related to the Miocene rift system. Since pre-existing faults within the ancient rift system are weak due to thermal softening (Hansen and Nielsen, 2003) and over-pressurized fluids beneath the seismogenic zone (Sibson, 2007), these faults are mechanically easy to reactivate as

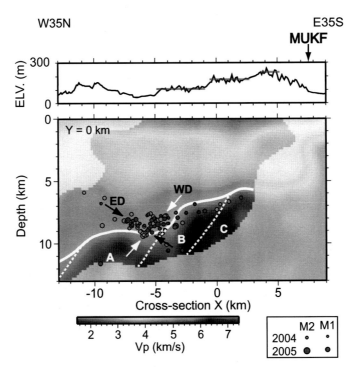

Figure 9.3 Depth section through the V_p model obtained at the southwest edge of the source area of the 2004 Chuetsu earthquake, with superimposed relocated aftershocks distributed within ±4 km from the section (light gray circles observed in 2004, and dark gray circles in 2005) (Kato *et al.*, 2010b). Masked areas correspond to low model resolution. White curves denote iso-velocity contours of $V_p = 5.7$ km/s, and white broken lines show faults suggested from aftershock alignments and the top surface geometries of the basement. Three tilted blocks are labeled as A, B, and C. Eastward- and westward-dipping alignments of aftershocks near the top surface of the basement are indicated by black (ED) and white (WD) arrows. The top figure shows a topography variation along the profile. Gray lines correspond to the three plateaus in the topography. MUKF is the surface trace of the Mui-kamachi fault.

reverse faults under compression. Indeed, the NW-dipping fault planes with high dip angles are far from an optimal orientation against the regional stress field (e.g., Kato *et al.*, 2006a; Townend and Zoback, 2006; Terakawa and Matsu'ura, 2010), which means that those NW-dipping faults are mechanically weak.

The planar distribution of V_p at a depth of 9 km (Figure 9.4a) shows that the DCT between the two source areas extends continuously for over 20 km from south to north. The azimuth of the DCT filled with thick sediments lies almost parallel to the surface traces of major active faults and fold axes. In contrast, remarkable high-velocity bodies ($V_p = $ ~6.8–7.1 km/s) are imaged beneath the DCT (-15 km $< X < -5$ km) and the source region of the 2007 earthquake (deeper than 15 km, in Figures 9.2b and 9.4b). As a result, the velocity gradient of V_p with depth is very steep at the DCT. We consider that the DCT corresponds to the Miocene rift axis, because the dominant dip direction of each tilted

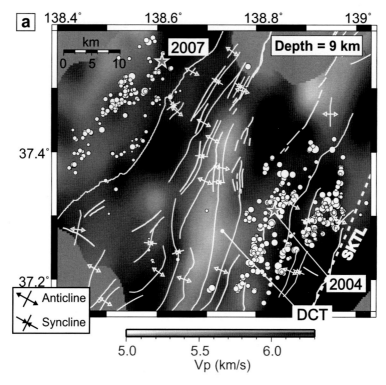

Figure 9.4 Map views of V_p structures at representative depths (Kato *et al.*, 2009). (a) 9 km. Relocated aftershocks distributed at depths from 7.5 km to 10.5 km (white circles) are superimposed. Gray stars denote epicenters of the 2004 and 2007 mainshocks. Major active faults are drawn as white lines. Anticlines and synclines are represented by white lines with arrows. A broken line indicates the SKTL.

block changes from westward to eastward across it. Based on laboratory measurements of exposed rocks (Christensen, 1996), we interpret the high-velocity bodies beneath the DCT to be a diabase intrusion into the Miocene rift axis. We propose that the diabase body intruded into the upper crust, acting as a magma source within the extensional rift system during the opening of the Japan Sea (Bjorklund *et al.*, 2002; White *et al.*, 2008; Figure 9.5). Indeed, andesitic and basaltic rocks are partially exposed on the surface in the study area (GSJ, 2002).

The spatial correlation of the DCT with the surface traces of major active faults and fold axes (Figure 9.4a) illustrates that the major active faults have slipped in response to upward movements of the stepwise and tilted blocks in the basement, which moved as a result of the horizontal compressional stress (Figure 9.5). These upward movements have led to the active growth of anticlinal folds. The elevations of topographic hills on the eastern side of the DCT are higher than those on the western side, due to a difference in dip angles of faults in the basement crossing the Miocene rift axis. Additionally, it is worth noting

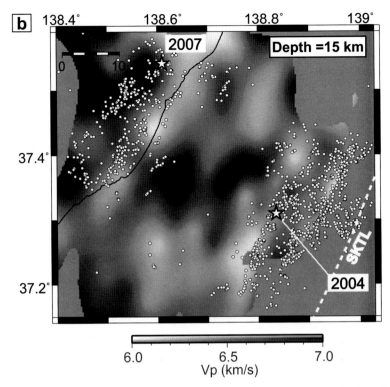

Figure 9.4 (b) 15 km. The center of the gray scale corresponds to an average velocity calculated at this depth (6.5 km/s). All of the relocated aftershocks (white circles) are superimposed.

that the eastward- or westward-dipping faults on the flanks of the DCT are situated along mechanically weak structural boundaries between the thick sediment and the basement (Figure 9.5).

The geometries of the block structures associated with the ancient rift system change along the fault strike in the vicinity of the cross-section at around $Y = 0$ km in Figure 9.2b. In the source region of the 2004 Chuetsu earthquake, the thickness of sedimentary layers in the hanging wall increases markedly on the southwest side of the mainshock hypocenter (e.g., Kato *et al.*, 2006a). For the case of the 2007 Chuetsu-Oki earthquake, a southeast-tilted basement structure dominates in the south, whereas a northwest-tilted basement structure gradually develops towards the north (e.g., Kato *et al.*, 2008a). These spatial variations indicate the presence of a segment boundary of the block structure near the cross-section of $Y = 0$ km in Figure 9.2b.

9.3.3 Stress loading processes to the reactivation of the ancient rift system

Even if several ancient weak fault zones in the upper crust exist within the buried rift system, a driving force to reactivate them is needed. We here discuss the mechanisms of stress loading into source faults of intraplate earthquakes in the Niigata region.

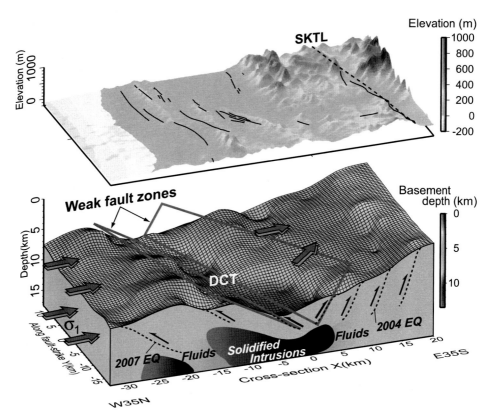

Figure 9.5 Three-dimensional perspective views of active tectonics in the study area (modified after Kato *et al.*, 2009). (Top) Topography, over which active faults are drawn as solid lines. The broken line indicates the SKTL. (Bottom) Perspective image of the depth to $V_p = 5.7$ km/s with interpretations. Gray rectangles denote potential weak fault zones along the flanks of the DCT. Dominant reactivated normal faults are drawn as broken lines with arrows showing slip directions during compressional inversion. Large arrows indicate the inferred direction of the maximum stress σ_1. Note that the σ_1 axis rotates from W30°N–E30°S in the central area of the source region of the 2004 Chuetsu earthquake to the E–W horizontal direction in the southwestern area. For color version, see Plates section.

The spatial variation of stress fields in the source area of the 2004 Chuetsu earthquake provides us with a valuable opportunity to understand the deformation of sedimentary layers in the hanging wall (Kato *et al.*, 2006a). Results of stress tensor inversions using dense aftershock data (focal mechanisms) show that the maximum principal stress rotates from W30°N–30°S in the central area of the source region to the E–W horizontal directions in the southwestern area (Figure 9.5). In short, the compressional stress axis in the southwest area is oblique to the fault strike of N35°E, even though the compressional stress axis in the central part is almost perpendicular to the fault strike. Near the hypocenter of the mainshock rupture, the thickness of sediment layers in the hanging wall increases towards E35°S, as well as increasing in the S35°W direction. As a result, the iso-velocity contour of

$V_p = 5.7$ km/s undergoes a marked change around the mainshock hypocenter (Figure 9.5). The structural coincidence between the stress axis distribution and the velocity structure raises the possibility that rotation of the compressional stress axis in the southwest area might be caused by lateral variation of sediment layers in the hanging wall. Sediments in the hanging wall with low elastic modulus can potentially allow ductile flow along the fault zone when compressional shear stress is applied. The ductile deformation of the sediments can partially accumulate elastic strain in the brittle parts of the fault zone, and may play a role in stress loading. Although it is difficult to directly demonstrate the stress loading process by the ductile deformation of the sediments into source faults, there is geodetic evidence showing ongoing ductile deformation of the sediments. For example, following the 2007 Chuetsu-Oki earthquake, episodic growth of fault-related folds in the shallow sedimentary layer was clearly detected by SAR interferometry, and did not accompany any seismicity (Nishimura *et al.*, 2008). The long-term leveling measurement also supports the episodic growth of folds therein.

In addition to the deformation of sediments, it has been proposed that ductile creeping of the weak lower crust could cause stress loading into seismogenic faults (Iio *et al.*, 2002). Note that slow anomalies in the lower crust ($V_p = {\sim}6.1$–6.3 km/s) are localized around deep extensions of mainshock faults for the 2004 and 2007 Niigata earthquakes (Figures 9.4b and 9.5). Furthermore, a highly conductive body was found in the lower crust (deeper than 15 km) beneath the source region of the 2004 earthquake by wideband magnetotelluric survey (Uyeshima *et al.*, 2005). Thus, these slow anomalies in the lower crust probably represent crustal fluids that might be exsolved from the solidified intrusions beneath the rift axis. According to a regional (larger-scale) tomography study conducted in the Niigata region (Nakajima and Hasegawa, 2008), these slow anomalies appear to extend to the deeper part of the crust and connect to a distinct low-velocity zone beneath the Moho (${\sim}30$–50 km depth) under source areas of the two Niigata earthquakes. Similarly, from the backbone mountain range to the fore-arc side of northeast Japan, several low-velocity zones are recognized just below the source areas of large intraplate earthquakes (e.g., Okada *et al.*, 2010). In the cases of the 2003 North Miyagi earthquake and the 2008 Iwate-Miyagi Nairiku earthquake, the upper crustal structures implying an ancient rift system were imaged by temporary dense seismic observations conducted after those mainshocks, as well as the low-velocity zones below the source areas (Okada *et al.*, 2007, 2012). Several seismic reflection profiles also confirmed that reverse-fault reactivation of pre-existing Miocene normal faults occurred across the entire seismogenic zone in the fore-arc side of northeast Japan (Kato *et al.*, 2006b).

Since the strength of ductile creep, which is a dominant deformation process in the lower crust, is weakened by crustal fluids (e.g., Carter and Tsenn, 1987), the slow anomalies in the lower crust correspond to locally weak zones (Iio *et al.*, 2002). Concerning the deformation process in the weak zone, Iio *et al.* (2002) assumed that the weak zone consists of numerous aseismic faults since the deformation is thought to be localized in ductile shear zones. Thus, local ductile creep of the weak zone within the lower crust causes stress loading, or stress transfer to the upper ancient rift system (Kenner and Segall, 2000), leading to the reactivation of this system in the form of intraplate earthquakes. Since the width and depth of the weak zones provide constraints on the process of stress loading to intraplate earthquakes, detailed knowledge of the heterogeneity within the lower crust is quite important.

9.3.4 Numerical modeling of development of fault zones

Using finite element modeling incorporating visco-elasto-plasticity, Shibazaki and Kato (2012) numerically simulated the development of fault zones in a geological setting with thick sedimentary layers and weak zones in the basement inferred from the tomographic study in the Niigata region (Figure 9.2b; Kato *et al.*, 2009). It is assumed that values of elastic constants and frictional coefficients in the sedimentary layer are smaller than those in the basement. Furthermore, they assumed that the frictional coefficient in the basement is low in areas of low P-wave velocity, examining how the present rheological structures inferred from P-wave velocity structure affect the development of fault zones. Figure 9.6 shows the equivalent total strains defined for the deviatoric components of the total strains after 8.0×10^4 years (i.e., the amount of contraction is 1.2 km). The equivalent total strains include viscous, plastic, and elastic strains. Fault zones are created just above concavities along the boundary between the lower sedimentary layer and the basement. For example, along all sections, fault zones are well developed at locations where the effective frictional coefficient is low in the basement and above the DCT, confirming the development of a NW-dipping reverse fault zone, which was activated by the mainshock of the 2004 Chuetsu earthquake. The model also shows the development of NW-dipping faults that were parallel to the fault of the mainshock. These faults may correspond to the faults of large aftershocks follow the 2004 Chuetsu earthquake. In addition, the model shows the development of SE-dipping reverse fault zones that were activated by the mainshock of the 2007 Chuetsu-Oki earthquake. Thus, the numerical results indicate that buried rift and weak zones in the basement caused the development of the complex fault configuration observed in the Niigata region.

9.4 The 2007 Noto-Hanto earthquakes

A shallow M_{JMA} 6.9 inland earthquake occurred on the west coast of the Noto Peninsula in Japan on March 25, 2007. The focal mechanism estimated by moment tensor inversion (NIED) revealed a dominant reverse slip component with a strike of approximately N55 °E (Figure 9.7a). The Noto Peninsula is situated close to the Yamato Basin in the Japan Sea, which is a large rift basin formed during the extension stage of the Japan Sea (e.g., Shimazu *et al.*, 1990). Based on geological studies of early Miocene syn-rifting succession in the northern regions of the Noto Peninsula (e.g., Kano *et al.*, 2002), it is proposed that small-scale rift structures were formed along the northern coast of the Noto Peninsula during the extension stage of the Japan Sea. Following the 2007 Noto-Hanto earthquake, we deployed a total of 89 temporary seismic stations on land in the source region. The seismic network was densely deployed with station spacing averaged less than 2 km (Figure 9.7a).

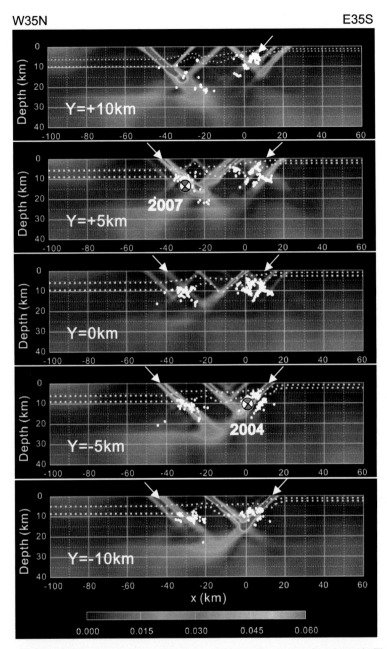

Figure 9.6 Numerical modeling of equivalent total strain in each cross-section shown in Figure 9.2b (Shibazaki and Kato, 2012). Viscous, plastic, and elastic strains are included in the total strain. Strain weakening in plastic deformation is incorporated. White arrows indicate the fault zones corresponding to the 2007 Chuetsu-Oki (left) and 2004 mid-Niigata Prefecture (right) earthquakes. The upper and lower dotted lines indicate the boundaries between the upper and lower sedimentary layers and between the lower sedimentary layer and the basement. White circles indicate the locations of aftershocks from the 2004 and 2007 earthquakes, as determined by Kato *et al.* (2009).

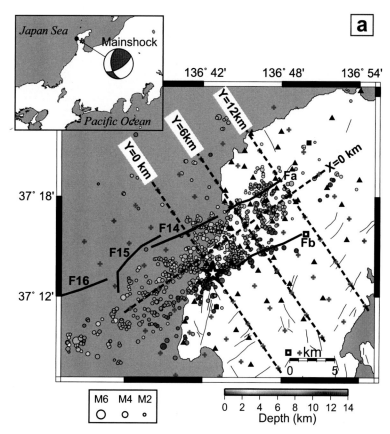

Figure 9.7 Dense seismic observations and V_p model in the source area of the 2007 Noto-Hanto earthquake (modified after Kato *et al.*, 2008b). (a) Map of the relocated aftershock hypocenters with the circle size scaled to earthquake magnitude and tones scaled to depth. The star denotes the epicenter of mainshock. The inset indicates the location of the area investigated with moment tensor for the mainshock (NIED). Filled triangles and squares denote temporary seismic stations and online stations, respectively. Open squares are temporary online stations operated by the Japanese University Group of the Joint Seismic Observations at NKTZ (2005). The grid used in the tomography is shown by crosses. Active (F14–F16) and geological faults (Fa, Fb) associated with the present earthquake are drawn as thick solid lines, and the other major active faults are drawn as thin solid lines.

9.4.1 Aftershock distribution and pre-existing structures within ancient rift system

Figure 9.7b shows the P-wave velocity (V_p) along five cross-sections perpendicular to the fault strike along relocated aftershocks within ±1.5 km of the section. Based on the aftershock distributions, it is estimated that the mainshock occurred on the approximately 55° southeastward-dipping plane. Surface extension of the mainshock fault roughly corresponds to the surface traces of several active faults beneath the ocean (F14–F16) and a

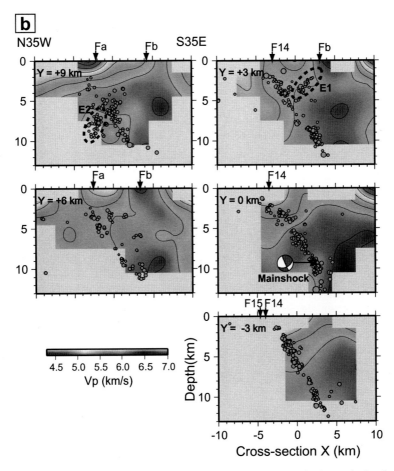

Figure 9.7 (b) Depth sections through the V_p model with superimposed relocated aftershocks distributed within ± 1.5 km of each cross-section. The white-masked areas correspond to the low-resolution model. The contour line interval is 0.3 km/s. Dashed ellipses (E1, E2) are explained in text. Arrows at the top of each section denote the approximate surface locations of faults.

geological fault (defined as a fault for which activity history has not been recognized since the Quaternary period) F_a (Katagawa *et al.*, 2005).

P-wave velocities of the hanging wall in the southeast appear to be higher than those of the footwall in the northwest. In addition, the high-V_p body of the hanging wall has a slightly high V_p/V_s ratio (~ 1.75–1.8), while the low-V_p body in the footwall appears to have a low V_p/V_s ratio (~ 1.6–1.7) at depths greater than 3 km in the cross-sections (see Figure 2 in Kato *et al.*, 2008b). The aftershocks associated with the mainshock fault are distributed approximately along this SE-dipping velocity boundary. Based on these observations, we propose that the rupture of the Noto-Hanto earthquake was likely to have propagated along these pre-existing faults. In addition to the mainshock faults, NW-dipping alignment of aftershocks with an angle of 50° is approximately distributed on a velocity boundary at shallow and moderate depths (E1 and E2 in Figure 9.7b). The surface extension of this NW-dipping velocity boundary almost coincides with geological fault F_b (Figure 9.7a).

Near-surface thin layers with significantly low V_p (<5.0 km/s) and high V_p/V_s (>1.9) were imaged to the northwest of the mainshock epicenter (X < ~0 km). This slow layer corresponds to the sediments that have piled over the half-grabens formed by crustal stretching at the time of expansion of the Japan Sea. Since the thickness of the sediments is less than approximately 3 km, the extent of the crustal stretching is considered to be less significant compared with the Niigata region (Figure 9.2b).

Geological studies show that the crustal shortening that began in the late Miocene (e.g., Itoh and Nagasaki, 1996) has continued to the present and led to the reactivation of a fault that formed as a normal fault under an extensional stress field. The SE-dipping fault plane of the mainshock with high dip angle for a thrust-type event is unfavorably oriented in the direction of the regional maximum stress, which is almost horizontally compressional to the WNW–ESE strike (e.g., Terakawa and Matsu'ura, 2010). This implies that the mainshock fault plane is mechanically weak. Therefore, a likely explanation for the Noto-Hanto earthquake involves reactivation of a normal fault as a reverse fault in terms of inversion tectonics. The tomographic images associated with the inverted normal fault were also observed in the source region of the 2004 and 2007 Niigata earthquakes, as mentioned in the previous sections.

9.4.2 Crustal fluid beneath the mainshock hypocenter

Interestingly, the dip angle of the aftershock alignment changes markedly from 55° to nearly 90° beneath the mainshock hypocenter in the cross-section of Y = 0 km (Figure 9.7b). In addition, the epicenter distributions of these aftershocks beneath the mainshock hypocenter were linearly concentrated along the fault strike of N55°E (Figure 9.7a) and the size of the vertical aftershock alignment was approximately 3 km × 3 km. Interestingly, both the mainshock hypocenter and the vertical alignment of aftershocks beneath it are located in the relatively low-V_p zones where the low-V_p/V_s values were imaged. According to Takei (2002), the relatively low V_p/V_s and low V_p can be explained by the presence of water-filled pores with high aspect ratios. The study of the stress field using high-quality aftershock data showed that aftershocks that occurred above 4 km in depth indicated a strike-slip stress regime (Kato *et al.*, 2011; Figure 9.8). In contrast, aftershocks in deeper parts (>7 km) indicated a thrust-faulting stress regime. In addition to this depth variation of the stress field, the maximum principal stress (σ_1) axis was stably oriented approximately W20°N down to the depth of the mainshock hypocenter, largely in agreement with the regional stress field, but below that depth the σ_1 axis had no definite orientation, indicating horizontally isotropic stress. One likely cause of these drastic changes in the stress regime with depth is the buoyant force of a fluid reservoir localized beneath the seismogenic zone. Indeed, the detection of a conductive layer beneath the mainshock hypocenter by a magnetotelluric

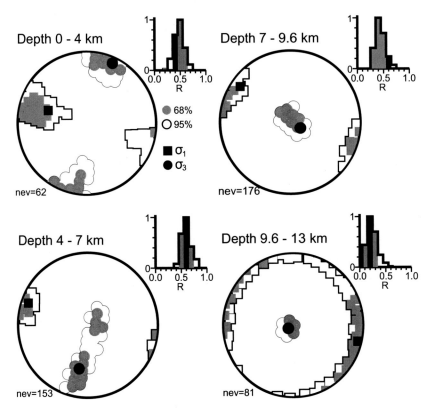

Figure 9.8 Results of stress tensor inversion for four depth ranges (Kato *et al.*, 2011): 0–4 km, 4–7 km, 7–9.6 km, and 9.6–13 km, showing lower-hemisphere equal-area projections of the orientations of σ_1 (squares) and σ_3 (circles), each with their marginal confidence limits. Black-filled symbols, optimal solutions. Gray shades, 68% confidence limits. Open contours, 95% confidence limits. In the top right margins, frequency histograms of the R-values, with 68% and 95% confidence intervals denoted by gray and open bars, respectively. In the bottom left margins, the number of events is shown.

survey (MT) conducted after the occurrence of the mainshock (Yoshimura *et al.*, 2008) supports the interpretation of the presence of water (Figure 9.9).

Given that σ_1 corresponds to the maximum horizontal stress (σ_h^{max}), the transition of the stress field with depth may be explained by an increase in magnitude of the minimum horizontal stress (σ_h^{min}). One simple candidate for the origin of such an increase is a hypothetical, upward flexure of the upper crust, with its hinge axis oriented parallel to the σ_1 axis (Figure 9.9). In shallow parts, σ_h^{min} remains smaller than σ_v (vertical stress) because of extensional stresses associated with the bending ($\sigma_v = \sigma_2$, $\sigma_h^{min} = \sigma_3$), resulting in a strike-slip regime. In deeper parts, by contrast, σ_h^{min} becomes larger than σ_v because of compressional stresses associated with the bending ($\sigma_h^{min} = \sigma_2$, $\sigma_v = \sigma_3$), which leads to a thrust-faulting regime. At great depths, σ_h^{min} ($= \sigma_2$) grows very close to σ_h^{max}

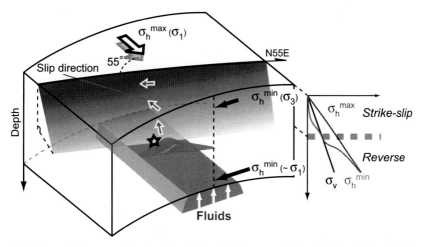

Figure 9.9 Schematic image of the depth variations in the stress field and a hypothetical fluid reservoir beneath the 2007 Noto-Hanto mainshock hypocenter (star) (Kato *et al.*, 2011). The lengths of vectors are scaled to the magnitudes of the principal stresses that they represent. The fault plane is shown as a shaded, inclined surface. Gray arrows, slip directions derived from a finite source model (Ozawa *et al.*, 2008).

($= \sigma_1$), resulting in a horizontally isotropic stress field. A possible support for this hypothesis comes from geomorphological data, where the height profile of a marine terrace formed about 120,000 years ago (Ozawa *et al.*, 2008) hints at the presence of a similar upward flexure in the earthquake source region.

We hypothesize that the buoyant force of a fluid reservoir beneath the mainshock hypocenter is causing such an upward flexure of the upper crust. Our study has only indicated localization of fluids just beneath the mainshock hypocenter, but regional (larger-scale) tomography (Hasegawa *et al.*, 2009) has suggested that fluids are apparently infiltrating into the seismogenic zone from a deeper and larger fluid reservoir, located at approximately 20 km in depth.

It has been postulated that fluids are involved in the initiation of mainshock ruptures (e.g., Miller *et al.*, 2004). According to the fault-valve model (e.g., Sibson, 2007), for example, high-pressured fluids intrude episodically into the fault region, reduce shear strengths, and induce mainshock ruptures. In fact, weak phases in the 2007 Noto-Hanto earthquake waveforms, observed at several seismic stations, have revealed the occurrence of an initial breakdown rupture close to the mainshock hypocenter (Sakai *et al.*, 2008). Furthermore, several foreshocks were identified in the vicinity of the mainshock hypocenter approximately 10 minutes prior to the rupture initiation. These observations lead us to propose the following scenario for the rupture propagation. The initiated rupture is likely to have propagated along the structural boundary between the hanging wall and the footwall, though it was oriented unfavorably for failure. It is likely that fluid migrations, which triggered the initial breakdown, continued to proceed into broader areas along the structural boundary, reducing shear strengths and facilitating ruptures there. Given these considerations, we hypothesize that fluid migrations along the fault, along with relative mechanical weaknesses within the fault zone, were the principal factors that caused the 2007 Noto-Hanto earthquake by reactivating a pre-existing normal fault created during the opening of the Japan Sea.

9.5 Conclusions

We imaged the entire crustal structure and stress field of buried ancient rift systems including heterogeneity within the crust through the seismic-tomographic analysis utilizing a dense seismic network deployed immediately after three recent large intraplate earthquakes along the eastern margin of the Japan Sea. We discovered that stepwise and tilted block structures of the basement, which are geophysical evidence of a Miocene rift system, are widely distributed beneath the thick sedimentary basin in the Niigata region. A similar structure associated with the ancient rift system was imaged in the source area of the Noto-Hanto earthquake. Mainshock fault planes with high dip angles are far from an optimal orientation against the regional stress field, which means that those faults are mechanically weak. Most aftershocks following recent intraplate earthquakes align roughly along the tilted block boundaries of the basement and are controlled by weaknesses associated with buried rift systems.

Furthermore, the structural coincidence between the stress axis distribution and the velocity structure observed in the Niigata region raises the possibility that ductile deformation of the sediments can partially accumulate elastic strain in the brittle parts of the fault zone. In addition, low-velocity anomalies are localized beneath the seismogenic zones, indicating that fluids may have locally weakened the crust. This study therefore suggests that reactivation of pre-existing faults within ancient rift systems by stress loading through ductile flow in the upper crust and creeping of the locally weakened lower crust is a plausible mechanical explanation for intraplate earthquakes. We believe that this mechanical explanation is applicable to other intraplate earthquakes in compressional inverted basins, beneath which ancient rift systems may exist, such as the New Madrid seismic zone (NMSZ) (Kenner and Segall, 2000), and the El Asnam fault zone in Algeria (Chiarabba *et al.*, 1997). In the NMSZ, a local low-velocity anomaly in the lower crust and upper mantle is imaged along the seismically active zone (Zhang *et al.*, 2009). These weak zones may transfer stress to the upper crust when loaded, thus leading to repeated shallow earthquakes in the NMSZ, which is a similar picture to that illuminated along the eastern margin of the Japan Sea.

In order to deepen our understanding of intraplate earthquake generation, a more quantitative approach such as numerical simulation is quite important. Although Shibazaki and Kato (2012) explored the possibility of clarifying the relationship between the configuration of fault zones and the heterogeneous crustal structure deduced from seismic tomography, more information is required on the distribution of pre-existing weak fault zones in order to perform more realistic modeling of fault development. Therefore, we need to implement numerical models by developing rheological models within and beneath seismogenic zones (crust and mantle) through a multi-disciplinary study consisting of dense seismic and electromagnetic surveys, geological observations, and laboratory measurements.

References

Aochi, H., and Kato, A. (2010). Dynamic rupture of crosscutting faults: a possible rupture process for the 2007 M_w 6.6 Niigata-ken Chuetsu–Oki earthquake. *Journal of Geophysical Research*, 115, B05310, doi:10.1029/2009JB006556.

Aoi, S., Enescu, B., Suzuki, W., *et al.* (2010). Stress transfer in the Tokai subduction zone from the 2009 Suruga Bay earthquake in Japan. *Nature Geoscience*, 3, 496–500, doi:10.1038/ngeo885.

Bjorklund, T., Burke, K., Zhou, H., and Yeats, R. S. (2002). Miocene rifting in the Los Angeles Basin: evidence from the Puente Hills half-graben, volcanic rocks, and P-wave tomography. *Geology*, 30, 451–454.

Carter, N. L., and Tsenn, M. C. (1987). Flow properties of the continental lithosphere. *Tectonophysics*, 136, 27–63.

Chiarabba, C., Amato, A., and Meghraoui, M. (1997). Tomographic images of the El Asnam fault zone and the evolution of a seismogenic thrust-related fold. *Journal of Geophysical Research*, 102, 24485–24498.

Chiarabba, C., De Gori, P., and Boschi, E. (2009). Pore-pressure migration along a normal-fault system resolved by time-repeated seismic tomography. *Geology*, 37, 67–70. doi:10.1130/G25220A.1.

Christensen, N. I. (1996). Poisson's ratio and crustal seismology. *Journal of Geophysical Research*, 101, 3139–3156.

Geological Survey of Japan, AIST (2002). *Geological Map of Japan*, CD-ROM Version, Digital Geoscience Map G-3.

Hansen, D. L., and Nielsen, S. B. (2003). Why rifts invert in compression. *Tectonophysics*, 373, 5–24.

Hasegawa, A., Nakajima, J., Uchida, N., *et al.* (2009). Plate subduction, and generation of earthquakes and magmas in Japan as inferred from seismic observations: an overview. *Gondwana Research*, 16, 370–400, doi:10.1016/j.gr.2009.03.007.

Iio, Y., Sagiya, T., Kobayashi, Y., and Shiozaki, I. (2002). Water-weakened lower crust and its role in the concentrated deformation in the Japanese Islands. *Earth and Planetary Science Letters*, 203, 245–253.

Iio, Y., Sagiya, T., and Kobayashi, Y. (2004). Origin of the concentrated deformation zone in the Japanese Islands and stress accumulation process of intraplate earthquakes. *Earth, Planets and Space*, 56, 831–842.

Itoh, Y., and Nagasaki, Y. (1996). Crustal shortening of southwest Japan in the late Miocene. *The Island Arc*, 5, 337–353, 1996.

JNOC (1988). *Report on basic geophysical exploration in onshore area "Kubiki-Tamugiyama"*. Tokyo: Japan National Oil Corporation (in Japanese).

Kano, K., Yoshikawa, T., Yanagisawa, Y., Ogasawara, K., and Danhara, T. (2002). An unconformity in the early Miocene syn-rifting succession, northern Noto Peninsula, Japan: evidence for short-term uplifting precedent to the rapid opening of the Japan Sea. *The Island Arc*, 11, 170–184.

Katagawa, H., Hamada, M., Yoshida, S., *et al.* (2005). Geological development of the west sea area of the Noto Peninsula district in the Neogene Tertiary to Quaternary, central Japan. *Journal of Geography*, 114, 791–810.

Kato, A., and the research team of aftershock observations for the 2004 mid-Niigata Prefecture Earthquake (2007). High-resolution aftershock observations in the source region of the 2004 mid-Niigata Prefecture earthquake. *Earth, Planets and Space*, 59, 923–928.

Kato, A., and group for the aftershock observations of the 2007 Niigataken Chuetsu-oki Earthquake (2008a). Imaging heterogeneous velocity structures and complex aftershock distributions in the source region of the 2007 Niigataken Chuetsu-oki Earthquake by a dense seismic observation. *Earth, Planets and Space*, 60, 1111–1116.

Kato, A., and group for the aftershock observations of the 2007 Noto Hanto Earthquake (2008b). Three-dimensional velocity structure in the source region of the Noto Hanto earthquake in 2007 imaged by a dense seismic observation. *Earth, Planets and Space*, 60, 105–110.

Kato, A., Sakai, S., Hirata, N., *et al.* (2006a). Imaging the seismic structure and stress field in the source region of the 2004 mid-Niigata Prefecture earthquake: structural zones of weakness, seismogenic stress concentration by ductile flow. *Journal of Geophysical Research*, 111, B08308, doi:10.1029/2005JB004016.

Kato, A., Kurashimo, E., Igarashi, T., *et al.* (2009). Reactivation of ancient rift systems triggers devastating intraplate earthquakes. *Geophysical Research Letters*, 36, L05301, doi:10.1029/2008GL036450.

Kato, A., Miyatake, T., and Hirata, N. (2010a). Asperity and barriers of the 2004 Mid-Niigata Prefecture earthquake revealed by highly dense seismic observations. *Bulletin of the Seismological Society of America*, 100(1), 298–306, doi:10.1785/0120090218.

Kato, A., Iidaka, T., Iwasaki, T., Hirata, N., and Nakagawa, S. (2010b). Reactivations of boundary faults within a buried ancient rift system by ductile creeping of weak shear zones in the overpressured lower crust: the 2004 mid-Niigata Prefecture earthquake. *Tectonophysics*, 486, 101–107.

Kato, A., *et al.* (2011). Anomalous depth dependency of the stress field in the 2007 Noto Hanto, Japan, earthquake: potential involvement of a deep fluid reservoir. *Geophysical Research Letters*, 38, L06306, doi:10.1029/2010GL046413.

Kato, N., Sato, H., and Umino, N. (2006b). Fault reactivation and active tectonics on the fore-arc side of the back-arc rift system, NE Japan. *Journal of Structural Geology*, 28, 2011–2022.

Kenner, S. J., and Segall, P. (2000). A mechanical model for intraplate earthquakes; application to the New Madrid seismic zone. *Science*, 289, 2329–2332.

Loveless, J. P., and Meade, B. J. (2010). Geodetic imaging of plate motions, slip rates, and partitioning of deformation in Japan. *Journal of Geophysical Research*, 115, B02410, doi:10.1029/2008JB006248.

Maruyama, T., Fusejima, Y., Yoshioka, T., Awata, Y., and Matsu'ura, T. (2005). Characteristics of the surface rupture associated with the 2004 Mid Niigata Prefecture earthquake, central Japan and their seismotectonic implications. *Earth, Planets and Space*, 57, 521–526.

Michael, A. J., and Eberhart-Phillips, D. (1991). Relations among fault behavior, subsurface geology, and three-dimensional velocity models. *Science*, 253, 651–654.

Miller, S. A., Collettini, C., Chiaraluce, L., *et al.* (2004). Aftershocks driven by a high-pressure CO_2 source at depth. *Nature*, 427, 724–727, doi:10.1038/nature02251.

Nakajima, J., and Hasegawa, A. (2008). Existence of low-velocity zones under the source areas of the 2004 Niigara-Chuetsu and 2007 Niigarta-Chuetsu-Oki earthquakes inferred from travel-time tomography. *Earth, Planets and Space*, 60, 1127–1130.

Nishimura, T., Tobita, M., Yarai, H., *et al.* (2008). Episodic growth of fault-related fold in northern Japan observed by SAR interferometry. *Geophysical Research Letters*, 35, L13301, doi:10.1029/2008GL034337.

Obara, K., Kasahara, K., Hori, S., and Okada, Y. (2005). A densely distributed high-sensitivity seismograph network in Japan: Hi-net by National Research Institute for Earth Science and Disaster Prevention. *Review of Scientific Instruments*, 76, 021301, doi:10.1063/1.1854197.

Okada, T., Hasegawa, A., Suganomata, J., *et al.* (2007). Imaging the heterogeneous source area of the 2003 M6.4 northern Miyagi earthquake, NE Japan, by double-difference tomography. *Tectonophysics*, 430, 67–81.

Okada, T., Umino, N., and Hasegawa, A. (2010). Deep structure of the Ou mountain range strain concentration zone and the focal area of the 2008 Iwate-Miyagi Nairiku earthquake, NE Japan: Seismogenesis related with magma and crustal fluid. *Earth, Planets and Space*, 62, 347–352.

Okada, T., and group for the aftershock observations of the Iwate-Miyagi Nairiku Earthquake in 2008 (2012). *Earth, Planets and Space*, 64, 717–728.

Okamura, Y., Ishiyama, T., and Yanagisawa, Y. (2007). Fault-related folds above the source fault of the 2004 mid-Niigata Prefecture earthquake, in a fold-and-thrust belt caused by basin inversion along the eastern margin of the Japan Sea. *Journal of Geophysical Research*, 112, B03S08, doi:10.1029/2006JB004320.

Ozawa, S., Yarai, H., Tobita, M., Une, H., and Nishimura, T. (2008). Crustal deformation associated with the Noto Hanto earthquake in 2007 in Japan. *Earth, Planets and Space*, 60, 95–98.

Sagiya, T., Miyazaki, S., and Tada, T. (2000). Continuous GPS array and present-day crustal deformation of Japan. *Pure and Applied Geophysics*, 157, 2303–2322.

Sakai, S., and the group for the joint aftershock observation of the 2007 Noto Hanto Earthquake (2008). Highly resolved distribution of aftershocks of the 2007 Noto Hanto earthquake by a dense seismic observation. *Earth, Planets and Space*, 60, 83–88.

Sato, H. (1994). The relationship between late Cenozoic tectonic events and stress field and basin development in northeast Japan. *Journal of Geophysical Research*, 99, 22261–22274.

Sato, H., and Kato, N. (2005). Relationship between geologic structure and the source fault of the 2004 mid-Niigata Prefecture earthquake, central Japan. *Earth, Planets and Space*, 57, 453–457.

Shibazaki, B., and Kato, A. (2012). Modeling the development of a complex fault configuration in the source region of two destructive intraplate earthquakes in the mid-Niigata region. *Tectonophysics*, 562–563, 26–33, doi:10.1016/j.tecto.2012.06.046.

Shimazu, M., Yoon, S., and Takeishi, M. (1990). Tectonics and volcanism in the Sado-Pohang Belt from 20 to 14 Ma and opening of the Yamato Basin of the Japan Sea. *Tectonophysics*, 181, 321–330.

Shinohara, M., Kanazawa, T., Yamada, T., *et al.* (2008). Fault geometry of 2007 Chuetsu-oki Earthquake inferred from aftershock distribution by ocean bottom seismometer network. *Earth, Planets and Space*, 60, 1121–1126.

Sibson, R. H. (2007). An episode of fault-valve behaviour during compressional inversion? The 2004 MJ6.8 Mid-Niigata Prefecture, Japan, earthquake sequence. *Earth and Planetary Science Letters*, 257, 188–199.

Takei, Y. (2002). Effect of pore geometry on *VP/VS*: from equilibrium geometry to crack. *Journal of Geophysical Research*, 107(B2), 2043, doi:10.1029/2001JB000522.

Takenaka, H., Yamamoto, Y., and Yamasaki, H. (2009). Rupture process at the beginning of the 2007 Chuetsu–oki, Niigata, Japan, earthquake. *Earth, Planets and Space*, 61, 279–283.

Terakawa, T., and Matsu'ura, M. (2010). The 3-D tectonic stress fields in and around Japan inverted from centroid moment tensor data of seismic events. *Tectonics*, 29, TC6008, doi:10.1029/2009TC002626.

The Japanese University Group of the Joint Seismic Observations at NKTZ (2005). The Japanese University Joint Seismic Observations at the Niigaka-Kobe Tectonic Zone (in Japanese with English abstract), *Bulletin of the Earthquake Research Institute, University of Tokyo*, 80, 133–147.

Townend, J., and Zoback, M. D. (2006). Stress, strain, and mountain building in central Japan. *Journal of Geophysical Research*, 111, B03411, doi:10.1029/2005JB003759.

Usami, T. (2003). *Materials for Comprehensive List of Destructive Earthquakes in Japan*. University of Tokyo Press (in Japanese).

Uyeshima, M., Ogawa, Y., Honkura, Y., *et al.* (2005). Resistivity imaging across the source region of the 2004 Mid-Niigata Prefecture earthquake (M6.8), central Japan. *Earth, Planets and Space*, 57, 441–446.

White, R. S., and the iSIMM Team (2008). Lower-crustal intrusion on the North Atlantic continental margin. *Nature*, 452, 460–464.

Williams, G. D., Powell, C. M., and Cooper, M. A. (1989). Geometry and kinematics of inversion tectonics. In *Inversion Tectonics*, ed. M. A. Cooper and G. D. Williams, Geological Society, London, Special Publication 44, pp. 3–15.

Yoshimura, R., Oshiman, N., Uyeshima, M., *et al.* (2008). Magnetotelluric observations around the focal region of the 2007 Noto Hanto earthquake (Mj 6.9), central Japan. *Earth, Planets and Space*, 60, 117–122.

Zhang, H., and Thurber, C. H. (2003). Double-difference tomography: the method and its application to the Hayward fault, California. *Bulletin of the Seismological Society of America*, 93, 1875–1889.

Zhang, Q., Sandvol, E., and Liu, M. (2009). Lithospheric velocity structure of the New Madrid Seismic Zone: a joint teleseismic and local P tomographic study. *Geophysical Research Letters*, 36, L11305, doi:10.1029/2009GL037687.

Zhao, D., Huang, Z., Umino, N., Hasegawa, A., and Kanamori, H. (2011). Structural heterogeneity in the megathrust zone and mechanism of the 2011 Tohoku-Oki earthquake (M_w 9.0). *Geophysical Research Letters*, 38, L17308, doi:10.1029/2011GL048408.

10

Deep controls on intraplate basin inversion

SØREN B. NIELSEN, RANDELL STEPHENSON, AND CHRISTIAN SCHIFFER

Abstract

Basin inversion is an intermediate-scale manifestation of continental intraplate deformation, which produces earthquake activity in the interior of continents. The sedimentary basins of central Europe, inverted in the Late Cretaceous–Paleocene, represent a classic example of this phenomenon. It is known that inversion of these basins occurred in two phases: an initial one of transpressional shortening involving reverse activation of former normal faults and a subsequent one of uplift of the earlier developed inversion axis and a shift of sedimentary depocentres, and that this is a response to changes in the regional intraplate stress field. This European intraplate deformation is considered in the context of a new model of the present-day stress field of Europe (and the North Atlantic) caused by lithospheric potential energy variations. Stresses causing basin inversion of Europe must have been favourably orientated with respect to pre-existing structures in the lithosphere. Furthermore, stresses derived from lithospheric potential energy variations as well as those from plate boundary forces must be taken into account in order to explain intraplate seismicity and deformation such as basin inversion.

10.1 Introduction

The focus of this chapter is intraplate stress and deformation in the European continent and the adjacent North Atlantic area. By "deep controls" we understand the involvement of the whole crust and mantle lithosphere, i.e., the lithosphere-scale processes involved in intraplate deformations such as basin inversion, which includes dynamic interactions with the underlying mantle. We will not be concerned with the relationship between basement faulting and the related folding and faulting of the overlying sediments. An extensive literature covers this subject from an observational, experimental, and modelling point of

Intraplate Earthquakes, ed. Pradeep Talwani. Published by Cambridge University Press. © Cambridge University Press 2014.

view, and we refer to the review by Turner and Williams (2004) and the references quoted therein.

One of the fundamental problems of geoscience is to link cause and effect on a regional scale and many million years back in time. What are the causative events that result in intraplate deformation? The North Atlantic realm and the European continent furnish excellent present and past examples of the existence and action of intraplate stresses originating from different sources.

In the past, some 65+ Ma ago, the generally north–south convergence of Africa and Europe during the Late Cretaceous (e.g., Rosenbaum *et al.*, 2002) furnishes an example of stresses transmitted into the interior of the European plate causing compressional shortening of sedimentary basins and rifts in the Alpine foreland, and hence their inversion (Ziegler, 1987, 1990). These relatively mild intraplate continental deformations have in Europe become known as "basin inversion" and the locations of the main examples are sketched out in Figure 10.1. The term "basin inversion" describes the process when an elongate zone of a former area of subsidence – a sedimentary basin or a continental rift – reverses its vertical direction of movement and becomes uplifted and eroded (Ziegler, 1987). The concept was first considered on a regional scale in the European continent by Voigt (1962). With the work of Ziegler (1987, 1990) the concept was placed in a plate tectonic framework involving a causal relationship between stress-producing processes at plate boundaries (the Africa–Europe collision) and stress-induced deformation in the interior of the European continent.

At the present day, the North Atlantic depth Anomaly (NAA), related to the magmatic opening of the North Atlantic around 56 Ma (Tegner *et al.*, 1998) and still visible in the melt anomaly of Iceland and the anomalously shallow North Atlantic Ocean, is a source of excess lithospheric potential energy and anomalous mantle pressure, which causes stresses that propagate through the lithosphere into the surrounding continental plates.

In this chapter, keeping in mind our goal to link cause and effect, we start with an overview of the present-day stress field of the North Atlantic–European realm placed in the context of the NAA (Section 10.2), something reasonably well known, in order to make inferences about present-day lithosphere processes that "cause" intraplate stresses and, possibly, deformation. This is followed by an overview of past (~65 Ma) intraplate deformation in Europe and how it occurred (the "effect"), as expressed by the geological record and predictive models of "basin inversion" (Section 10.3), and a discussion on how this might inform us as regards the link between intraplate forces and strain in continents in general (Section 10.4).

10.2 Present-day intraplate stress in the Europe–North Atlantic area

A range of stress sources contribute to the stress state of the lithosphere. These include (e.g., Ranalli, 1995) slab pull, shear resistance at subduction zones and strike slip faults, convection drag at the base of the lithosphere, stresses transferred to the interior of plates from plate boundary processes, horizontal gradients of lithospheric potential energy, and

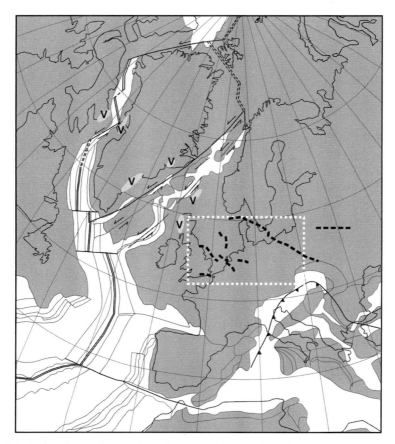

Figure 10.1 North Atlantic plate configuration around 62 Ma, simplified from Nielsen *et al.* (2007), showing the axes of European inversion structures formed in the Late Cretaceous–Paleocene (thick black dashed lines). Grey background represents continental and continental shelf areas, white background oceanic (with the double line being the Atlantic spreading centre) and deep marine areas. Light grey areas labelled "V" represent areas of major magmatism around this time, which was also the time of "secondary inversion" of the inversion structures. White dotted box shows the location of the more detailed map of Figure 10.3.

horizontal gradients of pressure variations at the base of the lithosphere. The last gives rise to dynamic topography.

10.2.1 Model of lithospheric stress from potential energy variations

Lateral variations in the density structure of the lithosphere, and lateral pressure variations in the mantle below the lithosphere due to density contrasts and related convection, mean that the lithostatic pressure obtained as the weight of the rock column above a certain depth below sea level depends on location. These differences in lithostatic pressure are balanced by (mainly) horizontal stresses within the lithosphere. It can be demonstrated that the resulting "swell push force" (Sandwell *et al.*, 1997), F_s, is proportional to the vertical integral of the first moment of the anomalous density, and given by

$$F_s = g \int_0^D \Delta\rho(z)z\,dz \qquad (10.1)$$

In Eq. (10.1), g is the gravitational acceleration at the Earth's surface, $\Delta\rho$ is the deviatoric lithospheric density with regard to a reference density at depth z, and D is the depth of isostatic compensation.

Jones *et al.* (1996) presented the classical derivation of how the fundamental entities of vertical density profiles and lithospheric potential energy lead to a vertically averaged, horizontal stress balance equation where horizontal gradients of the potential energy and the basal pressure become sources of stress in the lithosphere. For the vertically averaged deviatoric stresses the set of equations reads

$$\frac{\partial \tau_{xx}}{\partial x} + \frac{\partial \tau_{yx}}{\partial y} = \frac{1}{L}\left(\frac{\partial E}{\partial x} + L\frac{\partial \tau_{zz}}{\partial x}\right)$$

$$\frac{\partial \tau_{xy}}{\partial x} + \frac{\partial \tau_{yy}}{\partial y} = \frac{1}{L}\left(\frac{\partial E}{\partial y} + L\frac{\partial \tau_{zz}}{\partial y}\right) \qquad (10.2)$$

In Eq. (10.2), x and y are local horizontal coordinates, τ_{xx}, τ_{yy}, and τ_{xy} are the horizontal deviatoric stresses, E is the potential energy of the lithosphere of thickness L, and τ_{zz} is the average vertical deviatoric stress caused by deviations of the mantle pressure from a reference pressure.

A number of studies with different focuses on the major stress sources have investigated lithospheric stresses. Gosh *et al.* (2009) calculated the geopotential stress field of a mainly crustal model (based on CRUST2.0) in different isostatic states. Lithgow-Bertelloni and Guynn (2004) took a similar approach to the crustal contribution to the geopotential stresses and introduced vertical and horizontal mantle tractions. The approach of Bird *et al.* (2008) included geopotential, plate boundary, and basal stresses. They compared their results to observed seafloor spreading rates, plate velocities, anisotropy measurements, and principal stress directions. In general, the main conclusion of all these approaches was that one main driving force is not sufficient to explain the observations. A geopotential stress component is as important as basal mantle tractions and boundary forces to form the Earth's lithospheric stress field.

The present approach is similar but not identical to Jones *et al.* (1996) and differs from those of Lithgow-Bertelloni and Guynn (2004) and Bird *et al.* (2008) by considering only lithospheric potential energy and radial tractions. Plate velocities, shear tractions, and plate boundary forces are not considered.

Using CRUST2.0 (Bassin *et al.*, 2000; http://igppweb.ucsd.edu/~gabi/rem.html) we determine the potential energy of the lithosphere by isostatically balancing one-dimensional lithospheric columns in the presence of lateral pressure variations causing dynamic topography. In other words, we transfer some of the isostatic imbalance of the CRUST2.0 model to a basal pressure that supports topography. In the oceans we use the standard plate model

(Stein and Stein, 1992). The space between the reference depth (250 km) and the base of the lithosphere is filled with asthenosphere with a temperature that decreases upward along the adiabatic gradient of 0.6 °C/km at a reference potential temperature of 1315 °C (e.g., McKenzie *et al.*, 2005). This determines the temperature at the base of the lithosphere, the a-priori depth of which is obtained from a seismologically based model (Gung *et al.*, 2003; Conrad and Lithgow-Bertelloni, 2006). The CRUST2.0 model includes suggestions for the thicknesses and densities of sediment, upper crust, middle crust, and lower crust, and densities of the upper mantle although these values are not perfectly known. Furthermore, the pressure and temperature variations at the compensation depth are unknowns for which additional data in the form of topography and heat flow (Pollack and Chapman, 1977) are required to constrain the system. We consider this as an inverse problem and take the parameters of the lithospheric units as given by CRUST2.0 as tightly constrained inversion variables, the lithospheric thickness, the radiogenic heat production rate of the crust and the basal pressure as less constrained variables, and invoke a fixed coupling between basal pressure and temperature. By allowing the latter to exhibit up to ± 50 °C variations around the reference potential temperature, this inverse problem is sufficiently constrained by topography and heat flow. The basal pressure variation is parameterised using a spherical harmonic polynomial of degree 16 with a total of 153 variable parameters. The resulting model satisfies topography and (largely) heat flow by means of isostatically compensated lithospheric columns of almost known structure and basal pressure (and temperature) variations. This in turn determines the global lithospheric potential energy in the presence of a basal pressure, i.e., the source terms of Eq. (10.2).

To obtain the stresses of Eq. (10.2) we use a three-dimensional, spherical, global finite element mesh of flat, thick, elastic triangles each with 15 degrees of freedom. Each triangle has three corner nodes, each with three spatial coordinates, yielding 9 degrees of freedom. Each node is furthermore bestowed with a vertical axis with 2 angular degrees of freedom, pointing initially towards the centre of the sphere, but which upon loading can deviate slightly from the vertical by pivoting around the mid plane of the element as measured by the (small) angles. This accounts for the remaining 6 degrees of freedom. The relationship between strains and stresses for this thick element is given by Zienkiewicz (1977, Chapter 16). Each element furthermore has material parameters in the form of Young's modulus, Poisson's ratio and thickness, *h*.

10.2.2 Predicted lithospheric stress from potential energy variations in the Europe–North Atlantic area

The results of the global three-dimensional stress calculation (Figure 10.2a) are presented in terms of principal horizontal stress directions and magnitudes at the centre of the triangles. Effective stress, which is the square root of the second invariant of the deviatoric stress tensor, is also displayed. Our computed principal stresses compare favourably to previous models (e.g., Lithgow-Bertelloni and Guynn, 2004; Bird *et al.*, 2008) even though we

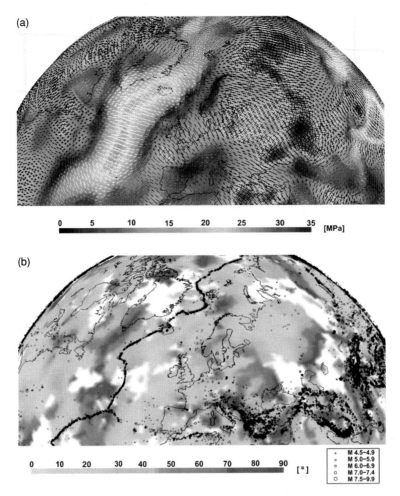

Figure 10.2 Lithospheric stress from potential energy variations. (a) Principal horizontal stress (white is extension and magenta is compression, relative to the lithostatic state of stress, where vector length represents stress magnitude) on a background of effective stress (i.e., the square root of the second invariant of the deviatoric stress tensor; e.g., Ranalli, 1995) in MPa (colour bar). (b) Difference between model predicted and observed principal stress directions (Heidbach *et al.*, 2007a, b) in degrees (colour bar: blue is a good fit and orange is a bad fit), also showing all magnitude greater than or equal to 4.5 earthquake epicentres since 1973 taken from the National Earthquake Information Centre. For colour version, see Plates section.

have not considered plate velocities, shear tractions, and plate boundary processes. The explanation could be that in the North Atlantic realm it is lithospheric potential energy and radial mantle tractions that exert the governing control on geopotential stresses. Indeed, we find that the combined stress field from radial tractions and lithospheric structure agrees better with the observed stress directions (Global Stress Map; Heidbach *et al.*, 2007) than if each source is considered individually.

The principal horizontal stresses are the vertically averaged deviatoric stresses relative to the lithostatic stress state, in which stresses (or pressure) at any depth are equal in any direction and equal to the weight of the overburden. To obtain the associated tectonic forces (N/m), the stresses should be multiplied by the thickness of the elastic shell, which is 100 km. The principal horizontal stress directions are compared to observed values (where available, from the Global Stress Map; Heidbach *et al.* [2007a, b], spatially averaged and extrapolated) and seismicity since 1973 in Figure 10.2b.

In the present context we wish to highlight the relationship between the stress field of oceans and that of the adjacent continental areas. Mature oceanic areas such as the central Atlantic Ocean clearly demonstrate the existence of ridge push in the form of relative compression of the older and deep parts of the oceanic lithosphere. Ocean ridges exhibit relative extension perpendicular to the spreading axis, but in the North Atlantic the Icelandic melt anomaly and the associated anomalous elevation of the ocean floor bear witness to a much more active spreading system. This anomaly produces a SE-directed maximum horizontal compressional stress field, which radiates from the Icelandic area through the British Isles and into central Europe, a model prediction that is in excellent agreement with observed directions of maximum compressional stress in this region (Figure 10.2b). However, it appears that the effect of the high potential energy and basal pressure in the North Atlantic around Iceland is not sufficient to place the highlands of southern Norway under significant compression. Rather, there is slight extension, which was also the conclusion of Pascal and Cloetingh (2009) using a one-dimensional approach.

We note that the conjugate Norwegian and Greenland margins exhibit very different stress states in the present model. While the Norwegian coastal areas generally are in a neutral to slightly compressive state of stress, the east Greenland coastal areas are in a state of relative extension. Apparently, this pattern correlates with the occurrence of extensive North Atlantic breakup magmatism (<62 Ma), which profoundly affected the central East Greenland oceanic and continental areas, but was far offshore on the continental shelf of western Norway. The dominant NW–SE direction of the axes of the intrusions of the British Tertiary Igneous Province (England, 1988) delivers further evidence of a correlation between the present-day potential energy related stress field and Paleocene North Atlantic magmatism. As dyke emplacement preferentially occurs within planes perpendicular to the direction of the minimum principal stress, σ_3 (Anderson, 1951), it appears that the predicted stress field of the British Isles (Figure 10.2a) with NE–SW relative extension has changed little from the stress field that furnished the overriding control on dyke emplacement in the Paleocene.

10.3 Past intraplate basin inversion in Europe

10.3.1 Style of Late Cretaceous–Paleocene basin inversion in Europe

Common to the west-central European structures are that the zones of inversion are Late Paleozoic–Mesozoic sedimentary basins and rifts that formed during the breakup of Pangaea

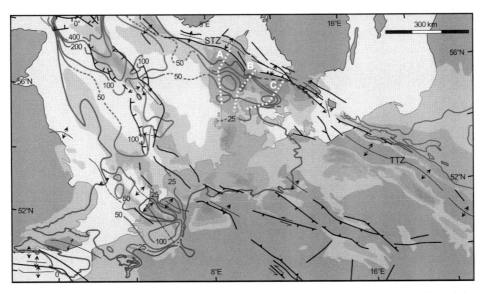

Figure 10.3 Inversion structures on the European continent (axes with anticline symbols) and the thickness of Late Cretaceous–Danian depocentres (yellow to brown colours increasing from 0–500 m, 500–1000 m, 1000–1500 m and >1500 m, respectively) and Middle–Late Paleocene depocentres (blue contours – labels in metres – with the red line indicating the depositional limit). The asymmetrical depocentres of the former are related to the primary, compressional, inversion and the more symmetrical and shallower Paleocene depocentres are related to secondary, relaxation, inversion. The depocentre east of the TTZ in Poland bounded by the green line represents Paleocene deposits <50 m. STZ is the Sorgenfrei–Tornquist Zone; TTZ is the Tornquist–Teisseyre Zone. The dotted white lines are the approximate locations of geological cross-sections (A, B, and C) shown in Figure 10.4. (The figure has been adapted from Nielsen *et al.* (2005).) For colour version, see Plates section.

(Ziegler *et al.*, 1995; Gutiérrez-Alonso *et al.*, 2008). The shallow manifestations of basin inversion are eroded ridges of up to some hundred kilometres length and of (order of magnitude) 50 km width. Flanking such zones are wider sedimentary depocentres (marginal troughs), which show a characteristic deepening towards the border faults of the uplifted zone as seen in the Upper Cretaceous–Paleocene thickness map of Europe (Figure 10.3). The internal structure of the uplifted area is characterised by reversely activated faults and thrusts, and erosion depths of up to some kilometres. It is characteristic that it is the same structures that have been reactivated over and over again.

The inversion movements occurred in phases. The earliest phase dates back to the Turonian of the early Late Cretaceous (Vejbæk and Andersen, 1987, 2002; Ziegler, 1987, 1990; Dadlez *et al.*, 1995; Ziegler *et al.*, 1995), and is believed to have heralded the change of the transtensional stress regime responsible for the breakup of Pangaea to an over-all transpressional stress regime produced by the African–European convergence and, indeed, the onset of continent–continent collision in the Alpine Orogeny (Eo-Alpine orogenic phase; e.g., Ziegler, 1990). The Late Cretaceous inversion phases were characterised by

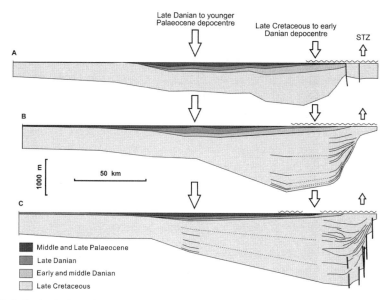

Figure 10.4 Geological profiles constructed from detailed subsurface data across the Sorgenfrei–Tornquist Zone (STZ) in Denmark and the associated sedimentary deposits (modified from Nielsen *et al.*, 2005). The northwestern part (profiles A and B) was mildly inverted and the southeastern part (profile C) was strongly inverted. Late Cretaceous deposition occurred during the primary, compressional, inversion. The shallow and symmetrical Paleocene sequence was deposited during the secondary, relaxation, inversion. For colour version, see Plates section.

transpressional shortening involving reverse activation of former normal faults and the creation of thrusts. During this phase the asymmetric marginal troughs formed (Figure 10.3). In the middle Paleocene, the European inversion structures experienced a distinct tectonic event that differed in style from the Late Cretaceous convergence-related phases. Now the inversion ridges experienced a domal, generally non-ruptural uplift, which involved both the Late Cretaceous inversion ridge and the proximal areas of the marginal troughs. Simultaneously, secondary, shallow and more symmetrical marginal troughs formed in more distal positions.

The occurrence of the two structurally distinct inversion phases is particularly well documented in the Sorgenfei–Tornquist Zone (STZ) of the Danish Basin (Figure 10.4), but their existence can also be inferred from the results of detailed studies of other structures (Vejbæk and Andersen, 2002; de Lugt *et al.*, 2003; Kockel, 2003; Lamarche *et al.*, 2003; Worum and Michon, 2005). Along the STZ the first inversion phase is visible in the asymmetrical chalk depocentre that formed mainly during the Campanian and Maastrichtian. This is evidenced by the thinning of internal chalk structures onto the inversion ridge and the embedded sandstone body that was shed into the chalk basin from source areas along the inversion ridge of southern Sweden (Erlström *et al.*, 1997). The detailed stratigraphic resolution reveals that the primary phase of inversion continued through the early and middle Danian. The onset of the secondary inversion phase occurred during the late Danian (lasting approximately from 62 to 61 Ma) when the depocentre shifted away from the inversion ridge to a more distal position and became more symmetrical. Simultaneously, the inversion ridge and the proximal areas of the asymmetrical marginal trough experienced a gentle doming, the erosion of which is revealed by the occurrence of Late Cretaceous cocolites in early Selandian sediments (Clemmensen and Thomsen, 2005; Nielsen *et al.*, 2005; Steuerbaut, 1998). It is apparent from Figure 10.3 how the secondary phase of inversion has exerted control on the Late Paleocene (and Eocene) deposits along the margins of the more prominent European inversion structures, although some of the depocentres that formed (e.g., along the eastern margin of the Middle Polish Swell and the Weald-Boulonnais area) initially were in a non-marine setting.

10.3.2 Modelling intraplate basin inversion

The typical width of inversion zones is on the order of a couple of lithosphere thicknesses, i.e., 200–250 km, when the extent of the marginal troughs is included. This order of magnitude wavelength points to a whole lithosphere involvement. The challenge is that a quantitative model of basin inversion must address both the relatively narrow localisation of shortening in the deeper parts of former sedimentary basins, the formation of marginal troughs, and the change of inversion style seen in the middle Paleocene in Europe.

The process of basin formation modifies the overall thermal and rheological structure of the lithosphere and it is obvious that this modification somehow is relevant to the later localisation of structural inversion. One fundamental question is simply why sedimentary basins are readily reactivated in compression. Although the mere existence of a sedimentary basin indicates the presence of a structural weakness in the continental lithosphere, it is not trivial how sedimentary basins can be reactivated a long time after formation, as is the case in Europe where the inversion structures generally are associated with Paleozoic and Mesozoic rift systems (Ziegler *et al.*, 1995).

The thermal and structural changes implied by extensional basin formation involve processes that both reduce and increase the load-carrying capacity of the lithosphere. For example, Braun (1992), Ziegler *et al.* (1995), van Wees and Beekman (2000), and Sandiford *et al.* (2003) pointed out that the formation of a rift basin elevates and strengthens (over time) the mantle beneath the basin because of the long-term cooling effected by the shallowing of the mantle and the attenuation of crustal heat production. This mainly thermal aspect should work against a later reactivation of the basin centre. Indeed, analysis of the subsurface temperature field in thermally equilibrated rifts has revealed (Sandiford, 1999; Sandiford *et al.*, 2003; Hansen and Nielsen, 2003) that a wide range of plausible values for the controlling parameters (thermal conductivities, heat production rates and their distribution, and the basin aspect ratio) result in a cooler upper mantle beneath the rift, while the mantle is warmer and therefore weaker beneath the margins of the rift. The fundamental mechanism here is refraction of heat around the relatively poorly conducting sediments and reduced crustal heat generation where the crust has been thinned.

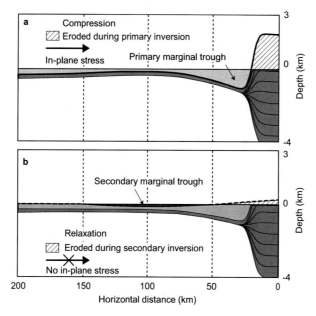

Figure 10.5 The primary and secondary inversion mechanisms: (a) basin fill after primary inversion, which occurs as a response to in-plane compression of a pre-existing rift zone; (b) basin fill after secondary inversion, which occurs as a response to relaxation of the earlier in-plane compression. Details of the modelling can be found in Nielsen *et al.* (2005).

On the other hand, the thinning of the crust during rifting, and the filling of the rift with sediments reduce the overall load-carrying capacity of the lithosphere because (1) less competent sediments have replaced more competent crustal rocks; (2) sediments are less dense than crustal rocks, reducing the confining pressure and thereby the strength of the crust below the sediments as compared to crust at the same depth outside the basin; (3) the crust beneath the sediments is likely to be warmer and therefore weaker than crust at the same depth outside the basin because of sediment blanketing; and (4) formation of crustal scale extensional faults that can be reactivated potentially have strong implications for the possibility of later inversion. Furthermore, the thermal refraction aspect of thermally equilibrated rift basins can in itself promote a strain energy favourable mode of basin inversion (Stephenson *et al.*, 2009).

Numerical modelling allows for investigating the relative importance of such oppositely directed mechanisms. Thus, assuming that a rift remains weak after its formation because of the wealth of faults that are produced during rifting, Nielsen and Hansen (2000) constructed a numerical thermo-mechanical model of compressional basin inversion that reproduced the fundamental observational features of inversion structures on the European continent (Figure 10.5). The model response to compression was localised shortening, thickening, and uplift of the upper crust and sediments within the rift, while the lower crust and upper mantle became slightly depressed. Simultaneously, syn-compressional and asymmetric marginal troughs formed on the former rift flanks. The model suggested that the shortening of crust and sediments is accommodated by shortening also in the lithospheric mantle in the vicinity of the rift, where the upper mantle is slightly warmer and therefore weaker. The occurrence of the asymmetrical marginal troughs was explained in this model by flexural loading of the lithosphere by the internal lithospheric load (as compared to the pre-inversion situation) that formed along the inversion axis because of thickening of the crust in the inversion zone and the replacement of less consolidated near-surface sediments with deeper and denser sediments. The most compelling proof for the existence of this lithospheric load along the inversion axis is the longevity of the asymmetric sedimentary depocentres flanking the European inversion structures (e.g., Figure 10.3). In some cases the axial load is also visible in a slightly elevated Bouguer gravity anomaly along the inversion trend (Wybraniec *et al.*, 1998).

Nielsen and Hansen's (2000) model also predicts the occurrence of "secondary inversion" (Figure 10.5). The mechanism here is that the compression during the primary inversion phase over-deepens the lithospheric flexure that is produced by the axial load of the inversion structure. During the convergence phase the lithosphere is relatively stiffer because the ongoing straining works against the continuous viscous relaxation of the stresses that are generated. When compressional stresses cease or decrease, the lithosphere performs a vertical, elastic flexural adjustment to the new boundary condition (not dissimilar to the mechanism described in Braun and Beaumont [1989]) in the form of an upward doming (order of magnitude 10^2 m) of the central inversion zone and the proximal areas of the primary marginal troughs, and a flexural down warp of smaller amplitude at a greater distance. The undulation continues beyond the elastic flexural adjustment, driven by viscous relaxation of stresses in the softer parts of the lithosphere and the associated regional isostatic adjustment.

10.4 Discussion

The computed stresses described in Section 10.2 and illustrated in Figure 10.2 are those derived from models of lateral variations in the present-day density structure of the lithosphere and lateral pressure variations in the mantle below the lithosphere due to density contrasts and related convection (i.e., variations of potential energy of the lithosphere and basal pressure). We will refer to these stresses as the "potential energy" stresses in the ensuing discussion. Figure 10.2b shows that these, expressed in terms of principal stress orientation, are largely incompatible with observed principal stresses, where available (e.g., Heidbach *et al.*, 2007a, b), in zones where the bulk of present-day seismicity occurs, which is along or close to active plate boundaries, as exemplified by the Tethyan convergence zone from the Mediterranean through central Asia.

This incompatibility demonstrates quite succinctly that interplate deformation (as expressed here as seismicity near a plate boundary) is driven by stresses heavily dominated by forces developed by plate interactions at plate boundaries. However, away from active plate boundary zones, Figure 10.2b shows that there is generally a good fit between predicted and observed stress directions where observed stress data are available. There is even the possibility of a vague correlation between intraplate seismicity (cf. central Europe) and regions where this good fit is seen, although we have made no attempt to determine any quantitative significance of this. In any case, it is fair to infer that intraplate stresses in Europe (that is, away from zones of convergence) are those being generated by potential energy effects rather than plate boundary effects. And, although there is some seismicity in such regions, there is no evidence, nor do our results suggest, that geologically significant intraplate deformation (i.e., basin inversion) is occurring at this time.

These correlations (good fit between modelled and observed stresses at the present day in intraplate settings and bad fit in tectonic plate boundary settings) in the context of the geological record of Late Cretaceous–Paleocene basin intraplate inversion in Europe allow us to make some inferences about cause and effect in intraplate deformation.

First, the potential energy stress field (Figure 10.2) is not in itself sufficient (for typical continental lithosphere composition, heterogeneity, and thermal structure) to produce geologically significant deformation, as recorded by what Nielsen and Hansen (2000) called "primary inversion" of basins in Europe (Figures 10.3–10.5). Rather, the main source of intraplate stress in driving intraplate strain expressed as such is that derived at plate boundaries. These stresses are of course superimposed upon the stresses derived from potential energy variations within the lithosphere. However, primary inversion occurs only when very high plate boundary forces combine favourably with potential energy "background" intraplate stresses (e.g., Figure 10.2a) in combination with the existence of pre-existing lithosphere-scale weaknesses that are also favourably orientated. The crustal geometries that are behind this are to some extent themselves a consequence of the processes that led to basin formation in the region in the first place. Nevertheless, intraplate primary basin inversion in Europe results not only from stresses building up – mainly from forces developed at a plate boundary convergent setting – but also depends on finding favourable rift-like structures amenable to reactivation as well as a favourable interaction (interference) of those stresses with the intrinsic "background" potential energy stress field.

Second, there is clear evidence to link the dissipation (relaxation) of the extraordinary stress field that caused primary inversion, derived mainly from plate boundary forces, to the occurrence of what Nielsen and Hansen (2000) called "secondary" basin inversion. This indirectly supports the inference that it was indeed the plate boundary derived stresses causing the primary inversion in the first place. Nielsen *et al.* (2007), following discussion by Nielsen *et al.* (2005), demonstrated that the relaxation of plate convergence derived stresses was the key factor explaining very precisely dated shifts in deposition patterns within the Danish Basin in the middle Paleocene (62 Ma). Possible causes for such a sudden plate-wide stress change suggested by Nielsen *et al.* (2005) were the Paleocene slowing down of the African–European convergence (e.g., Rosenbaum *et al.*, 2002) or the creation of a new plate margin in the North Atlantic (e.g., Harrison *et al.*, 1999) at the time of arrival of the proto Icelandic plume at the base of the North Atlantic lithosphere during the late Danian, slightly prior to the Danian–Selandian boundary (~62 Ma). However, the potential energy derived lithosphere stresses thought to be developed at this time with the postulated arrival of the Iceland plume and the initiation of the NAA, are actually directly incompatible with the observed secondary inversion deposition patterns (Nielsen *et al.*, 2007). These authors went on to point out that their stress modelling argued against a plume model for the tectonic evolution of the North Atlantic in the Paleocene. Rather, they proposed that magmatism and other North Atlantic tectonic events around this time were the consequence of changes to plate boundary interactions, specifically an embryonic strike-slip initiation of the North Atlantic plate boundary between Greenland and Scandinavia.

Third, we might presume that structures formed during pre-inversion rifting must also play a role in primary basin inversion, specifically that they should be favourably orientated with respect to the build-up of the causative tectonic convergence stress field. The importance of compressional reactivation of faults formed during extensional basin formation was noted by Buiter and Pfiffner (2003) and investigated in more detail in numerical experiments by Hansen and Nielsen (2003). By adjusting the degree of strain softening in three models they found that faults produced during rifting that remain weak, and thereby easily reversible, have the potential to influence profoundly the load-carrying capacity of the lithosphere and hence the future deformation history. The overall structural elements of the inverted rifts (a central inversion ridge flanked by asymmetrical marginal troughs) remained robust features of their models, similar therefore to the model of Nielsen and Hansen (2000). Regarding the orientation of reactivated structures in central Europe, it may simply be an intraplate setting fortuitous for basin inversion, with basement trends basically formed during late Paleozoic orogenesis (e.g., Ziegler *et al.*, 2006) being amenable to the particular plate boundary force derived stresses related to later collision with Africa.

An interesting result of the stress model presented here is that the potential energy stresses, expressed as effective stress (Figure 10.2a), are very small in the region of Europe that tends to have been most affected by intraplate basin inversion. This is the zone running from the North Sea and Denmark southeastwards to the Black Sea, corresponding to the "Trans-European Suture Zone" (TESZ; e.g., Pharaoh *et al.*, 2006) between cratonic lithosphere to the east and mostly younger accreted terranes to the west. The present-day stress field in this area is not one dominantly reflecting the NAA (which in any case was not yet in existence at the time of basin inversion) but has more to do with intrinsic lithosphere structure in the TESZ (which was already in place at that time). Though pre-existing structure and thermal refraction effects must have played a role, it is possible that the absence of a potentially counteracting "background" potential energy stress field in this area (i.e., similar to what is computed for the present day) may have been a factor in allowing the plate boundary derived stress field to be highly effective in causing intraplate deformation in this part of Europe.

10.5 Summary and conclusions

According to plate tectonics, lithospheric plates are essentially rigid with deformation resulting from interactions along (or very near) plate boundaries. In this respect basin inversion is an intermediate-scale manifestation of continental intraplate deformation, which, together with the occurrence of aseismic creep, distributed earthquake activity and the large-scale, large-amplitude deformations such as the Indian–Asian collision, testifies to the well-established deviation of large areas of the interior of continents from rigid plate tectonics. The inverted sedimentary basins of central Europe represent a classic example of this phenomenon. We have reviewed the dynamics of how inversion has occurred in these basins, where tightly constrained observations of the timing of vertical motions reveals that it occurs in two phases. The first phase, which in Europe begins in the early Late Cretaceous, displays transpressional shortening involving reverse activation of former normal faults and the creation of thrusts driven by stresses primarily derived from forces developed at the convergent southern margin of the Europe plate. The second phase, which in Europe occurs in the middle Paleocene, displays generally non-ruptural uplift of the earlier developed inversion ridge and the formation of shallow marginal troughs in more distal positions and is related to the lithosphere's response to the relaxation of the stress field responsible for the first inversion phase. The cause and effect relationship of intraplate stresses and intraplate deformation in Europe in the Late Cretaceous and Paleocene has been indirectly illuminated by considering a model of that part of the present-day stress field of Europe (and the North Atlantic) caused by potential energy variations in the present-day lithosphere. An important feature of intraplate basin inversion in Europe is that the causative stress field must have been favourably orientated with respect to pre-existing structures in the lithosphere and, further, that the stresses derived from plate boundary forces have not been destructively interfered with by the stresses derived from potential energy variations.

References

Anderson, E. M. (1951). *The Dynamics of Faulting and Dyke Formation, with Applications to Britain*. Edinburgh: Oliver and Boyd.

Bassin, C., Laske, G., and Masters, G. (2000). The current limits of resolution for surface wave tomography in North America. *Eos, Transactions, American Geophysical Union*, 81, F897.

Bird, P., Liu, Z., and Rucker, W. K. (2008). Stresses that drive the plates from below: definitions, computational path, model optimization, and error analysis. *Journal of Geophysical Research*, 113, 1406.

Braun, J. (1992). Post-extensional mantle healing and episodic extension in the Canning Basin. *Geophysical Research*, 97, 8927–8936.

Braun, J., and Beaumont, C. (1989). A physical explanation of the relationship between flank uplifts and the breakup unconformity at rifted continental margins. *Geology*, 17, 760–764.

Buiter, S. J. H., and Pfiffner, O. A. (2003). Numerical models of the inversion of half-graben basins. *Tectonics*, 22, 1057, doi:10.1029/2002TC001417.

Clemmensen, A., and Thomsen, E. (2005). Palaeoenvironmental changes across the Danian-Selandian boundary in the North Sea Basin. *Palaeogeography, Palaeoclimatology, Palaeoecology*, 219, 351–394.

Conrad, C. P., and Lithgow-Bertelloni, C. (2006). Influence of continental roots and asthenosphere on platemantle coupling. *Geophysical Research Letters*, 33, L05312, doi:10.1029/2005GL025621.

Dadlez, R., Narkiewicz, M., Stephenson, R. A., Visser, M. T. M., and Van Wees, J.-D. (1995). Tectonic evolution of the Mid-Polish Trough: modelling implications and significance for central European Geology. *Tectonophysics*, 252, 179–195.

de Lugt, I. R., van Wees, J. D., and Wong, Th. E. (2003). The tectonic evolution of the southern Dutch North Sea during the Paleogene: basin inversion in distinct pulses. In *Dynamics of Sedimentary Basin Inversion: Observations and Modelling*, ed. S. B. Nielsen and U. Bayer. *Tectonophysics*, 373, 141–159.

England, R. W. (1988). The Early Tertiary stress regime in NW Britain: evidence from the patterns of volcanic activity. In *Early Tertiary Volcanism and the Opening of the NE Atlantic*, ed. A. C. Morton and L. M. Parson, Geological Society, London, Special Publication 39, pp. 381–389.

Erlström, M., Thomas, S. A., Deeks, N., and Sivhed, U. (1997). Structure and tectonic evolution of the Tornquist Zone and adjacent sedimentary basins in Scania and the southern Baltic Sea area. *Tectonophysics*, 271, 191–225.

Ghosh, A., Holt, W. E., and Flesch, L. M. (2009). Contribution of gravitational potential energy differences to the global stress field. *Geophysical Journal International*, 179, 787–812.

Gung, Y., Panning, M., and Romanowicz, B. (2003). Global anisotropy and the thickness of continents. *Nature*, 422, 707–711, doi:10.1038/nature01559.

Gutiérrez-Alonso, G., Fernández-Suárez, J., Weil, A. B., *et al.* (2008). Self-subduction of the Pangaean global plate. *Nature Geosciences*, 1, 549–553.

Hansen, D. L., and Nielsen, S. B. (2003). Why rifts invert in compression. *Tectonophysics*, 373, 5–24.

Harrison, J. C., Sweet, A. R., McIntyre, D. J., *et al.* (1999). Correlation of Cenozoic sequences of the Canadian Arctic region and Greenland: implications for the tectonic history of northern North America. *Bulletin of Canadian Petroleum Geology*, 47, 223–254.

Heidbach, O., Fuchs, K., Muller, B., *et al.* (2007a). The World Stress Map. *Episodes*, 30, 197–201.

Heidbach, O., Reinecker, J., Tingay, M., *et al.* (2007b). Plate boundary forces are not enough: second- and third-order stress patterns highlighted in the World Stress Map database. *Tectonics*, 26, TC6014, doi:10.1029/2007TC002133.

Jones, C. H., Unruh, J. R., and Sonder, L. J. (1996). The role of gravitational potential energy in active deformation in the southwestern United States. *Nature*, 381, 37–41.

Kockel, F. (2003). Inversion structures in Central Europe: expressions and reasons, an open discussion. *Netherlands Journal of Geosciences*, 82, 367–382.

Lamarche, J., Scheck, M., and Lewerenz, B. (2003). Heterogeneous inversion of the Mid-Polish Trough related to crustal architecture, sedimentary patterns and structural inheritance. *Tectonophysics*, 373, 141–159.

McKenzie, D., Jackson, J., and Priestley, K. (2005). Thermal structure of oceanic and continental lithosphere. *Earth and Planetary Science Letters*, 233, 337–349.

Lithgow-Bertelloni, C., and Guynn, J. H. (2004). Origin of the lithospheric stress field. *Journal of Geophysical Research*, 109, B01408, doi:10.1029/2003JB002467.

Nielsen, S. B., and Hansen, D. L. (2000). Physical explanation of the formation and evolution of inversion zones and marginal troughs. *Geology*, 28, 875–878.

Nielsen, S. B., Thomsen, E., Hansen, D. L., and Clausen, O. R. (2005). Plate-wide stress relaxation explains European Palaeocene basin inversions. *Nature*, 435, 195–198.

Nielsen, S. B., Stephenson, R. A., and Thomsen, E. (2007). Dynamics of Mid-Palaeocene North Atlantic rifting linked with European intra-plate deformations. *Nature*, 450, 1071–1074.

Pascal, C., and Cloetingh, S. (2009). Gravitational potential stresses and stress field of passive continental margins: insights from the south-Norway shelf. *Earth and Planetary Science Letters*, 277, 464–473.

Pharaoh, T. C., Winchester, J. A., Verniers, J., Lassen, A., and Seghedi, A. (2006). The Western accretionary margin of the East European Craton: an overview. In *European Lithosphere Dynamics*, ed. D. G. Gee and R. A. Stephenson. Geological Society, London, Memoirs, 32, pp. 291–311.

Pollack, H. N., and Chapman, D. S. (1977). On the regional variation of heat flow, geotherms, and lithospheric thickness. *Tectonophysics*, 38, 279–296.

Ranalli, G. (1995). *Rheology of the Earth*. London, New York: Chapman and Hall.

Rosenbaum, G., Lister, G. S., and Duboz, C. (2002). Relative motions of Africa, Iberia and Europe during Alpine orogeny. *Tectonophysics*, 359, 117–129.

Sandiford, M. (1999). Mechanics of basin inversion. *Tectonophysics*, 305, 109–120.

Sandiford, M., Frederiksen, S., and Braun, J. (2003). The long-term thermal consequences of rifting: implications for basin reactivation. *Basin Research*, 15, 23–43.

Sandwell, D. T., Johnson, C. L., Bilotti, F., and Suppe, J. (1997). Driving forces for limited tectonics on Venus. *Icarus*, 129, 232–244.

Stein, C. A., and Stein, S. (1992). A model for the global variation in oceanic depth and heat flow with lithospheric age. *Nature*, 359, 123–129.

Stephenson, R., Egholm, D. L., Nielsen, S. B., and Stovba, S. M. (2009). Thermal refraction facilitates 'cold' intra-plate deformation: the Donbas foldbelt (Ukraine). *Nature Geosciences*, 2, 290–293.

Steurbaut, E. (1998). High-resolution holostratigraphy of Middle Paleocene to Early Eocene strata in Belgium and adjacent areas. *Palaeontographica A*, 247, 91–156.

Tegner, C., Duncan, R. A., Bernstein S., *et al.* (1998). 40Ar-39Ar geochronology of Tertiary mafic intrusions along the East Greenland rifted margin: relation to flood basalts and the Iceland hotspot track. *Earth and Planetary Science Letters*, 156, 75–88.

Turner, J. P., and Williams, G. A. (2004). Sedimentary basin inversion and intra-plate shortening. *Earth-Science Reviews*, 65, 277–304.

van Wees, J. D., and Beekman, F. (2000). Lithosphere rheology during intraplate basin extension and inversion: inferences from automated modelling of four basins in western Europe. *Tectonophysics*, 320, 219–242.

Vejbæk, O. V., and Andersen, C. (1987). Cretaceous Early Tertiary inversion tectonism in the Danish Central Trough. *Tectonophysics*, 137, 221–238.

Vejbæk, O. V., and Andersen, C. (2002). Post mid-Cretaceous inversion tectonics in the Danish Central Graben. *Bulletin of the Geological Society of Denmark*, 49, 129–144.

Voigt, E. (1962). Über Randtröge vor Schollenrändern und ihre Bedeutung im Gebiet der Mitteleuropäischen Senke und angrenzender Gebiete. *Zeitschrift der Deutschen Geologischen Gesellschaft*, 114, 378–418.

Worum, G., and Michon, L. (2005). Implications of continuous structural inversion in the West Netherlands Basin for understanding controls on Palaeogene deformation in NW Europe. *Journal of the Geological Society, London*, 162, 73–85.

Wybraniec, S., Zhou, S., Thybo, H., *et al.* (1998). New map compiled of Europe's gravity field. *Eos, Transactions, American Geophysical Union*, 79, 437–442.

Ziegler, P. A. (1987). Late Cretaceous and Cenozoic intra-plate compressional deformations in the Alpine foreland: a geodynamic model. *Tectonophysics*, 137, 389–420.

Ziegler, P. A. (1990). *Geological Atlas of Western and Central Europe*. Bath, UK: Geological Society of London Publishing House.

Ziegler, P. A., Cloetingh, S., and van Wees, J.-D. (1995). Dynamics of intraplate compressional deformation: the Alpine foreland and other examples. *Tectonophysics*, 252, 7–59.

Ziegler, P. A., Schumacher, M. E., Dèzes, P., van Wees, J.-D., and Cloetingh, S. (2006). Post-Variscan evolution of the lithosphere in the area of the European Cenozoic Rift System. In *European Lithosphere Dynamics*, ed. D. G. Gee and R. A. Stephenson. Geological Society, London, Memoirs, 32, pp. 97–112.

Zienkiewicz, O. C. (1977). *The Finite Element Method*, third edition. Maidenhead, UK: McGraw-Hill.

11

Unified model for intraplate earthquakes

PRADEEP TALWANI

Abstract

After the development of plate tectonic theory, understanding the genesis of
intraplate earthquakes has been the focus of many studies. I combine the
results of these studies with recently improved seismological and other data
to formulate a unified model for intraplate earthquakes. Intraplate earthquakes
occur within continental interiors in response to a generally uniform, compressional stress field associated with large-scale tectonic forces. The global
pattern of seismic energy release occurs preferentially in failed and passive
rifts, and less frequently on the edges of cratons. Thermo-mechanical modeling shows that during basin inversion, rifts preferentially utilize inherited zones
of crustal weakness. As a result, pockets of elevated strain rate and consequently
local stress accumulations occur on discrete structures, which I identify as local
stress concentrators (LSC). These are located in both the upper and lower crust
within the rift, and their reactivation occurs in the present-day compressional
stress field in the form of intraplate earthquakes. Commonly observed LSCs are
favorably oriented (relative to the regional stress field) fault bends and intersections, compressional stepovers, flanks of shallow plutons and buried rift
pillows. Stress build-up associated with one or more LSCs interacts with, and
produces a potentially detectable local rotation of the regional stress field with
wavelengths of tens to hundreds of kilometers. A local rotation of the regional
stress field provides evidence of local stress increase, and thus potentially
suggests the location of future intraplate earthquakes.

11.1 Introduction

Although intraplate earthquakes (IPE) are associated with only about 5% of global seismic energy release, historically these earthquakes account for a disproportionate number

Intraplate Earthquakes, ed. Pradeep Talwani. Published by Cambridge University Press. © Cambridge University Press 2014.

of deaths and destruction (www.earthquake.usgs.gov/earthquake/world/most_destructive. php). The earliest attempt to explain IPEs on a global scale was by Sykes (1978), who noted that due to a sparsity of seismic stations and short historical records the nature of IPEs was not well understood. Over the past four decades, since the development of the theory of plate tectonics, which explains nearly 95% of the Earth's seismic energy release near or along plate boundaries, there have been continuing improvements in the quality and quantity of data associated with IPEs. These include the development of improved techniques in locating IPEs with better instrumentation and denser seismic networks; paleo-seismology to document prehistoric earthquakes; global positioning systems to detect minute earth movements; *in situ* stress measurements; improved geophysical techniques, especially seismic tomography; mathematical models to explore various hypotheses; and improved analytical techniques to process various types of data. Improvements in observational data have been accompanied by the formulation of explanatory models which fall into two groups. The first were aimed at explaining the seismicity at a particular location, the New Madrid seismic zone (NMSZ) in the United States in particular. Most of these were conceptual in nature and are generally not testable. The second were based on global observations of geological and mechanical similarities, and spatial association between geological features and seismicity. However, a unified model for the genesis of IPEs has been missing.

Among the global observations are the apparent affinity of IPEs to rifted regions (Johnston and Canter, 1990), and the association of IPEs with identifiable geological features, e.g., with buried plutons (Long, 1976), fault bends (King, 1986) and rift pillows (Zoback and Richardson, 1996). These earthquakes occur due to reactivation caused by a regional, uniform compressional stress field (Zoback, 1992a). Simultaneously, sedimentologists developed basin inversion models to show how reactivation of elongate rift structures produces the sedimentary basins in central Europe (Ziegler, 1987; Nielsen *et al.*, this volume). Finally, *in situ* stress measurements and analysis of seismicity data revealed that in the vicinity of some of these earthquakes the local maximum horizontal stress field was rotated (Zoback and Richardson, 1996). These different observations and explanations were akin to the proverbial description of an elephant by blind men, each examining one part of the elephant's anatomy. I show that these observations, hypotheses, and ideas are not mutually exclusive, rather they are all different parts of a synoptic view. In this chapter I consider these earlier observations and integrate them with newer observations, data analyses, and mathematical models to develop a unified model for IPEs. In this testable model, the main idea is that IPEs are associated with stress build-up at local stress concentrators (LSC) due to a uniform, far-field regional stress field associated with plate tectonic forces. These LSCs were formed in and are preferentially located in former rift zones, and their reactivation occurs by the present-day compressive stress fields in the mid-continental regions. I will refer to this phenomenon as "reactivation by stress inversion." This stress accumulation causes detectable changes in the nature and direction of the regional stress field in their vicinity over wavelengths of tens to hundreds of kilometers.

In the first part I describe the results of earlier studies, which are then integrated to arrive at the new model. I start with a description of ideas about the present-day lithospheric stress field in continental regions (Section 11.2), and how it is perturbed on regional (Section 11.3) and local (Section 11.4) scales. Perturbation on a local scale and the resulting local rotation of the stress field attest to the presence of the LSCs (Section 11.5). The magnitudes of the local stress perturbations are discussed in Section 11.6. In Section 11.7, I address the global distribution of IPEs, and in Section 11.8 I use insights from basin inversion modeling to develop the unified model, presented in Section 11.9. Some consequences and uses of this model are discussed in Section 11.10, and the conclusions are presented in the final section.

11.2 Lithospheric stress field

In her seminal study, Mary Lou Zoback (1992a) identified two orders of stress in the continental lithosphere. The first-order mid-plate stress field, extending uniformly over thousands of kilometers, S_T, is associated with plate tectonic forces. Ziegler (1987) showed that collision-related major stresses can be transmitted over great distances through continental and oceanic lithosphere. This continental stress is generally compressional with one or both horizontal stresses (S_{Hmax} and S_{hmin}) greater than the vertical stress, S_V. In a compressional regional stress field, the maximum horizontal stress, S_{Hmax}, can be determined by direct measurements at shallow depths and its direction inferred from focal mechanisms at seismogenic depths. Zoback (1992a) showed that in continental interiors the direction of S_{Hmax} is the same at both depth ranges. Superimposed on S_T are second-order stress fields with wavelengths of hundreds to thousands of kilometers associated with specific geological and tectonic features. Following Zoback (1992a), the near-surface perturbing horizontal deviatoric stress above the LSC is labeled S_L (compression is assumed positive). Locally, this superposition of S_T with S_L can cause a local rotation of the regional stress field. The vertical stress due to the regional or local stress perturbation does not cause stress rotation, but can change the relative stress magnitude (stress regime, Zoback, 1992a). Rotation of the horizontal stress field depends on the angle between S_T and (the axis of) the local structure as well as on the relative magnitudes of S_T and S_L. For a potentially detectable rotation of S_T ($\sim 15°$) the magnitude of the local horizontal uniaxial stress, S_L, must be greater than about half the magnitude of the horizontal stress difference ($S_{Hmax} - S_{hmin}$). Zoback (1992a) showed that discernible rotations of S_T implied that the magnitudes of both S_T and S_L were of the same order, hundreds of megapascals.

11.3 Regional perturbation of the stress field S_T

11.3.1 Early ideas about perturbing lithospheric stresses

Before the development of plate tectonics theory, different lithospheric stresses were identified as potential causes of IPEs. We recognize them now as the second-order perturbing stresses described by Zoback (1992a). These stresses have also been incorporated in more recent models.

Artyushkov's (1973) suggestion that lithospheric stresses were caused by inhomogeneities in regional and global thickness led Mareschal and Kuang (1986) to speculate on the role of topography and density heterogeneities in the genesis of IPEs. Turcotte and Oxburgh (1976) hypothesized that IPEs were associated with tensional failure in the lithosphere caused by intraplate stresses generated by different processes. These processes included changes in latitude of surface plates (membrane stresses), change in temperature with depth (thermal stresses), addition or removal of overburden, changes in crustal thickness, in addition to the forces associated with the driving mechanism for plate tectonics. The role of addition and removal of overburden and associated thermal stresses was emphasized further by Haxby and Turcotte (1976).

11.3.2 *Perturbation of S_T by surface and other processes*

The regional compressive stress field due to plate tectonic forces, S_T, can be perturbed on both a regional scale due to surface or other processes, and on a local scale due to stress build up near LSCs. At many locations of IPEs, S_T is perturbed by a regional stress field due to surface processes, such as deglaciation, erosion, and sedimentation (see, e.g., Talwani and Rajendran, 1991; Muir-Wood, 2000; Mazzotti *et al.*, 2005) or, as suggested by modeling, by stress transfer to the brittle upper crust of thermal and other stresses caused by thermal and compositional anomalies in the lower crust and upper mantle (Liu and Zoback, 1997; Kenner and Seagall, 2000; Pollitz *et al.*, 2001; Sandiford and Egholm, 2008). For example, Liu and Zoback (1997) suggested that the higher heat flow within the NMSZ resulted in thermal weakening of the lower crust and upper mantle and the higher ductile strain rates led to stress concentrations. The transmission of these stresses to the upper crust leads to the observed seismicity. Kenner and Segall (2000) speculated that sudden tectonic perturbations and viscous relaxation of a weak lower crust within an elastic lithosphere caused stress transfer to the overlying crust and the resulting seismicity. Perturbation on a regional scale can advance or delay the occurrence of an IPE and also alter the style of the deformation and its location. They further modeled that coseismic slip in the crustal faults in turn reloads the lower crust, causing cyclic stress transfer and thus accounting for large repeat earthquakes.

11.3.3 *Deglaciation and erosion*

Of these various suggestions for regional stress perturbation, the role of surface processes, especially deglaciation, has been investigated by many authors.

Deglaciation occurred most rapidly around 10,000 years ago and melting was largely complete by about 6,000 years ago. The crust is still reacting to glacial unloading of the former ice sheet as it rebounds to a state near isostatic equilibrium (e.g., Steffen and Kaufmann, 2005). This is manifested by uplift in the deglaciated regions in northern latitudes. Stein *et al.* (1979), Quinlan (1984), Hasagawa and Basham (1989), and Wu (1998), among others, suggested that as a result of glacial isostatic adjustment (GIA), stresses accumulated during loading can be released by a reduction of the normal stress due to the upward flexure of the lithosphere due to the removal of the surface load. Wu (1998) estimated that the contribution of GIA towards fault instability in Laurentia and Fennoscandia is a few MPa, in agreement with Zoback's (1992b) estimate of $\sim$10 MPa. Wu and Johnston (2000) estimated that, because the flexural stresses decay rapidly from the ice margin, deglaciation has no effect hundreds of kilometers away from the edge of the ice sheet. They estimated that its influence on the seismogenic regions in Eastern Canada was <0.5 MPa, and concluded that the magnitude of post-glacial stress is too small to trigger the M8 earthquake near New Madrid, contrary to the conclusion of Grollimund and Zoback (2001).

In a related study, Calais *et al.* (2010) suggested that the seismicity in the NMSZ results from stress changes caused by the upward flexure of the lithosphere associated with river incision in the northern Mississippi Embayment. In their model the removal of $\sim$6 m of sediments over a 60 km wide area between 16 and 12 kyr BP, and another 6 m of sediments over a 30 km wide area between 12 and 10 kyr BP imposed additional tension stresses (<0.5 MPa) at depths between 5 and 15 km. Calais *et al.* (2010) suggest that these stress changes were adequate to trigger the 1811–1812 sequence of M $\sim$ 7 earthquakes on critically stressed faults in the NMSZ.

Pollitz *et al.* (2001) proposed that the seismicity results from the downward pull of the brittle upper crust by a suddenly sinking mafic body. According to this hypothesis, sinking began several thousand years ago due to perturbations in the lower crust related to the last North American glaciation.

11.3.4 Unverifiable models

Most of the above models are generally not testable since they are based on unverifiable assumptions. For example, as Calais *et al.* (2010) point out, there is no evidence of a weak lower crust or a sinking high-density body, as was suggested by Pollitz *et al.* (2001). Also, the magnitudes of the perturbing stresses associated with these models are about an order of magnitude lower than those of the regional stress field. However, I will show that they do contain some plausible elements, and can play a role in altering the stress regime at higher latitudes. Those will be incorporated in the unified model.

11.4 Local perturbation of the regional stress field: local stress concentrator models

The observation of a spatial association of seismicity locations with structures delineated by geophysical and geological data led to the suggestion of a causal relation between the two. The seismicity locations were inferred to be associated with local stress build-up, S_L, on structures due to S_T. From elasticity theory (e.g., Jaeger and Cook, 1979) a far-field stress is concentrated by a heterogeneity with a different elastic modulus than the surrounding space. The same is true for faults, which are equivalent to planar heterogeneities. Three types of LSCs were identified and incorporated into a testable local stress concentrator model for intraplate earthquakes by Talwani and Gangopadhyay (2000). These are shallow plutons, rift pillows, and fault intersections and kinks. These are described below.

11.4.1 Stress amplification around plutons

A spatial association between seismicity and stress accumulation on the periphery of mafic and ultramafic plutons embedded in felsic country rocks was identified by Long (1976), Kane (1977), and McKeown (1978), and ascribed as being due to rigidity contrast between the two. Analytical modeling by Campbell (1978) showed that the accumulated differential stress was largest when an elliptical inclusion was parallel to S_T, and it scaled with the size of the pluton. McKeown (1978) also noted that ancient rift zones may be a primary control on the locations of mafic intrusions.

With increasingly better seismic, geophysical, and geological data, a spatial association between seismicity and stress accumulation on the periphery of plutons was observed in the NMSZ (Hildenbrand *et al.*, 2001), and in South Carolina (Stevenson *et al.*, 2006; Talwani and Howard, 2012). Agreement of calculated stresses on the periphery of plutons, and the predictions of the analytical model of Campbell (1978), with the observed locations of seismicity has further strengthened the suggestion of a causal association between the two (Stevenson *et al.*, 2006).

11.4.2 Seismicity associated with rift pillows

A rift pillow forms as a result of mafic magmatic intrusion into the lower crust during rift formation. In failed rifts this high-density rift pillow forms in the lower crust and the excess mass of the pillow must be supported by the strength of the cooled lithosphere, inducing deviatoric stresses in the plate (Zoback and Richardson, 1996). A causal association of seismicity with buried rift pillows has been inferred in Brazil, in the NMSZ, and at two locations in India.

The east–west trending Amazonas rift in central Brazil is one of the largest continental rifts in the world (Zoback and Richardson, 1996). Two deep (23 km and 45 km) moderate (M 5.1 and 5.5) earthquakes occurred on the northern margin of the rift in 1963 and 1983. Based on the theoretical models by Sonder (1990) and Zoback (1992a), Zoback and Richardson (1996) suggested that these earthquakes were associated with local stress concentration around the rift pillow at those depths. The depth of the rift pillow was inferred from gravity data. The observed direction of the maximum horizontal stress, S_{Hmax}, in the vicinity of the rift inferred from bore-hole data and from focal mechanisms was found to be rotated $\sim 75°$ counter-clockwise relative to the east–west direction of S_T. Based on modeling, they suggested that the induced rift normal stresses, S_L, associated with the rift pillow can be as large as the tectonic stress, and that its influence on S_T depends both on the orientation of the rift pillow relative to S_T and on the ratio of the magnitude of the two stresses.

Seismic refraction data indicate the presence of a buried high-density rift pillow in the lower crust beneath the NMSZ (Mooney *et al.*, 1983). Grana and Richardson (1996) modeled the local stress field associated with a rift pillow at a depth of 30 km beneath the NMSZ. They found that the direction of S_{Hmax} near the top of the rift axis was rotated clockwise $\sim 10°$ to $30°$ relative to S_T. They suggested that the perturbed stress field over this and other stress pillows was adequate to trigger earthquakes.

In their study of the M 5.8 1997 Jabalpur earthquake in central India, Rajendran and Rajendran (1998) concluded that this deep earthquake (36 ± 4 km) located within the Narmada rift was associated with a rift pillow at that depth. The presence of the rift pillow has been inferred from gravity, deep seismic sounding, and seismic reflection data (Rajendran and Rajendran, 1998).

Mandal (2013) analyzed 10 years of aftershocks of the 2001 Bhuj earthquake occurring in the Kutch rift zone He inverted more than 450 well-determined focal mechanisms in 10 km depth slices to obtain the directions of the local stress orientations. While the orientation of S_{Hmax} for the top 30 km was along the regional direction of S_T, $\sim$N–S, he found that for the depth range of 30–40 km it was rotated clockwise by $\sim 50°$. The results of seismic tomography suggested that the earthquakes at that depth were possibly associated with fluid-filled mafic intrusions (rift pillow?).

In an alternative explanation for the role of a rift pillow in contributing to seismicity in NMSZ, Stuart *et al.* (1997) suggested that the cause of the observed seismicity was the stress concentrated in a weak sub-horizontal detachment fault in the lower crust directly above the rift pillow.

11.4.3 Stress concentration associated with fault geometry: the (fault) intersection model

It has been known for a long time that fault bends play an important role in the generation and termination of seismicity (see, e.g., King, 1986). With improving seismicity data the role of fault intersections in generating earthquakes was recognized. In the intersection model (Talwani, 1988) intersecting faults form a locked volume where stress builds up in response to S_T. This stress build-up can be large enough to cause major earthquakes. Fault intersections provide locations for the initiation and cessation of rupture and for the local generation and accumulation of stress and the resulting earthquakes (Talwani, 1999). The observed pattern of seismicity is consistent with the results of numerical analyses of the stress fields in the immediate vicinity of fault intersections (Andrews, 1989; Jing and Stephansson, 1990).

Simple 2D numerical models have been used to investigate the response to intersecting faults corresponding to known geological features subjected to far-field loading (Gangopadhyay *et al.*, 2004; Gangopadhyay and Talwani, 2005). These studies focused on the NMSZ and the Middleton Place Summerville seismic zone near Charleston, South Carolina. The resultant stress patterns, sign and amplitude of the maximum shear stress were consistent with the observed locations and structural style of faulting. These numerical models thus support the hypothesis that in a localized volume of previously weak crust, fault intersections act as stress concentrators that give rise to seismicity in their vicinity. Parametric studies by Gangopadhyay and Talwani (2007) showed that a fault at 45° ± 15° relative to S_{Hmax}, with an intersecting fault at 90° ± 35°, to it are optimal directions for stress accumulation. This model is scale independent, and has been found applicable in explaining many cases of IPE. On a continental scale, Hildenbrand *et al.* (1996, 2001) observed that along the 400 km long Reelfoot Rift axis, the only seismically active zone, the NMSZ, occurs near the 100 km wide intersection zone of the Reelfoot Rift and the Missouri batholiths. Perhaps the first regional-scale field evidence of stress concentration or perturbation in the vicinity of intersecting tectonic features was provided by Ellis (1991). By mapping stress distribution from 1,500 hydraulically fractured wells in south-central Oklahoma, he found a spatial correlation between the intersection of major crustal fault zones, the resulting stress distribution, and the contemporary seismicity. In another example, Dentith and Weatherstone (2003) also found that the spatial and temporal distribution of seismicity associated with the M 6.9 Meckering earthquake in southwestern Australia closely correlated with the predictions of the intersection model.

Many strike-slip fault systems consist of numerous discrete en echelon segments. Another geometrical configuration that acts as a local stress concentrator is a restraining stepover in an en echelon strike-slip fault system. Two-dimensional quasi-static elastic analysis of the restraining stepovers by Segall and Pollard (1980) showed that they store elastic energy and may be the sites of large earthquakes. We recognize them as potential local stress concentrators. Association of these stopovers with seismicity has been observed in the Middleton Place Summerville seismic zone near Charleston, South Carolina (Talwani, 1999), the NMSZ (Figure 7.11) and the 2001 Kutch earthquake (Figure 6.1).

11.4.4 Local shear model

Iio *et al.* (2004) presented a conceptual "local shear" model to explain the large recurring IPEs in Japan. They assumed that a seismogenic fault in the brittle crust extends as a ductile fault zone with low viscosity into the viscoelastic lower crust. This assumption was based on their inference of aseismic slip (localized shear deformation) on the downward extensions of seismogenic faults following several earthquakes (Iio and Kobayashi, 2002). In their conceptual model, plate tectonic forces load the crustal fault and can cause (intraplate) earthquakes. The stress drop that follows subsequently loads the fault extension in the viscoelastic lower crust. When the stress in the lower crust relaxes, it reloads the fault in the upper crust, which is further loaded by S_T, leading to the next earthquake. The recurrence time of the earthquakes in the brittle crust is a function of the viscosity of the lower crust and the strength of the brittle fault. The presence of fluids can drastically reduce the strength of the fault. This model presents a simple mechanism for transmitting stresses from the ductile lower crust to the brittle upper crust, and may be applicable to the models discussed earlier.

11.4.5 Local stress concentrator model

Talwani and Gangopadhyay (2000) suggested that one or more of the three LSCs, fault intersections, shallow plutons, and buried rift pillows were associated with most intraplate earthquakes. Gangopadhyay and Talwani (2003) examined seismicity, geophysical, and geological data for 20 M $\geq$ 5.0 intraplate earthquakes to identify the responsible LSCs. They found that fault intersections were associated with 30% of the events, fault intersections and plutons together with 35%, i.e., fault intersections were associated with nearly two-thirds of the events. Plutons alone and rift pillows alone were associated with 20% and 15% of the events, respectively. In the following section I examine some examples that provide evidence for the presence of local stress anomalies within the uniform regional tectonic stress field.

11.5 Evidence of the presence of a local stress anomaly

Zoback (1992a) identified large-scale regional crustal features as sources of second-order stress fields. Because of superposition of S_T by these regional stresses, the resultant stress field with wavelengths of hundreds to thousands of kilometers is rotated relative to the direction of S_T. Now, with the availability of modern seismic networks, stress field pertur- bations associated with LSCs and the resulting rotations of S_T with wavelengths of tens to hundreds of kilometers are being recognized and thus providing evidence for their pres- ence. The orientation and magnitude of the anomalous stress build-up in a discrete volume around a LSC, S_L, and the resulting local rotation, γ, of the S_T depends on the kind of stress concentrator and its geometrical relationship with S_T (Figure 11.1).

For a rift pillow, Sonder (1990) and Zoback (1992a) showed that γ depends on the angle, α, between the strike of the rift and S_T, and the ratio of the differential horizontal stress to S_L, $(S_{Hmax} - S_{hmin}) / S_L$. The magnitude of S_L depends on the mass contrast with the surrounding volume (Figure 11.1b). For intersecting faults, S_L is oriented along the direction of the shorter of the two intersecting faults (BC in Figure 11.1c) and γ depends on the angle, α, between the longer fault AB and S_T and the angle between the two faults, β. The magnitude of S_L depends on the lengths of the two faults and the angles α and β (Gangopadhyay and Talwani, 2007). For plutons, the directions of S_L and γ depend on the orientation of the long axis of an elliptical pluton relative to S_T (Figure 11.1d). The magnitude of S_L depends on the size of the pluton and the ratio of its rigidity modulus to that of the surrounding volume (Campbell, 1978). Next we present some examples of stress rotation associated with LSCs (see also Table 11.1). Due to uncertainties in the determination of the orientation of S_L from seismicity data, only those cases with $\gamma \geq 15°$ are considered meaningful (Mazzotti and Townend, 2010).

11.5.1 France

In a reanalysis of 40 years (1962–2002) of shallow crustal seismicity ($\leq$12 km) data from western and central France, including 4,500 events, four with M 5.1 to 5.7, Mazarbaud *et al.*

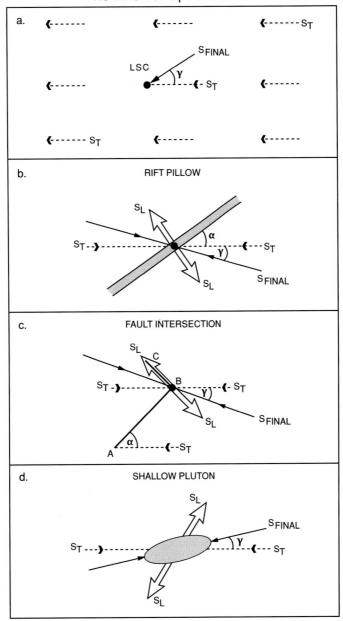

Figure 11.1 The stress field due to a local stress concentrator, S_L, causes a rotation of the regional stress field, S_T, by $\gamma°$ to S_{FINAL}. (a) In a continental setting, S_T is perturbed only in the vicinity of a LSC (solid circle). (b) The rift pillow (solid circle) lying within the rift (shaded area) oriented at an angle $\alpha°$ to S_T. (c) For intersecting faults AB and BC, S_L lies along shorter fault BC. The longer fault AB is oriented at an angle $\alpha°$ to S_T. (d) For a shallow pluton (shaded) S_T is rotated towards S_L.

Table 11.1 *Rotation of local S_{Hmax}*

Location seismic zone	Stress rotation angle (°) (+ cw)	Lateral extent km × km	References
Central-Western France			Mazarbaud *et al.* (2005)
N. Armorican Massif	39	~50–100	
S. Armorican Massif	5*	120 × 350	
E. Massif Central	24*	200 × 250	
Charente	17	~50–100	
Southeastern France			Baroux *et al.* (2001)
Rhone Valley	−8*	110 × 230	
Moyenne Durance FZ	0	70 × 210	
Digne Nappe reverse faulting domain	62	50 × 90	
Digne Nappe normal faulting domain	38*	30 × 70	
SE of Argentera Massif	−3*	80 × 120	
Ligurian basin	−43	50 × 200	
Eastern North America			Mazzotti and Townend (2010)
Lower St. Lawrence	44	100 × 300	
Charlevoix NW	1	20 × 40	
Charlevoix SE	47	20 × 40	
Gatineau	−5	150 × 200	
Ottawa	−8	100 × 450	
Montreal	14	100 × 150	
N. Appalacian	32	150 × 400	
C. Virginia	48	80 × 150	
E. Tennessee	−4	100 × 250	
NMSZ	6	100 × 300	
Northeastern Canada			Steffen *et al.* (2012)
Hudson Bay	~−45	~300 × 600	
Japan			Kawanishi *et al.* (2009)
Southwest Japan	20	35 × 250	
Brazil			Zoback and Richardson (1996)
Amazonas (rift pillow)	−75	~150 wide	
United States			Grana and Richarson (1996)
NMSZ (rift pillow)	10–30	~80 wide	
India			Mandal (2013)
Kutch (rift pillow)	~50	~10s	
NMSZ			Horton *et al.* (2005)
Bardwell, Kentucky	40	~60	

* Relative to direction of S_{Hmax} in local extensional stress regime.

(2005) found evidence of perturbation of S_T by local sources. They found that a regionally significant strike-slip regime with a northwest-trending S_{Hmax} was rotated at two locations and overprinted by local extensional perturbations at two others with wavelengths of tens to hundreds of kilometers. Mazarbaud *et al.* (2005) related the extensional deviatoric stress at the Eastern Massif Central (Table 11.1) to an underlying bulging of the crust at the apex of the hot mantle plume, inferred from seismic tomography.

In a similar study Baroux *et al.* (2001) determined the local stress regime near Provence in southeastern France by inverting focal mechanism data. They found evidence for short-scale (tens of kilometers) variations in the direction of S_{Hmax} close to the major faults in three locations. Significantly different stress regimes were detected, both compressional and extensional. The direction of the compressional stress was found to be both parallel and rotated relative to S_T.

11.5.2 Eastern North America

Mazzotti and Townend (2010) inverted focal mechanisms to determine the local state of stress in ten seismic zones in central and eastern North America. They compared the azimuth of the seismically determined S_{Hmax} with S_T obtained from relatively shallow boreholes within 250 km of the seismic zones. For four seismic zones the two azimuths were essentially parallel. However, a statistically significant clockwise rotation of ~30–50° was found for the Charlevoix, Lower St. Lawrence, and Central Virginia seismic zones, and to a lesser extent for the North Appalachian seismic zone (Table 11.1). The stress rotation occurs over distances of 50–100 km for the Lower St. Lawrence, Central Virginia, and northern Appalachian seismic zones and 20–40 km for the Charlevoix seismic zone. The Charlevoix seismic zone consists of two clusters, the northwest cluster with seismically determined S_{Hmax} parallel to S_T, and the southwest cluster where it is rotated clockwise 47°. The northwest cluster lies beneath the northern shore of the St. Lawrence River, with the southwest cluster beneath the river. Seismic tomography results show that the 5–10 km wide aseismic gap between the two clusters is associated with a high seismic velocity body (Vlahovic *et al.*, 2003).

The results of stress inversion of thrust focal mechanisms for five post-2007 (M 3.6–4.1) and three earlier (M 5.0–6.2) earthquakes in the Hudson Bay region of northeastern Canada show that locally the S_{Hmax} strikes NNW–SSE, in contrast with the regional NNE–SSW direction (Table 11.1; Steffen *et al.*, 2012). The authors attribute this counter-clockwise rotation of S_{Hmax} to the presence of a roughly E–W oriented fault zone and its combined effect with the regional perturbation due to glacial isostatic adjustment and the regional stress field S_T.

11.5.3 Japan

Kawanishi *et al.* (2009) examined earthquakes occurring over a 10-year period with M ≥ 1.0 and depths ≤ 30 km in the Chogochu district of southwest Japan. Using inversions of focal mechanisms obtained from a dense network of seismic stations, they found the azimuth of S_{Hmax} in the ~250 km long and ~35 km wide seismically active belt lying along the Japan Sea coast. In this seismic belt, where three M ≥ 7.0 earthquakes had occurred earlier, they found that the S_{Hmax} was rotated ~20° clockwise with respect to the S_T direction in the adjacent ~330 km long and 140 km wide area (Table 11.1). Based on modeling, the authors attribute the stress rotation to the deformation of an aseismic fault or a ductile fault zone in the lower crust beneath the seismic belt.

11.5.4 Continental rift zones

Some examples of stress rotation associated with buried rift pillows were described in Section 11.4b above. They vary from ~75° counter-clockwise for the Amazonas basin to ~50° clockwise for the Kutch rift zone (Table 11.1).

11.5.5 Bardwell, Kentucky, earthquake sequence

With increasing improvements in seismic monitoring, it now has become possible to detect stress rotations inferred from analysis of smaller earthquakes. The M 4 June 2003 Bardwell, Kentucky, earthquake and its aftershocks were recorded on a dense network (Horton *et al.*, 2005). Stress inversion of very accurately determined focal mechanisms reveal that locally (to ~60 km) S_T is rotated 40° clockwise (Table 11.1). The responsible LSC, however, was not identified by Horton *et al.* (2005).

11.6 Magnitude of local stress perturbations

In her analysis of the magnitude of various cases of regional secondary stresses with detectable rotations of S_T (i.e., $\gamma > 15°$), Zoback (1992a) showed that S_L must be greater than about half the regional horizontal stress difference, i.e., hundreds of megapascals. This suggests that, if a detectable stress field rotation is associated with a LSC, the magnitude of associated S_L needs to be of the same order, i.e. hundreds of megapascals. Mazzottii and Townend (2010) estimated the magnitudes of S_L associated with 30° to 50° rotation of S_T for the Lower St. Lawrence, Charlevoix, and Central Virginia seismic zones. Assuming that the responsible seismogenic structures were oriented perpendicular to S_L, and using Equation 8 of Zoback (1992a), they obtained an estimate of ~160 to 250 MPa for the differential stress in the horizontal plane at mid-crustal depths of ~8 km. This value was calculated based on an assumed coefficient of friction of $\mu = 0.8$, with near-hydrostatic pore pressures. The estimate of S_L reduces to 20 to 40 MPa for $\mu = 0.1$, or near-lithostatic pore pressure. However, the assumptions of such anomalous parameters are not compatible with recent analysis by Hurd and Zoback (2012), who suggest normal values of μ (0.6–0.8) and hydrostatic pore pressures, supporting the premise here that the stresses associated with LSCs that lead to moderate and large earthquakes are of the order of hundreds of megapascals.

In various models for the genesis of IPE discussed earlier, regional stress perturbations due to surface processes or due to anomalies in the lower crust and upper mantle (Section 11.3.3) are a few megapascals at best, at least an order of magnitude lower than the contribution of LSCs to the local stress perturbations. This magnitude difference suggests that, in the present-day stress field, local stress perturbations associated with LSCs are the likely cause of IPE rather than the smaller regional effect of surface processes, e.g., deglaciation and erosion. The latter could, however, provide a trigger for an earthquake. In the following section I examine the global distribution of IPEs and possible association with geological features.

11.7 Intraplate earthquakes and rifts

A possible spatial association between IPEs and rifts has been suggested by a number of earlier studies (see, e.g., Adams and Basham, 1989; Johnston, 1989). Johnston and Kanter (1990) found that globally most of the seismic energy release within intraplate regions occurs by the reactivation of weak pre-existing structures within failed rifts and passive margins in response to an ambient compressional stress field. Schulte and Mooney (2005) reevaluated this correlation by comparing an updated seismicity catalog with a compilation of "Rifts of the world" (Şengör and Natal'in, 2005). Using the latter's definition, " Rifts are fault-bounded elongate troughs, under and near which the entire thickness of the lithosphere has been reduced in extension during their formation," Schulte and Mooney (2005) found that failed rifts and passive margins (labeled "interior rifts and rifted continental margins" by them) together accounted for more than half of all IPEs and 90% of the seismic energy release. Twelve taphrogens, which are linked chains of rifts and grabens in rifted continental crust, account for 74% of all events and 98% of the total seismic energy release within failed rifts. Gangopadhyay and Talwani (2003) found that IPEs not associated with old rifts occur primarily in Pre-cambrian crust.

11.7.1 Correlation with deep mantle structure

Mooney *et al.* (2012) used shear-wave velocity perturbation, δV_S, at a depth of 175 km as a proxy for lithospheric temperature and composition. δV_S is the perturbation in the measured shear-wave velocity with reference to the Preliminary Reference Earth Model (Dziewonsky and Anderson, 1981) and was interpreted in terms of cratonic and non-cratonic lithosphere. They explored possible correlations between the gradient of lithospheric thickness, interpreted from δV_S anomalies at 175 km depth, and intraplate seismicity with magnitudes $\geq$ 4.5. They concluded that significant crustal intraplate seismicity is concentrated in rifted margins (in agreement with Schulte and Mooney, 2005) and the edges of cratons as defined by seismic tomography (Figure 11.2). As detailed knowledge of the structures associated with IPEs in the cratonic regions of Australia, Brazil, China, and India improves, we will be able to identify the nature of the responsible LSCs. In the absence of such detailed knowledge, the LSCs associated with rift structures will be the focus of this study. In this

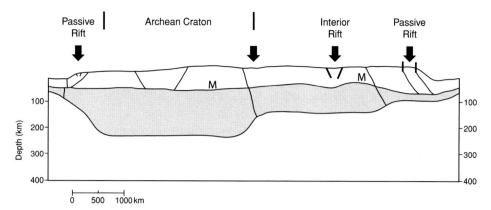

Figure 11.2 A schematic continental lithospheric cross-section. The primary locations of seismicity (solid arrows) are in rifts located along the passive margins and continental interior, and at the boundary of the deep Archean craton with the surrounding regions. The shaded area shows the lithosphere. M represents Moho. (Adapted from Mooney *et al.*, 2012.)

chapter I will focus on the larger earthquakes ($M \geq 4, 5$) with rift association, realizing that nearly 40% of $M \geq 4.5$ events (Schulte and Mooney, 2005) and a significantly higher percentage of smaller events may have other causes.

According to my hypothesis, large IPEs occur by reactivation of faults in the present-day compressional stress field. Whether a fault within a rift will be reactivated depends on whether or not it is favorably oriented relative to S_T, its strength, its tectonic history and geometry. An intracratonic rift forms in response to plate margin processes. The resulting geometry is a series of half-grabens with alternating polarity along strike, separated by accommodation zones (e.g., Ziegler, 1987). Thus, the present configuration of the rift is the result of its tectonic history, and reactivation of structures within it depends on their orientation with respect to S_T. Insight into reactivation of structures within a rift by stress inversion can be provided by modeling, which is described next.

11.8 Insights from basin inversion modeling

The term "basin inversion" is used to describe the tectonic process in which the deep part of a sedimentary basin or continental rift reverses its vertical direction of movement and becomes uplifted (Ziegler, 1987; Nielsen and Hansen, 2000). An inversion zone is an elongate structure that has deformed in response to compression. Among different intraplate discontinuities within the rigid crust, rifts with a thinner crust are most prone to inversion in response to compressional stresses emanating from plate boundaries (Ziegler, 1987). Ziegler described many examples of these inversion zones in central Europe and speculated on their genesis. To investigate the mechanism of these inversion zones, Nielsen and Hansen (2000) developed a numerical thermo-mechanical model of basin inversion in response to a compressional intraplate stress field. The results of their model showed that a pre-existing zone of structural and rheological weakness in a rift is the optimal location for reactivation under compressional inversion. These results suggested that pre-existing faults in a rift would be likely locations for reactivation. In a more detailed study Hansen and Nielsen (2003) developed a two-dimensional thermo-mechanical continuum model to investigate the whole sequence of lithospheric rifting and subsequent basin reactivation and inversion by compression. In this model, rifting was assumed to initiate from a thermal anomaly imposed at the base of the crust, with mass flux from below and above. The thermal anomaly is created by elevating the Moho temperature in a small area around the model center (Figure 11.3a). Allowing for strain hardening, the rifting process is carried out for 10 Ma during which boundary faults and interior conjugate faults extending to the brittle–ductile transition develop. The mantle undergoes regional uplift to compensate for localized crustal thinning and the development of crustal-scale faults significantly weakens the lithosphere and influences the rift structural style. Compressional stress is applied after 60 Ma and basin inversion follows as a natural consequence. In their reactivation model, the inversion preferentially utilizes the inherited zones of crustal weakness. After compression and post-compressional relaxation, at 100 Ma the modeled compressive strains are preferentially located along the boundary faults, interior through-going and conjugate faults, and on top of the up-welled mantle in the lower crust (Figure 11.3b). A comparison of the locations of these pockets of elevated strain rates (Figure 11.3b) with models showing seismicity and associated structures in the Sea of Japan (Figure 4 in Kato *et al.*, 2009; and reproduced as Figure 9.5 in this volume) and the Kutch rift (Figure 4 in Biswas, 2005; modified and reproduced as Figure 6.3c in this volume) was used to infer the locations of LSCs within the rift (Figure 11.3c). The comparison led to the identification of a rift pillow (1), border faults and interior conjugate faults (2 and 3). A through-going fault to the lower crust based on the model by Iio *et al.* (2004) was added to the figure (4) together with a shallow pluton (5) based on the observation of the Osceola pluton in the NMSZ (Hildenbrand *et al.*, 2001). Additionally, the rifts are broken and displaced laterally along transfer faults and compressional stepovers, providing additional fault intersections. Shallow plutons are emplaced at or near these intersections, producing more LSCs.

Next I will compare the results of this model with earlier models proposed to explain IPEs and with the global distribution of IPEs (Schulte and Mooney, 2005; Mooney *et al.*, 2012) to formulate a unified model for intraplate earthquakes.

11.9 Unified model for intraplate earthquakes

11.9.1 The model

Some of the observations and conclusions presented in earlier sections can be summarized as follows:

- There is a generally uniform compressional stress field, S_T, in continental regions.
- Globally, IPEs occur primarily in rifts and at craton boundaries.
- Basin inversion models show how weak structures within a rift are preferentially reactivated by S_T.

a.

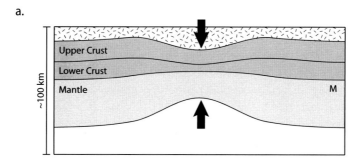

b.

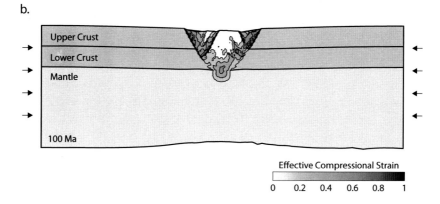

c.

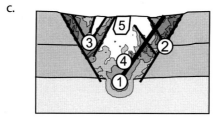

Figure 11.3 (a) The model set-up. A thermal anomaly is imposed at the base of the crust in the middle of the model. The stippled pattern shows sediments, and the arrows show mass flux. (b) Compressional strain-rate distribution after 100 Ma. The reactivation of weak rifted structures by stress inversion results in isolated locations of elevated strain rates (darker patterns). (c) Schematic figure showing interpreted locations of LSCs within the rift (this study). The regions of modeled high strain rates (darker pattern in (b)) were compared with locations of faults and rift pillows in the Kutch and Sea of Japan rift basins to infer the location of a rift pillow (1), border and interior conjugate faults (2 and 3). A through-going fault into the lower crust (4) is based on the model by Iio *et al.* (2004) and a shallow pluton (5) is based on the Osceola pluton within the NMSZ (Hildenbrand *et al.*, 2001). (Adapted from Hansen and Nielsen, 2003.)

- Within rifts, pockets of elevated modeled strains are identifiable with locations of LSCs.
- In response to S_T, around the LSC there is temporal accumulation of local stress, S_L, which can lead to IPEs.
- The influence of S_L extends over wavelengths of tens to hundreds of kilometers.
- An increase in S_L can cause a local rotation in the direction of S_T.
- The magnitude of S_L can be comparable to that of S_T, hundreds of megapascals.
- The structural style and timing of IPEs at any LSC can be modulated by the smaller second-order regional stresses such as those due to GIA, erosion, and lower crust–upper mantle thermal anomalies.

The observations and conclusions listed above address different aspects of IPEs and the factors that contribute to their genesis. Individually, most of these observations have been known for a long time. I combine them with the results of improved observational data to develop a testable, unified model for intraplate earthquakes.

Intraplate earthquakes occur in continental regions characterized by a uniform compressional stress field, S_T, extending over thousands of kilometers (Zoback, 1992a). This stress field is primarily associated with strike-slip faulting, i.e., $S_{Hmax} > S_V \geq S_{hmin}$ (e.g., in continental United States), unless perturbed by a secondary stress field due to regional or local features (Talwani and Rajendran, 1991; Zoback, 1992a, b). There is a global pattern of seismic energy release by IPEs in response to S_T. It preferentially occurs in failed (or interior) rifts and passive (or rifted) margins (see, e.g., Johnston and Kanter, 1990; Schulte and Mooney, 2005). Most of the seismic energy release not associated with failed rifts occurs on the edges of cratons (Mooney *et al.*, 2012). Nearly half of the smaller IPEs (M < 4.5) occur outside the failed rifts, and their contribution to the global seismic energy release from IPEs is negligible (Schulte and Mooney, 2005). Consequently, in this model I will focus on the larger (M ≥ 4.5) events.

Thermo-mechanical modeling by Hansen and Nielsen (2003) demonstrated that large strain accumulations are localized within rifts during their formation and reactivation. These high strain accumulations occur on discrete structures that act as local stress accumulators or concentrators. These LSCs are located in both the upper and the lower crust. Their reactivation in the form of IPEs occurs in the present-day compressional stress field. Commonly observed LSCs are favorably oriented, relative to S_T, fault bends and intersections, flanks of shallow plutons, and buried rift pillows (Talwani and Gangopadhyay, 2000; Gangopadhyay and Talwani, 2003). Local stresses build up on these discrete LSCs lying within the failed rifts or at the edge of cratons and cause IPEs. Next I explore some of the features of S_L and its bearing on the genesis of IPEs.

11.9.2 Build-up of S_L and sequential fault reactivation

Local stresses accumulate at the LSCs in response to the regional stress field S_T. At any LSC, the local stress, S_L, grows with time until it reaches a critical threshold, and interacts with S_T to generate an earthquake. At any time the stress accumulation can occur on

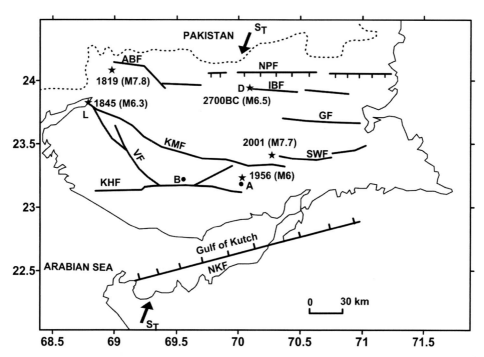

Figure 11.4 The Kutch Rift Basin lies between the Nagar Parkar and North Kathiawar faults (NPF and NKF). Historical earthquakes in 1819, 1845, and 2700 BC are inferred to be associated with the Allah Bund, Kutch Mainland, and Island Belt faults (ABF, KMF, and IBF) respectively. Instrumentally located 1956 Anjar (A) and 2001 Bhuj (B) earthquakes were associated with the Katrol Hill (?) and South Wagad faults (KHF and SWF) respectively. D and L show the locations of old historical towns at Dholavira and Lakhpat. S_T shows the direction of the regional S_{Hmax}. (Fault map courtesy B. K. Rastogi.)

one or more LSCs, and the release of this accumulated stress can occur sequentially. For example, within the Kutch Rift Basin destructive earthquakes have occurred on different faults (Figure 11.4). The 1819 M 7.8 earthquake occurred on the northern Allah Bund fault, the 1845 M 6.3 earthquake that destroyed Lakhpat on the Kutch Mainland fault, the 1956 M 6 Anjar earthquake on the Katrol Hill fault (?) and the 2001 M 7.7 Bhuj earthquake on the North Wagad fault (Rastogi *et al.*, this volume). Based on paleo-archeological data and historical reports of the destruction of Dholavira, the 2700 BC earthquake (estimated M 6.5) was associated with the Island Belt fault (Bisht, 2011). The absence of repeat earthquakes on any one of these faults possibly explains the absence of major topographic features associated with repeat earthquakes within the same rift.

Within the Reelfoot Rift in the Central United States, there is paleoseismic evidence of two large earthquakes hundreds of kilometers away from the NMSZ. These occurred ~6,100 yrs BP in the Wabash Valley (Munson *et al.*, 1997) and ~ 5,500 yrs BP in eastern Arkansas (Al-Shukri *et al.*, 2005). Using a ~2,000-year record of historical seismicity in

North China, Liu *et al.* (2011) documented the absence of a repeating major earthquake at any location in the Wiehe and Shanxi rifts. These observations of multiple events on different faults in the Kutch Rift, at distant locations in the Reelfoot Rift, and non-repeat earthquakes in the rifts in North China are in accord with the unified model wherein different LSCs can be sequentially activated within a rift.

11.9.3 Areal dimensions of local stress changes

Local stress concentrators and the associated surface area affected by S_L are of smaller areal dimensions, with wavelengths of tens to hundreds of kilometers (Table 11.1), that is, an order of magnitude smaller than the areas affected by perturbing secondary regional stresses associated with regional-scale features identified by Zoback (1992a). However, the local stresses S_L associated with these LSCs have magnitudes comparable to S_T as evidenced by the observed rotation of local S_T (next section).

11.9.4 Local rotation of the regional stress field S_T and magnitude of S_L

As discussed in Section 11.5, the interaction of S_L with S_T causes a local rotation in the direction of S_T which depends on the type of LSC, its orientation, and the ratio of the differential horizontal stress to S_L (Figure 11.1). S_L grows with time, and for optimal conditions the rotation γ is detectable from seismicity data. A stress rotation of $>15°$ associated with a perturbing stress suggests that its magnitude is hundreds of megapascals. The various examples of stress rotation (Section 11.5) associated with different LSCs thus suggest that the related stress accumulation on them is adequate to trigger IPEs. The actual magnitude of the stress accumulation will depend on the nature of the LSC, its geometry, and the duration of the stress accumulation.

11.9.5 Local deviatoric stress

Zoback (1992a) presented an example of a secondary deviatoric stress associated with lithospheric thinning in the East African Rift system. Assuming normal frictional and pore pressure distributions, she showed a regional extension in the upper brittle crust of $S_L = -180$ MPa. Mazarbaud *et al.* (2005) presented examples of local extensional deviatoric stress associated with Amorian Massif and Massif Central in northern and central France. Their results suggest that the local stress associated with these LSCs is extensional and of the order of hundreds of megapascals extending over wavelengths of a few hundred kilometers (Table 11.1).

11.9.6 Effect of deglaciation

Perturbing regional secondary stresses act over a large region and interact with local stress changes in the vicinity of LSCs. When this regional perturbing stress is due to elastic rebound due to deglaciation (GIA) and/or due to erosion, it opposes the ambient regional vertical stress. In some cases the relative value of the vertical stress changes from S_2 to S_3, promoting failure by thrust faulting (see, e.g., Talwani and Rajendran, 1991; Zoback, 1992b; Muir-Wood, 2000; Mazzotti *et al.*, 2005; Hurd and Zoback, 2012). Major reverse faulting observed in higher latitudes (e.g., in Fennoscandia) followed the onset of initial deglaciation (Wu, 1998). However, the contribution of GIA to the present-day stress field is less than 10 MPa (Zoback, 1992b; Wu, 1998; Wu and Johnston, 2000). Various authors have suggested that under the present-day stress field the local GIA may be adequate to trigger seismicity on critically stressed faults in Eastern Canada (see, e.g., Quinlan, 1984; Mazzotti *et al.*, 2005; Wu and Mazzotti, 2007; Mazzotti and Townend, 2010).

11.10 Discussion

Analysis of newly compiled stress data from Central and Eastern United States and south-eastern Canada "suggests that shear failure on the preferred nodal planes generally do not require reduced fault friction or elevated pore pressures" (Hurd and Zoback, 2012). Assuming the normal friction and pore pressure conditions recently confirmed by Hurd and Zoback (2012), Mazzotti and Townend (2010) estimated ∼160–250 MPa for the differential stress in the horizontal plane (S_L) for some seismogenic structures in the same study area. These results support the basic premise of the unified model that intraplate earthquakes result primarily from stress build-up of hundreds of megapascals on LSCs. They also argue against models wherein an order of magnitude lower stresses associated with deglaciation, erosion, or thermal and compositional anomalies in the lower crust and upper mantle were proposed to be the cause of intraplate earthquakes. However, these small secondary stresses can modulate S_L due to the LSCs and trigger seismicity.

Next, I discuss some observations related to IPEs and their possible explanations based on the unifying model. These relate to the absence of significant topography in some regions of IPEs, and the apparent absence of detection of anomalous strain accumulation in continental regions.

11.10.1 Absence of topography

In some rift basins there is evidence of uplift of tens to hundreds of meters along internal and border faults. Such is the case in the Kutch Rift Basin, where repeated thrusting along dipping faults that extend to great depths and are oriented almost orthogonally to S_T has led to the formation of local uplifts and grabens (Biswas, 2005; Figure 6.3, this volume). However, the deep-seated, near-vertical faults within the Reelfoot Rift are not favorably oriented relative to S_T, and there is a general lack of extensive topographic expression over extended distances. In the NMSZ thrusting is limited to cross faults and major deformation occurs by strike-slip faulting.

The absence of topography or significant offsets in sedimentary deposits within the Reelfoot Rift has been interpreted to suggest that present-day seismic activity in the NMSZ is of recent origin (e.g., Grollimund and Zoback, 2001). This interpretation is based on the assumption of repeated earthquakes on a single fault system and ignores the presence of other LSCs within the rift and the effect of erosion. In the presence of several LSCs within a rift system, seismicity is not restricted to any one particular fault, but can jump from one LSC to another (see, e.g., Section 11.9.2 and Figure 11.4). The effect of erosion can also be pronounced. If an earthquake is associated with a scarp, as was the case in 1819, when the M 7.8 earthquake in the Kutch Rift Basin produced a $\sim$4 m high, $\sim$90 km long scarp — the Allah Bund, evidence of it begins to fade due to erosion, and may eventually disappear. Commenting on the flat terrain south of the Allah Bund, where major upheaval had taken place during the earthquake, Frere (1870) wrote: "In some places we were told of small islets of raised ground, which formerly existed on the Runn, and which had been swallowed up, and we were taken to the spot, near Vingur, where such an islet used to stand out from the surface of the Runn. But in no such case did any hollow or chasm remain visible. The islet seemed to have melted down to the general level, and the description given by eye-witnesses of what they saw indicated the action one would expect, from the continued agitation of a mass of wet sand [from earthquakes] surrounded by water." The presence of multiple LSCs within the Reelfoot Rift (Section 11.9.2), and the possible obliteration of minor scarps by erosion over the past several millennia, suggest that tectonism within the Reelfoot Rift may not have been restricted to the Holocene period, but may have had a longer history.

11.10.2 Temporal growth of S_L as a predictor of earthquakes?

According to the unified model, for optimally oriented LSCs, and for any given horizontal stress difference, a temporal increase in S_L manifests itself as an increase in the associated rotation, γ. Conversely, after a major earthquake and a release of the accumulated stress there will be a decrease in both S_L and γ. Currently, with uncertainties in focal mechanisms, the accuracy with which γ can be detected is $\sim$15°. However, in the future, if the direction of the stress field at any location is continuously monitored, it may be theoretically possible to detect temporal changes in γ, and infer a potential growth in S_L as an earthquake precursor.

11.10.3 Apparent absence of strain accumulation

According to the two-dimensional thermo-mechanical continuum model (Hansen and Nielsen, 2003), as well as the unified model, the presence of local stress build-up in the vicinity of LSCs should be associated with local pockets of elevated strain. However, two decades of GPS measurements in the NMSZ have failed to provide a consensus that such pockets of elevated strain rate exist (see, e.g., Newman *et al.*, 1999). A possible reason may be that the instrument placement and "our geodetic approach implicitly focuses on motions due to plate wide rather than locally derived stresses" (Newman *et al.*, 1999). However, appreciable local pockets of horizontal strain were detected in Kutch, where a dense network of stations was deployed over the aftershock region of the 2001 earthquake and in the surrounding region (Reddy and Patil, 2008). After the coseismic strain had reduced to background levels, subsequent GPS and InSAR measurements led to the detection of up to ~13 mm/yr of vertical displacement in regions of continuing aftershock activity (Rastogi *et al.*, 2012). Another example of localized elevated coseismic epicentral strain was provided by the ~2 m shortening of railroad tracks in the 1886 Charleston earthquake (Talwani, 1999). Kato *et al.* (2009) reported that the 1964 Niigata (M 7.5) and the 1983 Sea of Japan intraplate earthquakes are located in a zone of contractural strain rates larger than 10^{-7}/yr. This elevated strain away from the plate boundary was detected along the Japan Sea coast, on a continuous GPS array (Sagiya *et al.*, 2000). The region had been host to the 1964 Niigata (M 7.5), the 1995 Kobe (M 7.2), and four other M > 7.0 earthquakes since 1847 (Sagiya *et al.*, 2000).

These observations suggest that in earthquake-prone regions it may be possible to detect preseismic increases in local strain rates, which may indicate potential locations of large earthquakes. A refocus of strain measurement strategy may be in order.

11.11 Conclusions

Intraplate earthquakes account for a very small fraction of the Earth's seismic budget. Although large intraplate earthquakes are much less frequent than their plate-boundary counterparts, when they do occur they can be associated with large-scale destruction. Because of their rarity, efforts to study them have been limited. As the results of these efforts began to accumulate, several features common to intraplate earthquakes were recognized and used to explain their genesis.

In this chapter I present a review of some earlier ideas and integrate them with more recent observations to develop a unified model of intraplate earthquakes. These include a review of both theoretical models and those based on a spatial association of intraplate earthquakes with identifiable geological features. These geological features, located within rigid plates where stresses accumulate in response to a regional compressional stress, S_T, were recognized as LSCs. Pockets of local stress accumulation, S_L, at any LSC extend over wavelengths of tens to hundreds of kilometers and were found to locally change the direction of S_T. At any LSC there is a build-up of S_L which interacts with S_T and may ultimately lead to an intraplate earthquake.

An inventory of M $\geq$ 4.5 intraplate earthquakes showed that they are preferentially located in old rifts and at boundaries of cratons. Possible reasons for this location preference were suggested by Ziegler (1987), who in his study of inverted basins in central Europe noted, "Amongst different types of intra-plate discontinuities, rifts with strongly thinned crust appear to be prone to early inversion in response to collision related intra-plate tangential stresses. Most wrench faults, which penetrate much of the crust and possibly extend into the upper mantle, are also prone to compressional activation". It thus appears that the location of intraplate earthquakes is not random, but that there is a solid mechanical basis for their preferred location in rifts. Models of basin inversion by Nielsen and Hansen (2000) and Hansen and Nielsen (2003) illustrate these mechanical underpinnings. Ziegler's

(1987) suggestion that deep faults that penetrate the crust can be locations for intraplate deformation could, perhaps, explain the incidence of intraplate earthquakes near craton boundaries. This speculation merits further research.

Detailed analyses of seismicity around LSCs show that the accumulated stress, S_L, affects S_T, locally rotating it over tens to hundreds of kilometers. This suggests that a search for potential locations of large future intraplate earthquakes could be carried out by monitoring evolving S_L on a dense network of seismic and GPS monitors.

Acknowledgements

I want to thank Sue Hough, Randell Stephenson, Tom Owens, and Bill Clendinen for their helpful reviews of various versions of the chapter. I also acknowledge with thanks Scott Howard and Erin Koch for help with the figures. Many of the ideas in this chapter were developed over the years and resulted from fruitful discussions with Kusala Rajendran and Abhijit Gangopadhyay.

References

Adams, J., and Basham, P. (1989). The seismicity and seismotectonics of Canada's eastern margin and craton. In *Earthquakes at North-Atlantic Passive Margins: Neotectonic and Post-Glacial Rebound*, ed. S. Gregersen and P. W. Basham. Boston, MA: Kluwer Academic, pp. 355–370.

Al-Shukri, H. J., Lemmer, R. E., Mahdi, H. H., and Connelly, J. B. (2005). Spatial and temporal characteristics of paleoseismic features in the southern terminus of the New Madrid seismic zone in eastern Arkansas. *Seismological Research Letters*, 76, 502–511.

Andrews, D. J. (1989). Mechanics of fault junctions. *Journal of Geophysical Research*, 94, 9389–9397.

Artyushkov, E. V. (1973). Stresses in the lithosphere caused by crustal thickness inhomogeneities. *Journal of Geophysical Research*, 78, 7675–7708.

Baroux, E., Béthoux, N., and Bellier, O. (2001). Analysis of the stress field in southeastern France from earthquake focal mechanisms. *Geophysical Journal International*, 145, 336–348.

Bisht, R. S. (2011). Major earthquake occurrences in archaeological strata of Harappan Settlement at Dholavira (Kachchh, Gujarat). Abstract No. S16_IGCP I1, *Proceedings of International Symposium on the 2001 Bhuj Earthquake and Advances in Earthquake Science*, AES 2011, ISR, Gandhinagar, p. 112.

Biswas, S. K. (2005). A review of structure and tectonics of Kutch basin, western India, with special reference to earthquakes. *Current Science*, 88, 1592–1600.

Calais, E., Freed, A. M., Van Arsdale, R., and Stein, S. (2010). Triggering of New Madrid seismicity by late-Pleistocene erosion. *Nature*, 466, 608–611.

Campbell, D. L. (1978). Investigation of the stress-concentration mechanism for intraplate earthquakes. *Geophysical Research Letters*, 5, 477–479.

Dentith, M. C., and Weatherstone, W. E. (2003). Controls on intra-plate seismicity in southwestern Australia. *Tectonophysics*, 376, 167–184.

Dziewonsky, A. M., and Anderson, D. L. (1981). Preliminary reference Earth model. *Physics of the Earth and Planetary Interiors*, 25, 297–356.

Ellis, W. L. (1991). Stress distribution in south-central Oklahoma and its relationship to crustal structure and seismicity. In *Rock Mechanics as a Multiple Science*, ed. J.-C. Rogiers. Rotterdam: Balkema, pp. 73–80.

Frere, H. B. E. (1870). Notes on the Runn of Cutch and neighbouring region. *Journal of the Royal Geographical Society London*, 40, 181–207.

Gangopadhyay, A., and Talwani, P. (2003). Symptomatic features of intraplate earthquakes. *Seismological Research Letters*, 74, 863–883.

Gangopadhyay, A., and Talwani, P. (2005). Fault intersections and intraplate seismicity in Charleston, South Carolina: insights from a 2-D numerical model. *Current Science*, 88, 1609–1616.

Gangopadhyay, A., and Talwani, P. (2007). Two-dimensional numerical modeling suggests preferred geometry of intersecting faults. In *Continental Intraplate Earthquakes: Science, Hazard and Policy Issues*, ed. S. Stein and S. Mazzotti. Geological Society of America Special Paper 425, pp. 87–99, doi:10.1130/2007.2425(07).

Gangopadhyay, A., Dickerson, J., and Talwani, P. (2004). A two-dimensional numerical model for current seismicity in the New Madrid Seismic Zone. *Seismological Research Letters*, 75, 406–418.

Grana, J. P., and Richardson, R. M. (1996). Tectonic stress within the New Madrid Seismic zone. *Journal of Geophysical Research*, 101, 5455–5458.

Grollimund, B., and Zoback, M. D. (2001). Did deglaciation trigger intraplate seismicity in the New Madrid seismic zone? *Geology*, 29, 175–178.

Hansen, D. L., and Nielsen, S. B. (2003). Why rifts invert in compression. *Tectonophysics*, 373, 5–24.

Hasagawa, H. S., and Basham, P. W. (1989). Spatial correlation between seismicity and postglacial rebound in eastern Canada. In *Earthquakes at North-Atlantic Passive Margins: Neotectonic and Post-Glacial Rebound*, ed. S. Gregersen and P. W. Basham. Boston, MA: Kluwer Academic, pp. 483–500.

Haxby, W. F., and Turcotte, D. L. (1976). Stresses induced by the addition or removal of overburden and associated thermal effects. *Geology*, 4, 181–184.

Hildenbrand, R. G., Grissom, A., van Schmus, R. S., and Stuart, W. (1996). Quantitative investigations of the Missouri gravity low: a possible expression of a large Late Precambrian batholith intersecting the New Madrid seismic zone. *Journal of Geophysical Research*, 101, 21,921–21,942.

Hildenbrand, T. G., Stuart, W. D., and Talwani, P. (2001). Geologic structures related to New Madrid earthquakes near Memphis, Tennessee, based on gravity and magnetic interpretations. *Engineering Geology*, 62, 105–121.

Horton, S. P., Kim, W.-Y., and Withers, M. (2005). The 6 June 2003 Bardwell, Kentucky earthquake sequence: evidence for a locally perturbed stress field in the Mississippi embayment. *Bulletin of the Seismological Society of America*, 95, 431–445.

Hurd, O., and Zoback, M. D. (2012). Intraplate earthquakes, regional stress and fault mechanics in the central and eastern U.S. and southeastern Canada. *Tectonophysics*, 581, 182–192.

Iio, Y., and Kobayashi, Y. (2002). A physical understanding of large intraplate earthquakes. *Earth, Planets and Space*, 54, 1001–1004.

Iio, Y., Sagiya, T., and Kobayashi, Y. (2004). What controls the occurrence of shallow intraplate earthquakes? *Earth, Planets and Space*, 56, 1077–1086.

Jaeger, J. C., and Cook, N. G. W. (1979). *Fundamentals of Rock Mechanics, 3*. London: Chapman and Hall, p. 591.

Jing, S., and Stephansson, O. (1990). Numerical modelling of intraplate earthquakes by 2-dimensional distinct element method. *Gerlands Beiträge zur Geophysik*, 99, 463–472.

Johnston, A. C. (1989). The seismicity of 'Stable Continental Interiors'. In *Earthquakes at North-Atlantic Passive Margins: Neotectonic and Post-Glacial Rebound*, ed. S. Gregersen and P.W. Basham. Boston, MA: Kluwer Academic, pp. 299–327.

Johnston, A. C., and Kanter, L. R. (1990). Earthquakes in stable continental crust. *Scientific American*, 262, 68–75.

Kane, M. F. (1977). Correlation of major earthquake centers with mafic/ultramafic basement masses. U.S. Geological Survey Professional Paper 1028-O, pp. 199–204.

Kato, A., Kurashimo, E., Igarashi, T., *et al.* (2009). Reactivation of ancient rift system triggers devastating intraplate earthquakes. *Geophysical Research Letters*, 36, L05301, doi:10.1029/2008GL036450.

Kawanishi, R., Iio, Y., Yukutake, Y., Shibutani, T., and Katao, H. (2009). Local stress concentration in the seismic belt along Japan Sea coast inferred from precise focal mechanisms: implications for the stress accumulation process on intraplate faults. *Journal of Geophysical Research*, 114, doi:10.1029/2008JB005765.

Kenner, S. J., and Segall, P. A. (2000). A mechanical model for intraplate earthquakes: application to the New Madrid seismic zone. *Science*, 289, 2329–2332.

King, G. C. P. (1986). Speculations on the geometry of the initiation and termination process of earthquake rupture and its relation to morphological and geological structure. *Pure and Applied Geophysics*, 124, 567–585.

Liu, L., and Zoback, M. D. (1997). Lithospheric strength and intraplate seismicity in the New Madrid seismic zone. *Tectonics*, 16, 585–595.

Liu, M., Stein, S., and Wang, H. (2011). 2000 years of migrating earthquakes in north China: how earthquakes in midcontinents differ from those at plate boundaries. *Lithosphere*, 3, 128–132.

Long, L. T. (1976). Speculations concerning southeastern earthquakes, mafic intrusions, gravity anomalies and stress amplification. *Earthquake Notes*, 47, 29–35.

Mandal, P. (2013). Seismogenesis of the uninterrupted occurrence of the aftershock activity in the 2001 Bhuj earthquake zone, Gujarat, India, during 2001–2010. *Nature Hazards*, 65, 1063–1083.

Mareschal J.-C., and Kuang, J. (1986). Intraplate stresses and seismicity: the role of topography and density heterogeneities. *Tectonophysics*, 132, 153–162.

Mazarbaud, Y., Béthoux, N., Guilbert, J., and Bellier, O. (2005). Evidence for short-scale stress field variations within intraplate central-western France. *Geophysical Journal International*, 160, 161–178.

Mazzotti, S., and Townend, J. (2010). State of stress in central and eastern North America seismic zones. *Lithosphere*, 2, 76–83.

Mazzotti, S., James, T. S., Henton, J., and Adams, J. (2005). GPS crustal strain, postglacial rebound, and seismic hazard in eastern North America: the Saint Lawrence valley example. *Journal of Geophysical Research*, 110, B11301, doi:10.1029/2004JB003590.

McKeown, F. A. (1978). Hypothesis: many earthquakes in central and southeastern United States are causally related to mafic intrusive bodies. *Journal of Research, U.S. Geological Survey*, 6, 41–50.

Mooney, W. D., Andrews, M. C., Ginzburg, A., Peters, D. A., and Hamilton, R. M. (1983). Crustal structure of the northern Mississippi embayment and a comparison with other rift zones. *Tectonophysics*, 94, 327–348.

Mooney, W. D., Ritsema, J., and Hwang, Y. K. (2012). Crustal seismicity and earthquakes catalog maximum moment magnitude (Mcmax) in stable continental regions (SCRs): correlation with the seismic velocity of the lithosphere. *Earth and Planetary Science Letters*, 357–358, 78–83.

Muir-Wood, R. (2000). Deglaciation seismotectonics: a principal influence on intraplate seismogenesis at high latitudes. *Quaternary Science Reviews*, 19, 1399–1411.

Munson, P. J., Obermeier, S. F., Munson, C. A., and Hajic, M. R. (1997). Liquefaction evidence for Holocene and latest Pliestocene seismicity in southern halves of Indiana and Illinois: a preliminary overview. *Seismological Research Letters*, 68, 521–536.

Newman, A., Stein, S., Weber, J., *et al.* (1999). Slow deformation and lower seismic hazard at the New Madrid seismic zone. *Science*, 284, 619–621.

Nielsen, S. B., and Hansen, D. L. (2000). Physical explanation of the formation and evolution of inversion zones and marginal troughs. *Geology*, 28, 875–878.

Pollitz, F. F., Kellogg, L., and Burgmann, R. (2001). Sinking mafic body in a reactivated lower crust: a mechanism for stress concentration at the New Madrid seismic zone. *Bulletin of the Seismological Society of America*, 91, 1882–1897.

Quinlan, G. (1984). Postglacial rebound and focal mechanisms of eastern Canada earthquakes. *Canadian Journal of Earth Science*, 21, 1018–1023.

Rajendran, K., and Rajendran, C. P. (1998). Characteristics of the 1997 Jabalpur earthquake and their bearing on its mechanism. *Current Science*, 74, 168–174.

Rastogi, B. K., Choudhury, P., Dumka, R., Sreejith, K. M., and Majumdar, T. J. (2012). Stress pulse migration by viscoelastic process for long-distance delayed triggering of shocks in Gujarat, India, after the 2001 M_w 7.7 Bhuj earthquake. In *Extreme Events and Natural Hazards: The Complexity Perspective*, ed. A. S. Sharma, A. Bundle, V. P. Dimri and D. N. Baker. Geophysical Monograph Series, 196, AGU, Washington, D.C., pp. 63–73, doi:10.10.1029/GM196.

Reddy, C. D., and Patil, P. S. (2008). Post-seismic crustal deformation and strain rate in Bhuj region, Western India, after the 2001 January 26 earthquake. *Geophysical Journal International*, 172, 593–606.

Sagiya, T., Miyazaki, S., and Tada, T. (2000). Continuous GPS array and present day crustal deformation of Japan. *Pure and Applied Geophysics*, 157, 2303–2322.

Sandiford, M., and Egholm, D. L. (2008). Enhanced intraplate seismicity along continental margins: some causes and consequences. *Tectonophysics*, 457, 197–208.

Schulte, S., and Mooney, W. (2005). An updated global earthquake catalogue for stable continental regions: reassessing the correlation with ancient rifts. *Geophysical Journal International*, 161, 707–721.

Segall, P., and Pollard, D. D. (1980). Mechanics of discontinuous faults. *Journal of Geophysical Research*, 85, 4337–4350.

Şengör, A. M. C., and Natal'in, B. A. (2005). Rifts of the world. In *Mantle Plumes: Their Identification Through Time*, ed. R. E. Ernst and K. L. Buchan, Geological Society of America Special Paper 352, pp. 389–482.

Sonder, L. J. (1990). Effects of density contrasts on the orientation of stress in the lithosphere: relation to principal stress directions in the Transverse Ranges, California. *Tectonics*, 9, 761–771.

Steffen, H., and Kauffman, G. (2005). Glacial isostatic adjustment of Scandinavia and northwestern Europe and the radial viscosity structure of the earth's mantle. *Geophysical Journal International*, 163, 801–812.

Steffen, R., Eaton, D. W., and Wu, P. (2012). Moment tensors, state of stress and their relation to post-glacial rebound in northeastern Canada. *Geophysical Journal International*, 189, 1741–1752.

Stein, S., Sleep, N., Geller, R. J., Wang, S. C., and Kroeger, C. (1979). Earthquakes along the passive margin of eastern Canada. *Geophysical Research Letters*, 6, 537–540.

Stevenson, D., Gangopadhyay, A., and Talwani, P. (2006). Booming plutons: source of microearthquakes in South Carolina. *Geophysical Research Letters*, 33, L03316, doi:1029/2005 GL 024679.

Stuart, W. D., Hildenbrand, T. G., and Simpson, R. W. (1997). Stressing of the New Madrid seismic zone by lower crustal detachment fault. *Journal of Geophysical Research*, 102, 27623–27633.

Sykes, L. R. (1978). Intra-plate seismicity, reactivation of pre-existing zones of weakness, alkaline magmatism, and other tectonics post-dating continental separation. *Reviews of Geophysics and Space Physics*, 16, 621–688.

Talwani, P. (1988). The intersection model for intraplate earthquakes. *Seismological Research Letters*, 59, 305–310.

Talwani, P. (1999). Fault geometry and earthquakes in continental interiors. *Tectonophysics*, 305, 371–379.

Talwani, P., and Gangopadhyay, A. (2000). Schematic model for intraplate earthquakes. *Eos, Transactions, American Geophysical Union*, 81(48), F 918.

Talwani, P., and Howard, C. S. (2012). January 1, 1913 Union county, South Carolina earthquake, revisited. *South Carolina Geology*, 48, 11–24.

Talwani, P., and Rajendran, K. (1991). Some seismological and geometrical features of intraplate earthquakes. *Tectonophysics*, 186, 19–41.

Turcotte, D. L., and Oxburgh, E. R. (1976). Stress accumulation in the lithosphere. *Tectonophysics*, 35, 183–199.

Vlahovic, G., Powell, C., and Lamontagne, M. (2003). A three-dimensional P wave velocity model for the Charlevoix seismic zone, Quebec, Canada. *Journal of Geophysical Research*, 108, 1–12.

Wu, P. (1998). Intraplate earthquakes and postglacial rebound in eastern Canada and Northern Europe. In *Dynamics of Ice Age Earth: A Nuclear Perspective*, ed. P. Wu. Uetikon-Zurich, Switzerland: Trans Tech Publications, pp. 603–628.

Wu, P., and Johnston, P. (2000). Can deglaciation trigger earthquakes in North America? *Geophysical Research Letters*, 27, 1323–1326.

Wu, P., and Mazzotti, S. (2007). Effects of a lithospheric weak zone on postglacial seismo-tectonics in eastern Canada and northeastern USA. In *Continental Intraplate Earthquakes: Science, Hazard and Policy Issues*, ed. S. Stein and S. Mazzotti. Geological Society of America Special Paper 425, pp. 113–128, doi:10.1130/2007.2425(09).

Zeigler, P. A. (1987). Late Cretaceous and Cenozoic intra-plate compressional deformations in the Alpine foreland: a geodynamic model. *Tectonophysics*, 137, 389–420.

Zoback, M. L. (1992a). First- and second-order patterns of stress in the lithosphere: The World Stress Map Project. *Journal of Geophysical Research*, 97, 11,703–11,728.

Zoback, M. L. (1992b). Stress field constraints on intraplate seismicity in eastern North America. *Journal of Geophysical Research*, 97, 11,761–11,782.

Zoback, M. L., and Richardson, R. M. (1996). Stress perturbation associated with the Amazonas and other ancient continental rifts. *Journal of Geophysical Research*, 101, 5459–5475.

12

Intraplate seismic hazard: Evidence for distributed strain and implications for seismic hazard

SUSAN E. HOUGH

Abstract

In contrast to active plate boundaries, the lack of a fundamental scientific framework to understand seismogenesis combines with the lower rates of earthquakes to make probabilistic seismic hazard assessment (PSHA) challenging in intraplate regions. Seismic hazard maps for intraplate regions tend to indicate hazard dominated by source zones that have produced large earthquakes during historical times. The level of hazard can also depend critically on the magnitudes of pre-instrumental earthquakes, which are clearly far less certain than earthquakes during the instrumental era. In this brief chapter I first summarize recently published results that address the long-standing paradox of low strain accrual and high moment release in the Central/Eastern United States (CEUS); I then summarize recent published studies from the CEUS and elsewhere suggesting that strain release is more spatially distributed over timescales of millennia to tens of millennia than over the past few centuries. I explore a simple quantitative model that has been discussed in numerous past studies, namely that strain is everywhere low, with weak modulation in the CEUS due to glacial isostatic adjustment and concentration within failed rifts that represent large-scale zones of weakness in the crust. Within this paradigm, one can consider an overall moment-release budget associated with low but spatially distributed strain. I show that even a conservative estimate of strain rate (0.5–1.0×10^{-9}/yr) can generate the rate of moment release observed in historical times, if the strain is distributed and large historical earthquakes were close to M_w 7.0. Basic considerations from both geological observations and statistical seismology support the assumption that, over decadal to century timescales, activity is likely to remain concentrated in areas that have seen elevated activity in the recent geological past. Significant rate changes are possible, however, over short timescales, and may be likely over millennial scales –

Intraplate Earthquakes, ed. Pradeep Talwani. Published by Cambridge University Press. © Cambridge University Press 2014.

as demonstrated, in addition to other evidence discussed in this paper, by abundant and widely accepted evidence that the New Madrid seismic zone itself "turned on" in the relatively recent geological past. One logic-tree approach to PSHA might therefore be to include a low probability that recent clusters have ended along with a non-zero probability that future activity within other rifted zones will exceed the rate predicted by an extrapolation of current background rates.

12.1 Introduction

Quantifying probabilistic seismic hazard remains a vexing problem in intraplate regions throughout the world. Along active plate boundary regions, expected long-term earthquake rates can be estimated from seismicity catalogs, geological evidence of past earthquakes, and GPS estimates of ongoing deformation. Hazard assessment remains challenging, and under active investigation, even in active regions (e.g., WGCEP, 2013); however, in a well-studied region such as Japan or California, geological results and deformation data provide at least a first-order constraint on seismic moment accrual and release rates. The magnitude of the maximum possible earthquake in a region, M_{max}, can also be estimated given an overall strain rate, paleoseismic constraints on the rate of surface-rupturing earthquakes, and the commonly accepted assumption that, in most regions if not for specific faults, regional earthquake magnitudes follow a Gutenberg–Richter distribution (Gutenberg and Richter, 1944; King, 1983; Felzer, 2006; Hough, 1996).

In intraplate regions, estimation of long-term earthquake rates is far more challenging. A fundamental difficulty for hazard assessment is that, unlike active plate boundary zones, no basic, integrated physical framework can explain the first-order processes controlling seismogenesis in intraplate regions. Given the lower rate of earthquakes, even relatively long historical catalogs are inadequate to capture long-term average rates; direct geological evidence of past earthquakes is generally scant, and hopes of constraining low strain rates from GPS data have not generally been borne out to date, with improved data often revealing little to no detectable surface deformation (e.g., Calais *et al.*, 2005; Frankel *et al.*, 2012). With no long-term constraint on fault slip rate or deformation from geological or geodetic data, estimates of earthquake rates and thus hazard are estimated from seismicity data; i.e., the earthquakes observed during historic times and/or documented in prehistoric times, and their estimated magnitudes. Even where the largest known earthquakes – the M_{max}, or characteristic earthquakes – occurred during pre-instrumental times, magnitude estimates can be uncertain by a full magnitude unit or more (see Hough and Page, 2011). Recurrence times for an assumed characteristic event can sometimes be estimated from geological evidence; however, as I discuss in a later section, even these long-term recurrence rate estimates are highly uncertain in intraplate regions.

In this chapter I focus initially on our current understanding of seismic hazards in the Central/Eastern United States (CEUS), where (1) large earthquakes have struck in historical

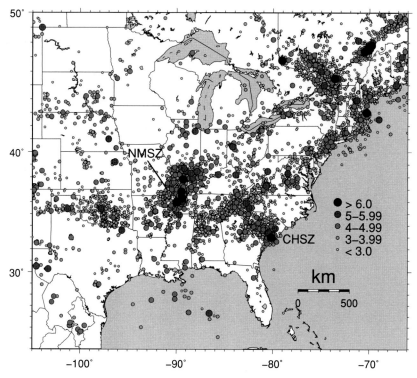

Figure 12.1 CEUS earthquake catalog determined for the CEUS SSC Project (Coppersmith *et al.*, 2012). The CEUS-SSC Catalog is modified to include a M5.2 inferred triggered earthquake near Louisville, Kentucky, which occurred at 10:40 LT on February 7, 1812 (Hough, 2001), and to revise the magnitude of the 1843 Marked Tree, Arkansas, earthquake from 6.0 to 5.4 (Hough, 2013).

times, (2) GPS results provide an increasingly tight (and low) bound on strain accrual rates, and (3) extensive research has focused on both the magnitudes and the recurrence rates of the (assumed) characteristic earthquakes. Moreover, efforts to understand intraplate seismogenesis, both in the CEUS and more generally, have focused largely on reconciling the apparent paradox posed by high rates of moment release versus a low rate of strain accrual. Research results to date as well as the conclusions of this chapter are expected to have general applicability for low strain rate regions, although, as I discuss later in the chapter, there appear to be some differences in the nature of seismicity characteristics between intraplate regions.

12.2 Seismic moment release in the Central/Eastern United States

12.2.1 Historical earthquakes: introduction and rupture scenarios

Moderate earthquakes have occurred in many regions in the CEUS during the ~300-year historical record (Figure 12.1). However, according to all recent characterizations, overall

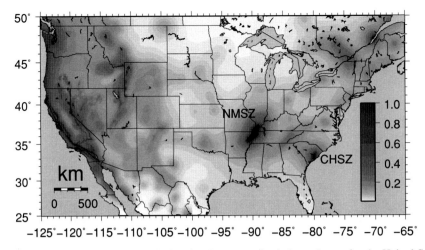

Figure 12.2 National Seismic Hazard Mapping Program seismic hazard map for the United States: PGA (%g) with 2% probability of exceedence in 50 years, rock site (grayscale) (results from Petersen *et al.*, 2008).

moment release in the region has been concentrated in two source zones: the New Madrid, Central United States, seismic zone (NMSZ) and the Charleston, South Carolina, seismic zone (CHSZ) (Figure 12.2). Because of low regional attenuation (e.g., Nuttli, 1973a), the NMSZ) contributes significantly to seismic hazard in relatively distant, large midwestern United States cities such as St. Louis, Missouri (Frankel *et al.*, 2002; Petersen *et al.*, 2008). I note that, in general, low attenuation will be a key factor contributing to seismic hazard estimates in intraplate regions; the focus of this chapter, however, is on issues related to source characterization. The CHSZ also contributes significantly to regional as well as local hazard estimates. Away from the NMSZ and CHSZ, the likelihood of a large (M ≥ 7) earthquake is low, typically estimated by extrapolating background seismicity rates assuming a Gutenberg–Richter distribution with an imposed M_{max}.

Given the importance of the NMSZ and CHSZ for probabilistic seismic hazard analysis (PSHA) in the CEUS, considerable effort has been devoted to understanding the large earthquakes that occurred in both regions during historical times. Decades of painstaking investigations have led to the development of detailed rupture scenarios for both the 1811–1812 New Madrid earthquake sequence and the 1886 Charleston earthquake. For the NMSZ, available data include (1) paleoliquefaction features preserved by the sediments within the Mississippi Embayment (e.g., Tuttle and Schweig, 1996); (2) the present-day distribution of seismicity in the NMSZ, which has commonly been assumed to be a long-lived aftershock sequence that illuminates the principal fault zones (e.g., Gomberg, 1993; Johnston, 1996; Mueller *et al.*, 2004; Ebel *et al.*, 2000; Stein and Liu, 2009); (3) first-hand reports ("felt reports") of the shaking and/or damage caused by the events over the CEUS (e.g., Nuttli, 1973b; Street, 1982, 1984).

The 1811–1812 New Madrid earthquake sequence comprised four principal earthquakes that occured over 3 months. By some accounts the principal events in this sequence are

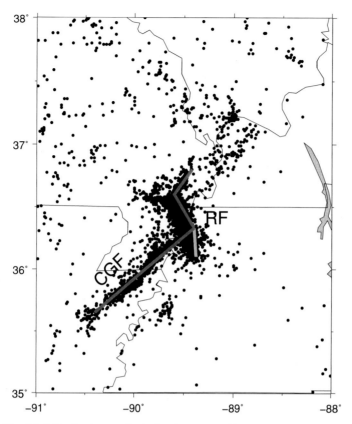

Figure 12.3 New Madrid seismic zone, including instrumentally recorded seismicity (black dots) and inferred traces of the Cottonwood Grove fault (gray line; CGF) and Reelfoot fault (gray line; RF). The location of the northern limb of the NMSZ is also indicated (gray line extending northeastward from the northern end of the RF).

among the largest – if not the largest – earthquakes to have ever occurred in a so-called Stable Continental Region (SCR; Johnston, 1996). Ground motions from the principal events were felt by individuals as far away as Canada, New England, and at a number of locations along the Atlantic coast (Mitchill, 1815; Fuller, 1912; Nuttli, 1973b; Street, 1982, 1984; Johnston, 1996). Contemporary accounts document three principal mainshocks: approximately 2:15 a.m. local time (LT) on December 16, 1811; around 8:00 a.m. LT on January 23, 1812, and approximately 3:45 a.m. LT on February 7, 1812 (henceforth NM1, NM2, and NM3, respectively). Additionally, a large aftershock to NM1 (NM1-A) occurred near dawn on December 16, 1811. There is some documentation of energetic aftershock sequences following each mainshock (Drake, 1815; McMurtie, 1819; Hough, 2009).

As noted, rupture scenarios have been developed for the four principal 1811–1812 events based on a number of lines of evidence (e.g., Johnston and Schweig, 1996; Figure 12.3). Of these, the association of NM3 with the Reelfoot thrust fault is the best supported by direct evidence, including the appearance of waterfalls along the Mississippi River (e.g., Russ, 1982; Odum *et al.*, 1998). NM1 is commonly associated with the strike-slip Cottonwood Grove fault, largely based on the prominent lineation illuminated by microseismicity as well as the distribution of liquefaction. Direct evidence from written archival accounts points only to a location somewhere south of the Reelfoot fault; i.e., unlike NM3, there are no accounts suggestive of surface rupture or accounts suggesting near-fault effects (e.g., localized extremely strong ground motions). The rupture scenario for NM1A is similarly uncertain. Two different scenarios have been proposed for NM1A: a rupture on the northern Cottonwood Grove strike-slip fault (Johnston and Schweig, 1996), or a rupture on the southeastern extension of the Reelfoot thrust fault (Hough and Martin, 2002). For this event direct evidence – the shaking distribution as inferred from written accounts – suggests only a location to the north of NM1. The rupture scenario of NM2 is the least well constrained: although associated by most authors with strike-slip motion on the northern limb of the NMSZ (e.g., Johnston and Schweig, 1996), Mueller *et al.* (2004) and Hough *et al.* (2005) summarize several lines of evidence from written accounts suggesting the event might have been centered in the Wabash Valley, well north of the NMSZ, in a region where a contemporaneous account describes "wagon loads" of sand erupting at the surface during the 1811–1812 sequence (although the account does specify when in the sequence the sand blows occurred).

Owing to the incomplete historical and paleoseismic records, the various 1811–1812 rupture scenarios and our general understanding of the NMSZ are heavily guided by the distribution and interpretation of present-day seismicity; I therefore now review this issue briefly. The long-lived aftershock hypothesis (e.g., Ebel *et al.*, 2000; Stein and Liu, 2009) has recently been called into question by Page *et al.* (2012), who used simulated sequences based on the epistemic triggering of aftershock (ETAS) model to show that no plausible set of sequence parameters can explain the robust known features of the early 1811–1812 aftershock sequence and the current rate of small earthquakes, assuming them to be aftershocks. It remains unclear, then, why present-day background seismicity does appear to illuminate the primary 1811–1812 ruptures. One might question this assumption; however, there is no question that present-day microseismicity illuminates the Reelfoot fault and, as noted, there is compelling evidence that this fault produced NM3. It reasonably follows, then, that microseismicity may well illuminate the overall sequence. An alternative hypothesis is that present-day microseismicity is driven by low levels of deep creep on continuations of faults below the brittle seismogenic crust, as Frankel *et al.* (2012) conclude is occurring on the Reelfoot and that perhaps continues at lower levels on the Cottonwood Grove fault. This would explain why the seismicity pattern of the NMSZ, with its well-defined side limbs, is consistent with predicted Coulomb stress increase associated with mainshock rupture on these two faults (Mueller *et al.*, 2004). (So far as we know, the temporal decay of aftershocks universally follows Omori's law (e.g., Omori, 1895). Since rigorous statistical tests reveal that NMSZ seismic activity since the early nineteenth century cannot be explained by Omori's law (Page *et al.*, 2012) and are therefore not aftershocks, a plausible alternative hypothesis is that ongoing creep at depth is not a consequence of post-seismic deformation, but rather is a steady-state process that drives both background seismicity as well as occasional large earthquakes.) In either case, any inference of rupture length for a historical earthquake based on present-day microseismicity patterns is clearly uncertain. A defensible alternative constraint on rupture length, for example, discussed in detail by Mueller *et al.* (2004) and Hough and Page (2011), is that NM3 rupture was bounded by the intersection with the northern and southern limbs of the NMSZ (Figure 12.3) rather than the significantly longer rupture length assumed by a number of other researchers (e.g., Johnston and Schweig, 1996; Frankel *et al.*, 2012). I discuss this issue at more length in a later section.

The Charleston earthquake of September 1, 1886 – 9:50 p.m. LT on August 31, 1886 – was the primary event in an apparently more typical earthquake sequence: a single large mainshock preceded by a small number of foreshocks and followed by a conventional, although spatially distributed, aftershock sequence (Dutton, 1889; Seeber and Armbruster, 1987). As discussed by Hough (2004), the overall felt extent of the mainshock was similar to that of the principal 1811–1812 New Madrid earthquakes, although far better sampled by extant archival accounts. As summarized by Talwani and Dura-Gomez (2009) and Dura-Gomez and Talwani (2009), available geophysical data combined with precise locations of present-day seismicity reveals a complex fault system including the northeast-striking Woodstock fault, an oblique right-lateral strike-slip fault with a ~6 km long antidilational left step through which the Sawmill Branch fault is the most active. As discussed by Dura-Gomez and Talwani (2009), detailed contemporary accounts point to the involvement of multiple faults, likely including the Woodstock fault, in the 1886 Charleston mainshock. Proposed rupture scenarios for this earthquake have been less detailed than those discussed above for the 1811–1812 New Madrid sequence. Recent microseismicity does not appear to delineate the historical mainshock rupture, and extensive liquefaction, as documented initially by Dutton (1889) and later by Talwani and Schaeffer (2001) is distributed over a broad source zone.

12.2.2 Historical earthquakes: magnitudes

The magnitudes of the principal 1811–1812 earthquakes are of critical importance for hazard assessment and efforts to understand intraplate seismogenesis. As discussed by numerous past studies (see Johnston and Schweig, 1996, for summary), inferred rupture scenarios and the size and distribution of liquefaction features provide some constraint on magnitude. There is no question that both the 1811–1812 mainshocks and the 1886 Charleston earthquake generated extensive liquefaction (e.g., Fuller, 1912; Obermeier *et al.*, 1990; Saucier *et al.*, 1991; Tuttle *et al.*, 2002; Dutton, 1889; Talwani and Cox, 1985). Widespread and enormous sand blows such as those in the NMSZ provide prima facie evidence for large magnitudes. However, such magnitude estimates are not well constrained (e.g., Pond and Martin, 1997; Stein and Newman, 2004). Recently, Holzer *et al.* (2011) presented preliminary results from a new method to estimate peak ground acceleration (PGA) from cone penetration test soundings, concluding that PGA values in the liquefaction zone of the 1811–1812 sequence were in the range 0.2–0.46 g. Holzer *et al.* (2011) state that these moderate-to-high values are consistent with $M_w \approx 7.5$ values and inconsistent with magnitudes smaller than 7. However, predicted accelerations at deep-sediment sites for earthquakes of M_w 7.0 and larger are uncertain in intraplate regions owing to a lack of direct observation. For example, the preliminary results of Ramirez-Guzman *et al.* (2011a), based on ground motion simulations, reveal perhaps surprising insensitivity of predicted intensities within the embayment, at distances over which significant liquefaction occurred, to magnitude values ranging from 7.0 to 7.7. That is, over this range of magnitudes, predicted intensities do not vary enough to be useful as a discriminant given the uncertainties in observed intensity values. Noting this uncertainty, Ramirez-Guzman *et al.* (2011b) suggest that predicted shaking intensities within the Mississippi Embayment will not be useful to constrain historical earthquake magnitudes within the range 7.0 to 7.7, and conclude that additional constraints, "such as those provided by paleoliquefaction analyses," will be needed to reduce uncertainties in magnitude estimates. However, while the distribution and size of liquefaction features might conceivably shed further light on magnitudes, the method developed by Holzer *et al.* (2011) uses paleoliquefaction analysis to derive ground motions, the interpretation of which will be plagued by the same uncertainties documented by Ramirez-Guzman *et al.* (2011a, b).

In light of these uncertainties, determination of magnitudes for the 1811–1812 mainshocks has thus hinged critically on the felt reports and the interpretation of associated modified Mercalli intensity (MMI) values by various investigations. Magnitude estimates have varied enormously, from ~7 (Hough and Page, 2011) to as high as 8¾ (Nuttli, 1979). Values close to M_w 7 have also been suggested based on constraints on the overall present-day strain rate (e.g., Newman *et al.*, 1999), although inference of magnitude from strain rate is highly model-dependent (e.g., Kenner and Segall, 2000). While different analyses of macroseismic data from the New Madrid sequence have yielded widely varying magnitude estimates, underscoring the inherent difficulty in assessing historical earthquake records, in all studies the four principal events are found to have roughly comparable magnitudes, with a range of at most 1 magnitude unit. For the maps released by the U.S. Geological Survey National Seismic Hazard Mapping Project (NSHMP) in 2002, values of 7.3–8.0 were considered using a logic-tree approach, with highest weight assigned to 7.7 (Frankel *et al.*, 2002; Petersen *et al.*, 2008).

Investigation of the Charleston earthquake dates back to the immediate post-earthquake investigations led by Clarence Dutton, an Army officer detailed to the U.S. Geological Survey. This effort culminated in the publication of one of the earliest comprehensive, scientific reports of a large earthquake (Dutton, 1889). The so-called Dutton Report includes thorough and consistent compilations of near-field geological effects of the earthquake and accounts of far-field macroseismic effects. Whereas about 100 or fewer intensity values are available for each of the New Madrid mainshocks, the Dutton Report provides the basis for assignment of over 1,000 intensity values. In a comprehensive interpretation of these accounts, Bollinger (1977) assigned almost 800 intensity values based on the 1,337 intensity reports tallied by Dutton (1889). Bollinger (1977) estimated an m_b value of 6.8–7.1

using the same techniques that Nuttli (1973a, b) used to estimate magnitudes for the New Madrid earthquakes. The intensity values determined by Bollinger (1977) have provided the basis for later investigation using increasingly modern methodology. Johnston (1996) estimated M_w 7.3 +/− 0.26 for the Charleston earthquake. Bakun and Hopper (2004) report a preferred M_w value of 6.9. Published M_w values for the Charleston earthquake have thus been relatively consistent: the U.S. National Hazards Mapping project currently assumes a range between M 6.8 and 7.5, with highest weight given to a value of 7.3 (Frankel *et al.*, 2002).

Comparing the intensity distributions of the 1886 mainshock with the best characterized intensity distribution from the 1811–1812 New Madrid sequence (the December main-shock), Hough (2004) concludes that these events had comparable magnitudes. Based on the results of Hough and Page (2011), this suggests $M_w \approx 6.8$ for the Charleston mainshock, consistent with the estimate of Bakun and Hopper (2004). This estimate is also consistent with the results of Leon *et al.* (2005), who analyze liquefaction observations and estimate magnitudes of Charleston paleoearthquakes of 6.5–7.0.

12.2.3 Historical earthquakes: recurrence rates

To estimate the seismic moment release rates for the NMSZ and CHSZ, it is necessary to estimate both the size of the large historical events, which are assumed to be characteristic earthquakes for their respective source zones, and their average recurrence rates. Sand blows generated by prehistoric earthquakes have been identified in both the NMSZ and Charleston regions. Although it is likely that future investigations will improve estimates of recurrence times, studies to date have produced compelling evidence of prehistoric earthquakes comparable in magnitude to those in historic times. Paleoseismic investigations suggest a repeat time on the order of 400–500 years over the past 2,000–3,000 years for both the New Madrid sequence and the Charleston earthquake (Talwani and Cox, 1985; Talwani and Schaeffer, 2001; Tuttle *et al.*, 2002); they also suggest that the NMSZ tends to produce prolonged sequences with multiple, distinct mainshocks, the magnitudes of which are comparable to those of the 1811–1812 events (e.g., Tuttle and Schweig, 1996; Tuttle *et al.*, 2002).

As summarized by Grollimund and Zoback (2001), this documented rate of Holocene NMSZ activity cannot be representative of a long-term rate (e.g., Schweig and Ellis, 1994), because extensive seismic reflection data reveal small cumulative offsets in post-Cretaceous embayment sediments (e.g., Hamilton and Zoback, 1981). Van Arsdale (2000) concluded from a combination of trench and reflection results that Holocene slip rates on the Reelfoot fault are at least four orders of magnitude higher than during the Pleistocene. Pratt (2012) suggests that pre-Holocene seismic activity involved faults other than those currently active in the NMSZ. However, in all attempts to model NMSZ seismogenesis, including that of Grollimund and Zoback (2001), it is taken as an incontrovertible constraint that NMSZ activity has "turned on" at some point in the relatively recent geological past, most likely within the Holocene.

12.2.4 Prehistoric earthquakes in other regions

Paleoseismic investigations in the CEUS have generally focused in the regions where large earthquakes occurred during historical times. However, paleoliquefaction features have been identified in other places in the CEUS, away from the NMSZ and the CHSZ. Significant paleoliquefaction features have been documented in the Wabash Valley (e.g., Munson *et al.*, 1997) associated with a large earthquake approximately 6,100 years BP. Green *et al.* (2004) reassessed the magnitude of this event, known as the Vincennes earthquake, using recently developed field and analytical techniques, estimating an approximate magnitude of M_w 7.5. The results of these investigations have been incorporated in the USGS hazard map. The Wabash Valley remains seismically active in modern times: since 1900 the region has produced a higher rate of $M \geq 5$ events than the NMSZ, which has not produced an earthquake larger than M 5 since 1895.

More recently, Al-Shukri *et al.* (2005) documented evidence for a significant paleo-liquefaction feature near Marianna, Arkansas, more than 100 km south of the currently seismically active segments of the NMSZ. These sand blows, dated to approximately 5,500 years BP, have very large dimensions, comparable to those found in the NMSZ. As noted by Tuttle *et al.* (2006), initial findings from the Marianna site suggest that seismicity might vary in space and time within the Reelfoot fault system (i.e., beyond the NMSZ proper), although more field investigation is needed to assess the history of activity at the Marianna site. Additionally, a number of geological investigations, including seismic reflection profiling and trenching, within or adjacent to the greater NMSZ have pointed towards Holocene activity on faults other than those apparently involved with the 1811–1812 sequence (e.g., Cox *et al.*, 2001; Magnani *et al.*, 2011). Several studies have presented evidence for Holocene faulting along the Commerce Geophysical lineament, which trends northeastward through central Arkansas, southeastern Missouri, southern Illinois, and south/central Indiana (Baldwin *et al.*, 2006; Harrison *et al.*, 1999; Figure 12.4). Conceivably, further investigation of Holocene faults adjacent to the central NMSZ could provide additional evidence for localized hazard, but without magnitude estimates or recurrence rates it is impossible to predict the implications for hazard.

Recent NSHMP maps assume a characteristic earthquake distribution for the NMSZ and the CHSZ; i.e., the rate of large events estimated from paleoseismic evidence is significantly higher than an extrapolation of background seismicity rates assuming a Gutenberg–Richter distribution (Gutenberg and Richter, 1944). (For the NMSZ, the rate of characteristic events is higher than the extrapolated Gutenberg–Richter rate by a factor of ~3 to a factor of 10 or more, depending on the assumed magnitude of the characteristic earthquakes [see Stein and Newman, 2004]). Apart from these two source zones and the Cheraw and Meers faults in Colorado and Oklahoma, respectively, earthquake rates for the United States national seismic hazard maps are estimated from background seismicity rates assuming a Gutenberg–Richter distribution, with an imposed M_{max} of 7.5 in rifted crust and 7.0 in SCR regions apart from rift zones (e.g., Frankel *et al.*, 2002).

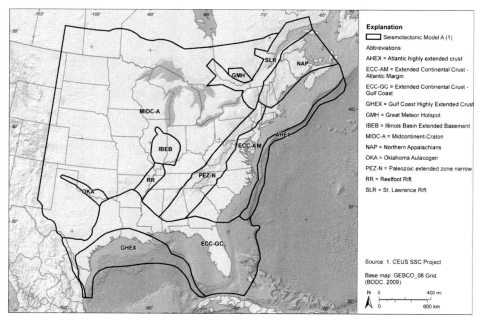

Figure 12.4 Seismotectonic zonation map developed for CEUS–SSC project (modified from Copper-smith *et al.*, 2012). The aim of this project was not to identify all failed rifts within the mid-continent region, but only those considered to have an elevated earthquake potential relative to stable continental regions. The Commerce Geophysical Lineament tracks the northwestern edge of the Reelfoot Rift (RR), extending from roughly the southwestern corner of the RR, continuing through the Illinois Basin Extended Basin (IBEB).

12.3 Strain rate

12.3.1 Observed strain rate

The interiors of tectonic plates are assumed to be rigid, with strain rates 3–4 orders of magnitude lower than active plate boundaries. In the NMSZ, increasingly precise estimates of strain rate from GPS data have revealed an increasingly tight upper bound on the level of strain accrual in the region. Although the initial investigation of Liu *et al.* (1992) estimated a strain rate of 5–7 mm/yr across the Reelfoot fault based on a combination of GPS data and triangulation data dating back to the 1950s, Calais *et al.* (2005) show that no statistically significant site motions are found within the NMSZ using GPS data collected through 2005. Using GPS data recorded through 2005, Calais *et al.* (2005) estimate an upper bound on residual velocities relative to stable North America of 1.4 mm/yr, at a 95% confidence level. Frankel *et al.* (2012) analyze GPS data recorded through 2010 and conclude that there is statistically significant motion of 0.37 +/− 0.07 mm/yr across the Reelfoot fault at the surface, which they show is consistent with an interseismic deep creep rate of 4 mm/yr on the Reelfoot fault at depths between 12 and 20 km. Their model predicts that sufficient strain will accumulate across the zone to produce a $M_w \approx 7.3$ earthquake every 500 years, which is roughly consistent with the Holocene paleoseismic record if characteristic earthquakes are low magnitude 7.

Any model derived from surface measurements of strain is non-unique, with generally poor resolution at depth (e.g., Page *et al.*, 2009). The assumed rupture length for a Reelfoot fault event is moreover uncertain. To estimate a magnitude for a Reelfoot fault event, Frankel *et al.* (2012) assume a rupture length of 60 km, with rupture extending over a full 24 km down-dip width. This rupture length is based primarily on the distribution of present-day microseismicity. Based on analog and box modeling, Pratt (2012) concludes that thrust faulting does extend over the full length of the central NMSZ limb as defined by microseismicity (see Figure 12.3.) However, as noted earlier, several other lines of evidence suggest predominant moment release associated with NM3 involved a shorter rupture length, bounded between the intersections of the Reelfoot and the northern and southern limbs of the NMSZ (Figure 12.3): (1) an extension of thrust motion south/southeast of the intersection of the Cottonwood Grove and Reelfoot faults is kinematically inconsistent, as right-lateral strike-slip movement on the Cottonwood Grove fault would lower the likelihood of failure of the thrust fault extending south/southeast of the junction with the Reelfoot fault (e.g., Mueller *et al.*, 2004); (2) the scarp/surface flexure associated with the Reelfoot fault is less well expressed geomorphically to the south/southeast of the intersection (e.g., Champion *et al.*, 2001); (3) Mueller *et al.* (2004) show that the "side limbs" on the NMSZ are consistent with off-fault lobes of increased Coulomb stress generated by ruptures on two master faults, the Cottonwood Grove and Reelfoot faults. Further, it is possible that the southeastern extension of the Reelfoot fault ruptured not during the February mainshock but rather during the dawn aftershock (NM1-A), as proposed by Hough and Martin (2002). Thus, while a plausible case can be made for a longer rupture length, a length of 40 km is also defensible. If one assumes a rupture length of 40 km and a depth of 20 km, the moment of a "500-year event" inferred by Frankel *et al.* (2012) would be reduced by nearly a factor of 2 (i.e., the ratio of fault area: $800 \text{ km}^2/1440 \text{ km}^2$), implying a M_w of 7.1 rather than 7.3. It has sometimes been suggested that the magnitudes of the 1811–1812 earthquakes could be higher than predicted from standard scaling relationships by virtue of high stress drop and/or unusually high shear modulus (e.g., Johnston, 1996). However, as discussed by Hanks and Johnston (1992), there is a strong dependence of high-frequency strong ground motions on stress drop. It is thus not possible to appeal to high stress drop as an explanation for high magnitudes, if the magnitudes are determined from high-frequency ground motions (i.e., intensities). In fact Hanks and Johnston (1992) themselves conclude that both the 1811–1812 mainshocks and the 1886 Charleston earthquake might be "no larger than $M = 6.5$ to 7, provided their stress drops are higher than average by a factor of 2 or so." (Note that this estimate was based on the original intensity assignments for the 1811–1812 earthquakes.)

The strain-rate observations from the NMSZ are thus consistent with two interpretations: (1) As argued by Hough and Page (2011), and corroborated by Frankel *et al.* (2012), a localized low but non-zero surface strain rate (e.g., 10^{-9}/yr), within the bounds imposed by the most recent GPS studies, could be sufficient to account for a sequence with an equivalent

$M_w \approx 7$ event every 500 years, a rate that is roughly consistent with the historic and prehistoric moment release rate if the principal 1811–1812 events and earlier mainshocks were in the range M_w 6.7–7; or (2) the principal 1811–1812 events were significantly larger than M_w 7, in which case deep-seated strain accrual processes in the NMSZ give rise to very low rates of surface deformation that are close to the level of detectability with available GPS data (e.g., see Kenner and Segall, 2000). The latter interpretation could be effectively untestable, at least in the foreseeable future, although increasingly precise results from GPS data will continue to improve an upper bound on strain accrual. I suggest the former interpretation is more likely, and better supported by available evidence.

12.3.2 Mechanism of strain accrual

The mechanism for localized strain accumulation in the NMSZ further remains enigmatic. Several mechanisms have been proposed, including (1) stress concentration associated with a Paleozoic mafic rift pillow (Grana and Richardson, 1996; Stuart et al., 1997); (2) glacial isostatic adjustment (GIA) following the removal of the Laurentide ice sheet (e.g., Grollimund and Zoback, 2001); (3) localized mantle flow driven by descent of the ancient Farallon slab (e.g., Forte et al., 2007); and (4) isostatic adjustment following the rapid incision of the Mississippi River Valley (Van Arsdale et al., 2007; Calais et al., 2011). As discussed by Talwani (this volume), the predicted magnitudes of stresses generated by the second and fourth mechanisms are low, which poses a problem for appealing to these mechanisms as the explanation for Holocene NMSZ activity. However, the explanations are attractive because they provide a qualitative and to some extent quantitative explanation for why NMSZ activity has been concentrated in the late Holocene (e.g., Kelson et al., 1996; Tuttle et al., 2002). There is no question that GIA contributes a significant level of present-day strain in eastern North America, not only over the region formerly covered by the Laurentide ice sheet but also well to the south, in the forebulge region (e.g., James and Bent, 1994; Grollimund and Zoback, 2001). Glacial isostatic adjustment is sometimes discounted as the likely cause of seismic activity because the strain signal is spatially distributed whereas significant moment release is apparently localized. Also, Wu and Johnston (2000) discounted the importance of the mechanism for NMSZ activity because earthquakes triggered by GIA in this location are predicted to have predominantly thrust mechanisms. However, unlike earlier interpretations of the 1811–1812 sequence (e.g., Johnston, 1996), the sequence is characterized by a predominantly thrust mechanism in the interpretations of Hough et al. (2000) and Hough and Page (2011). It is moreover plausible that GIA is a necessary but not sufficient condition for seismic activity: it has long been proposed that pre-existing zones of weakness such as failed rifts concentrate strain in intraplate crust (e.g., Sykes, 1978). Along the St. Lawrence Seaway in Canada, where the present-day GIA signal is higher than in the New Madrid region and has been measured with GPS data, Mazzotti et al. (2005) show that Holocene seismic moment release is consistent with the strain accrual associated with GIA (within the limitations of the short historical earthquake catalog).

Although the association of elevated modern-day seismicity with failed rifts dates back to the earliest efforts to develop models for CEUS (and intraplate) seismogenesis, some key questions clearly remain unanswered. For example, the Precambrian Mid-Continent (Keweenawan) Rift (e.g., Keller *et al.*, 1983) through the upper Midwest has not been characterized by elevated seismicity during the instrumental era. One possibility is that GIA and the presence of a failed rift together are a necessary but not a sufficient condition for elevated activity. Another possibility is that, as I discuss more in a later section, the historical catalog is simply too short to reveal all active source zones if recurrence intervals are long. I note, for example, that the largest historical earthquake in the state of Nebraska, an estimated M 5.1 event on November 15, 1877, is inferred to be associated with the western flank of the Keweenan mafic belt; it is possible, given the uncertainties in the event location, that this event was within the rift.

12.3.3 Distributed strain release: a simple model

I now explore simple calculations to consider the possibility that CEUS moment release is driven by strain accrual that is low but more spatially distributed than current hazard maps suggest. The precise level of strain rate is unknown. Although attempts have been made to estimate strain rates from historical moment release rates (e.g., Anderson, 1986), these are not expected to be reliable given the short catalog and the enormous uncertainties associated with the magnitude estimates of the largest events. Somewhat better constraint is available from GPS investigations. Calais *et al.* (2011) estimate only an upper bound of 1.3×10^{-9}/yr in the NMSZ, while Galgana and Hamburger (2011) recently estimated a value of 1–2×10^{-9}/yr in the Wabash valley. The model of Grollimund and Zoback (2001), which includes an imposed zone of weakness centered on the NMSZ, predicts a strain rate from GIA on the order of 10^{-9}/yr.

As an initial conservative estimate, I assume a strain rate of 0.5–1.0×10^{-9}/yr is distributed over the failed rifts that have elevated seismic potential (Coppersmith *et al.*, 2012), and that these zones cumulatively make up approximately 10% of the CEUS–SCR crust (Figure 12.4). A moment rate corresponding to distributed strain accumulation can be estimated from

$$d\varepsilon/dt = dM_o/dt/(2.67\mu Ah) \tag{12.1}$$

where h is the thickness of the brittle layer, μ is shear modulus taken to be 3×10^{11} dyne/cm^2, and A is area (Anderson, 1986). As an illustrative simple calculation, I consider a total CEUS area of 2800×2400 km^2 in which there is an elevated strain rate of 0.5–1.0×10^{-9} within failed rifts that make up 10% of the total area. Assuming $h = 15$ and no aseismic strain release, Equation (12.1) predicts an overall moment accrual rate of $(4$–$8) \times 10^{24}$ dyne cm/yr. If this rate were released solely in M_{max} events, it would be sufficient to produce one M_w 7.0 earthquake somewhere within failed rifts every 45–90 years. Assuming that 10% of the moment rate is released by smaller earthquakes, the result is 50–100 years. Further assuming a *b*-value of 1, this corresponds to a M_w 6 or greater earthquake approximately one every 5–10 years.

The above calculation, while highly simplified, can be compared with the observed rate of large earthquakes in the CEUS. The rate of $M_w \geq 7$ events is difficult to estimate, so I instead consider the rate of $M_w \approx 6$ events. For example, since 1850 there have been approximately six CEUS earthquakes with magnitudes close to M_w 6, as well as the 1886 mainshock: the largest aftershock of the 1886 Charleston earthquake (Talwani and Sharma, 1999), the 1895 Charleston, MO, earthquake (Bakun and Hopper, 2004), the 1897 Giles County, VA, earthquake (Bollinger and Hopper, 1971), the 1931 Valentine, TX, earthquake (Doser, 1987), the 1944 Massena, NY, earthquake (Bent, 1996), and the 2011 Mineral, VA, earthquake (e.g., http://neic.usgs.gov/neis/eq_depot/2011/eq_110823_se082311a/se082311a_l.html, last accessed 19 Dec. 2012). The observed rate of $M_w \geq 6$ events between 1850 and 2013 has thus been one every 23 years, an obviously rough estimate of long-term rate but one that is within a factor of 2–4 of the ballpark estimate. If one instead considers the expected rate of $M_w \geq 5.8$ events based on the above model, since several of the historical earthquakes were slightly smaller than M_w 6, the expected rate is higher (one every 3–6 years) than the ballpark estimate. I note, however, that (1) the discrepancy is significantly smaller than any model that proposes to account for $M_w \geq 7.5$ earthquakes every $\sim$500 years given best-available constraints on strain rate, and (2) both the calculation and the estimation of observed rates are highly simplified and uncertain. For example, the list of observed earthquakes since 1850 excludes five events that occurred in Canada but within the region defined in Figure 12.4. Other complications abound. For example, as shown by Hough *et al.* (2003), even a low level of permanent deformation (e.g., folding, or pressure solution) could account for a higher percentage of overall strain release in a low strain rate region than an active plate boundary, reducing the level of strain released seismically. For example, using calculations based on a simple model, they show that, for a strain rate of 5×10^{-8}/yr, permanent deformation will reduce accrued strain on the order of 30% over 3,000 years. For this and other reasons, a quantitative assessment of strain accrual is highly uncertain. The key point, however, is that even a low level of distributed strain accrual can generate a significant (and distributed) overall hazard. That is, even a strain rate as low as 0.5×10^{-9}/yr would be sufficient to produce as many or more large earthquakes as have been observed in historical times, if the strain is distributed over a region covering roughly 10% of the CEUS. (While significant non-zero strain rates have only barely been observed to-date in the NMSZ and Wabash region, a rate on the order of 0.5×10^{-9}/yr would be below the current level of detection in other CEUS regions.)

12.3.4 Other intraplate regions

A growing body of geological evidence reveals that many faults in low strain rate regions generate large earthquakes that are clustered on timescales of a few centuries to perhaps a few millennia, with much longer intervening periods when the fault might be considered dormant. Examples (see Coppersmith [1988] and Crone *et al.* [2003] for summaries) include the Meers fault in Oklahoma (Crone and Luza, 1990), the clear scarp of which is inferred to have been generated by two M $\approx$ 7 earthquakes during the past ~3,000 years, with no evidence of prior events over 100,000 years or more (see Crone *et al.*, 2003), and the Cheraw fault in Colorado, which has produced three large earthquakes between 8 and 25 ka BP.

Especially compelling geological evidence for temporally clustered seismic activity on individual faults has been documented in Australia. A number of active faults in Australia reveal behavior similar to that of the Meers fault (e.g., Crone *et al.*, 2003; Clark *et al.*, 2012), including the Roopena fault in South Australia, the Hyden fault in Western Australia, and the Lake Edgar fault in southwest Tasmania. To quote Clark *et al.* (2012), "A common characteristic of morphogenic earthquake activity in Australia appears to be temporal clustering. Periods of earthquake activity comprising a finite number of large events are separated by much longer periods of seismic quiescence." In eastern Utah, in a region east of the Basin and Range province, where a precise strain rate estimate is not available but thought to be low, Schwartz *et al.* (2012) summarize evidence that the entire neotectonic history of the Bear River fault consists of two M 6–6.5 surface-rupturing normal earthquakes during the late Holocene.

In low strain rate regions of China, the especially long historical record reveals that large earthquakes have occurred throughout a broad region, with no evidence for clustering or even repeated large events in any individual source zone over the past ~700 years (Liu *et al.*, 2011). One key caveat is that paleoseismological investigations that might document recurrence intervals have apparently not been undertaken at the source zones discussed by Liu *et al.* (2011). However, with a historical record extending back to 1300, it is unlikely that individual source zones in low strain rate regions in China are characterized by the same type of clustering documented in the NMSZ and CHSZ. It is not clear why low strain rate regions in China do not (apparently) generate the same type of clustering that appears to be common in other regions.

One possible explanation for clustering is a feedback mechanism between an effectively embedded finite fault in the upper crust and a viscoelastic lower crust (e.g., Kenner and Segall, 2000). In such a setting, an initial episode of strain release can trigger a sequence of large events as post-seismic motions in a deep viscoelastic layer repeatedly reload an individual fault in the brittle crust. A simplified analytical model predicts that the recurrence interval will scale linearly with the viscosity of the weak lower crust (Kenner and Segall, 2000). One might conjecture, then, that differences in degree of clustering in different regions are due to differences in lower crustal viscosity. Lower crustal viscosity is poorly constrained on a global basis. However, absolute plate motions are well established, and are known to vary significantly among intraplate regions (Table 12.1.) It has been shown that plate motions are driven primarily by lateral forces (e.g., Forsyth and Uyeda, 1975; Conrad and Lithgow-Bertelloni, 2002), but drag on the bottom of plates influences plate motion as well, and is stronger under continents than oceans (Forsyth and Uyeda, 1975). A relatively fast plate motion might therefore be expected to correlate to some extent

Table 12.1 *Plate rate versus propensity for large event clustering for four tectonic plates*

Tectonic plate	Plate rate (cm/yr)	Short-term clustering?
Australian	5.8	Yes
Indian	2.6	No?
North American	4.1	Yes
Eurasian	2.3	No

with low viscosity and, therefore, with a shorter recurrence interval for individual source zones. This hypothesis is highly speculative; however, both Australia and North America are characterized by relatively fast plate rates and clustering, whereas peninsular India and Eurasia are characterized by significantly lower plate rates and no clustering.

In any case, in all of the intraplate regions discussed above, evidence points to earthquake activity that in some SCR regions tends to be clustered on timescales of centuries to a few millennia, and that in all SCR regions is distributed throughout a broad region over longer timescales.

12.4 Statistical considerations

In the absence of geological and geodetic constraints on fault slip rates or direct evidence for large characteristic earthquakes, PSHA calculations have long relied on the observed rate of small-to-moderate earthquakes in a region to estimate the expected rate of large earthquakes. This practice is motivated by good evidence that, through any given region, seismicity is characterized by a Gutenberg–Richter magnitude distribution (Gutenberg and Richter, 1944) with a *b*-value of 1.0 (e.g., King, 1983; Felzer, 2006). The practice is further justified by appealing to the assumption of stationarity; the same rationale is sometimes also used to estimate M_{max}. However, it has been shown that a stationary process, for example the Epistemic Triggering of Aftershock (ETAS) model (Kagan and Knopoff, 1981; Ogata, 1988), will give rise to a constant *b*-value but an apparent *a*-value that can vary considerably if estimated from a relatively short earthquake catalog (Page *et al.*, 2010). As shown by Stein and Newman (2004), within a short catalog large earthquakes can appear to be either characteristic or "uncharacteristic"; i.e., to not be represented in the catalog at the average long-term rate.

The ETAS model, which provides an integrated framework for seismicity including foreshock and aftershock probabilities, has been shown to match the degree of clustering observed in typical earthquake catalogs (e.g., Hardebeck *et al.*, 2008). An ETAS model predicts not only traditional foreshock/aftershock clustering, but also regional clustering of moderate earthquakes. In California, for example, such clustering can give rise to periods of relative high or low activity that extend for several decades (Hardebeck *et al.*, 2008). Using simulated ETAS catalogs, Page *et al.* (2010) show that, for any region, reliable estimation of a long-term *a*-value requires a catalog that is several times longer than the recurrence time of the M_{max} event in the region. Even in California, where strain rate is relatively high, the long-term *a*-value remains uncertain by 20% or more (e.g., WGCEP, 2013). Relative to California, the historical record in the CEUS is at most a factor of 2 longer, but the strain rate is a factor of 1,000 or more lower. Even without detailed statistical calculations, it is clear that a 100- or 300-year snapshot of seismicity cannot capture all potentially active source zones or provide a reliable estimate of long-term rates on timescales of millennia.

12.5 Discussion and conclusions

Quantification of seismic hazard in the CEUS, and by extension any similar low strain rate region, emerges as a daunting prospect. Probabilistic seismic hazard assessment remains challenging in a well-studied, high strain rate region such as California (e.g., WGCEP, 2013), where geological and geodetic constraints are each uncertain, but presumably allow hazard to be characterized to first order. In the low strain rate CEUS region, where the seismic catalog is at best a factor of 2–10 longer than that in California, earthquake rates are a factor of 1,000 or more lower, geological and geodetic constraints are limited or non-existent, and hazard is likely to be dominated by low-probability events. There is little question that the inputs, for example given published magnitude estimates of characteristic earthquakes that vary by nearly a full magnitude unit, to PSHA remain highly uncertain in a low strain rate region. And, again, to quote from Clark *et al.* (2012), "This apparent bimodal recurrence behavior poses challenges for probabilistic seismic hazard assessment."

In the absence of adequate constraint on long-term slip rates, deformation, and seismic rates, an alternative to the use of smoothed seismicity models involves consideration of tectonic zonation. Since 2005 the seismic hazard map of Canada has been determined in part based on an interpretation of tectonic zonation (e.g., Adams and Halchuk, 2003). Whereas the maps prior to 2005 had conspicuous "bull's-eyes" around the largest known late Holocene earthquakes, high hazard in the new maps is more distributed throughout broader zones, with less pronounced bull's-eyes corresponding to regions of historical activity.

As paleoseismologists continue to study the Central and Eastern United States, it also appears likely that geological investigations will identify other source zones that have produced large Holocene earthquakes. Inclusion of rates based on observed prehistoric earthquakes in other areas will likely serve to distribute hazard over broader zones. A reduction of the estimated magnitudes of the New Madrid and perhaps Charleston mainshocks would further serve to lessen the estimated hazard in proximity to these zones.

A fundamental challenge, however, will be to account for uncertainties associated with long-term rate estimates, not only in the NMSZ and CHSZ regions, but also in regions where no large earthquake has occurred during historical times. In the NMSZ and CHSZ, as discussed earlier, late Holocene recurrence rates of large earthquakes have been estimated, but it is not clear whether recent events are part of (relatively) short-term clusters that will eventually end. For these regions there is a consensus if not a unanimous view that PSHA treatments should assume that future rates will be comparable to those estimated for recent millennia.

As pointed out by Stein and Newman (2004), short catalogs can give rise to not only apparent characteristic earthquakes with shorter recurrence rates than the long-term rate, but also to earthquakes that appear less frequently than their long-term rates ("uncharacteristic earthquakes.") The possibility has been at least discussed, for example at the 2012 National Seismic Hazard Mapping Program CEUS workshop, of including a low-weight branch in a logic-tree approach corresponding to the possibility that no large (M > 7) earthquakes will occur in the future in the NMSZ. One possible treatment of uncertainties associated with the short historical catalog and limited geological record, then, would also include logic-tree branches corresponding to the possibility that other source zones might produce large earthquakes at a higher rate over the next few millennia than the rate observed during the late Holocene.

Acknowledgments

I thank Pradeep Talwani for the invitation to contribute to this volume, as well as for feedback on the chapter, and am grateful to Oliver Boyd, Rob Williams, and an anonymous reviewer for critical reviews that improved the manuscript. I also acknowledge Karen Felzer and Morgan Page for many helpful discussions.

References

Adams, J., and Halchuk, S. (2003). *Fourth Generation Seismic Hazard Maps of Canada: Values for Over 650 Canadian Localities Intended for the 2005 National Building Code of Canada*. Geological Survey of Canada.

Al-Shukri, H. J., Lemmer, R. E., Mahdi, H. H., and Connelly, J. B. (2005). Spatial and temporal characteristics of paleoseismic features in the southern terminus of the New Madrid seismic zone in eastern Arkansas. *Seismological Research Letters*, 76, 502–511, doi:10.1785/gssrl.76.4.502.

Anderson, J. G. (1986). Seismic strain rates in the central and eastern United States. *Bulletin of the Seismological Society of America*, 76, 273–290.

Bakun, W. H., and Hopper, M. (2004). Magnitudes and locations of the 1811–1812 New Madrid, Missouri and the 1886 Charleston, South Carolina earthquakes. *Bulletin of the Seismological Society of America*, 94, 64–75.

Baldwin, J. N., Witter, R. C., Vaughn, J. D., *et al.* (2006). Geological characterization of the Idalia Hill fault zone and its structural association with the Commerce Geophysical Lineament, Idalia, Missouri. *Bulletin of the Seismological Society of America*, 96, 2281–230, doi: 10.1785/0120050136.

Bent, A. (1996). Source parameters of the damaging Cornwall-Massena earthquake of 1944 from regional waveforms. *Bulletin of the Seismological Society of America*, 86, 489–497.

Bollinger, G. A. (1977). Reinterpretation of the intensity data for the 1886 Charleston, South Carolina, earthquake. In *Studies Related to the Charleston, South Carolina Earthquake of 1886: A Preliminary Report*, ed. D. W. Rankin, U.S. Geological Survey Professional Paper 1028, pp. 17–32.

Bollinger, G. A., and Hopper, M. (1971). Virginia's two largest earthquakes – December 22, 1875 and May 31, 1897. *Bulletin of the Seismological Society of America*, 61, 1033–1039.

Calais, E., Han, J. Y., DeMets, C., and Nocquet, J. M. (2005). Deformation of the North American plate interior from a decade of continuous GPS measurements. *Journal of Geophysical Research*, 111, B06402, doi: 10.1029/2005JB004253.

Calais, E., Freed, A. M., Van Ardsale, R., and Stein, S. (2011). Triggering of New Madrid seismicity by late-Pleistocene erosion. *Nature*, 566, 608-U2, doi: 10.1038/nature09258.

Champion, J., Mueller, K., Tate, A., and Guccione, M. (2001). Geometry, numerical models and revised slip rate for the Reelfoot fault and trishear fault-propagation fold, New Madrid seismic zone. *Engineering Geology*, 62, 31–49, doi: 10.1016/S0013-7952(01)00048-5.

Clark, D., McPherson, A., and Van Dissen, R. (2012). The long-term behavior of the Australian stable continental region (SCR) faults. *Tectonophysics*, 566, 1–30, doi: 10.1016/j.tecto.2012.07.004.

Conrad, C. P., and Lithgow-Bertelloni, C. (2002). How mantle slabs drive plate tectonics. *Science*, 298, 207–209, doi: 10.1126/science.1074161.

Coppersmith, K. (1988). Temporal and spatial clustering of earthquake activity in the Central and Eastern United States. *Seismological Research Letters*, 59, 299–304, doi:10.1785/gssrl.59.4.299.

Coppersmith, K. J., Salomone, L. A., Fuller, C. W., et al. (2012). *Central and Eastern United States (CEUS) Seismic Source Characterization (SSC) for Nuclear Facilities Project* (No. DOE/NE-0140). Electric Power Research Institute.

Cox, R. T., Van Arsdale, R. B., Harris, J. B., and Larsen, D. (2001). Reelfoot rift zone margin, central United States, and implications for regional strain accumulation. *Geology*, 29, 419–422, doi: 10.1130/0091-7613(2001)029<0419:NOTSTRR>2.0.CO;2.

Crone, A. J., and Luza, K. V. (1990). Style and timing of Holocene surface faulting on the Meers fault, southwestern Oklahoma. *Geological Society of America Bulletin*, 102, 1–17, doi:10.1130/0016–7606(1990).

Crone, A. J., De Martini, P. M., Machette, M. N., Okumura, K., and Prescott, J. R. (2003). Quiescent faults in Australia: implications for fault behavior in stable continental regions. *Bulletin of the Seismological Society of America*, 93, 1913–1934, doi: 10.1785/010000094.

Doser, D. I. (1987). The 16 August 1931 Valentine, Texas, earthquake: evidence for normal faulting in west Texas. *Bulletin of the Seismological Society of America*, 77, 2005–2017.

Drake, D. (1815). *Natural and Statistical View, or Picture of Cincinnati and the Miami County, Illustrated by Maps*. Cincinnati: Looker and Wallace.

Dura-Gomez, I., and Talwani, P. (2009). Finding faults in the Charleston Area, South Carolina: 1. Seismological data. *Seismological Research Letters*, 80, 883–900, doi: 10.1785/gssrl.80.5.883.

Dutton, C. (1889). The Charleston earthquake of August 31, 1886. *U.S. Geological Survey Ninth Annual Report, 1887–88*, pp. 203–528.

Ebel, J. E., Bonjer, K.-P., and Oncescu, M. C. (2000). Paleoseismicity: seismicity evidence for past large earthquakes. *Seismological Research Letters*, 71, 283–294, doi: 10.1785/gssrl.71.2.283.

Felzer, K. R. (2006). Calculating the Gutenberg–Richter *b*-value (abstract), American Geophysical Union Fall meeting, abstract S42C-08.

Forsyth, D., and Uyeda, S. (1975). On the relative importance of the driving forces of plate motion. *Geophysical Journal of the Royal Astronomical Society*, 43, 163–200.

Forte, A. M., Mitrovica, J. X., Moucha, R., Simmons, N. A., and Grand, S. P. (2007). Descent of ancient Farallon slab drives localized flow below the New Madrid seismic zone. *Geophysical Research Letters*, 34, L04308, doi: 10.1029/2006GL027895.

Frankel, A., Petersen, M.D., Mueller, C. S., *et al.* (2002). Documentation for the 2002 Update of the National Seismic Hazard Maps. U.S. Geological Survey Open File Report 02-420.

Frankel, A., Smalley, R., and Paul, J. (2012). Significant motions between GPS sites in the New Madrid region: implications for seismic hazard. *Bulletin of the Seismological Society of America*, 102, 479–489. doi: 10.1785/0120100219.

Fuller, M. L. (1912). The New Madrid earthquakes. *U.S. Geological Survey Bulletin*, 494.

Galgana, G. A., and Hamburger, H. M. (2011). Geodetic observations of active intraplate crustal deformation in the Wabash Valley seismic zone and the southern Illinois basin. *Seismological Research Letters*, 81, 699–714, doi:10.1785/gssrl.81.5.699.

Gomberg, J. S. (1993). Tectonic deformation in the New Madrid seismic zone: inferences from map view and cross-sectional boundary element models. *Journal of Geophysical Research*, 98, 6639–6664.

Grana, J. P., and Richarson, R. M. (1996). Tectonic stress within the New Madrid seismic zone. *Journal of Geophysical Research*, 101, 5445–5458.

Green, R. A., Obermeier, S. F., and Olson, S. M. (2004). The role of paleoliquefaction studies in performance-based earthquake engineering in the central-eastern United States. 13th World Conference on Earthquake Engineering, Vancouver, Canada, August 1–6, Paper 1643.

Grollimund, B., and Zoback, M. D. (2001). Did glaciation trigger intraplate seismicity in the New Madrid Seismic Zone? *Geology*, 29, 175–178, doi:10.1130/0091–7613.

Gutenberg, B., and Richter, C. F. (1944). Frequency of earthquakes in California. *Bulletin of the Seismological Society of America*, 34, 185–188.

Hamilton, R. M., and Zoback, M. D. (1981). Tectonic features of the New Madrid seismic zone from seismic reflection profiles. In *Investigations of the New Madrid Earthquake Region*, ed. F. A. McKeown and L. C. Pakiser, U.S. Geological Survey Professional Paper 1236, pp. 55–82.

Hanks, T. C., and Johnston, A. C. (1992). Common features of the excitation and propagation of strong ground motion for North American earthquakes. *Bulletin of the Seismological Society of America*, 82, 1–23.

Hardebeck, J. L., Felzer, K. R., and Michael, A. J. (2008). Improved tests reveal that the accelerating moment release hypothesis is statistically insignificant. *Journal of Geophysical Research*, 113, B08310, doi:10.1029/2007JB005410.

Harrison, R. W., Hoffman, D., Vaughn, J. D., *et al.* (1999). An example of neotectonism in a continental interior: Thebes Gap, midcontinent, United States. *Tectonophysics*, 305, 399–417.

Holzer, T. L., Noce, T. E., and Burnett, M. J. (2011). Implications of liquefaction caused by the 1811–1812 New Madrid earthquakes for estimates of ground shaking and earthquake magnitudes (abstract). *Seismological Research Letters*, 82, 274.

Hough, S. E. (1996). The case against huge earthquakes. *Seismological Research Letters*, 67, 3–4.

Hough, S. E. (2001). Triggered earthquakes and the 1811–1812 New Madrid, central United States, earthquake sequence. *Bulletin of the Seismological Society of America*, 91, 1574–1581.

Hough, S. E. (2004). Scientific overview and historical context of the 1811–1812 New Madrid earthquake sequence. *Annals of Geophysics*, 47, 523–537.

Hough, S. E. (2009). Cataloging the 1811–1812 New Madrid, Central U.S. earthquake sequence. *Seismological Research Letters*, 80, 1045–1053.

Hough, S. E., and Martin, S. (2002). Magnitude estimates of two large aftershocks of the 16 December, 1811 New Madrid earthquake. *Bulletin of the Seismological Society of America*, 92, 3259–3268.

Hough, S. E., and Page, M. (2011). Towards a consistent model for strain accrual and release for the New Madrid Seismic Zone. *Journal of Geophysical Research*, 116, doi: 10.1029/2010JB007783.

Hough, S. E. (2013). Spatial variability of "Did You Feel It?" intensity data: insights into sampling biases in historical earthquake intensity distributions. *Bulletin of the Seismological Society of America*, 103, 2767– 2781.

Hough, S. E., Armbruster, J. G., Seeber, L., and Hough, J. F. (2000). On the modified Mercalli intensities and magnitudes of the 1811–1812 New Madrid, Central United States earthquakes. *Journal of Geophysical Research*, 105, 23, 839–823, 864.

Hough, S. E., Seeber, L., and Armbruster, J. G. (2003). Intraplate triggered earthquakes: observations and interpretation. *Bulletin of the Seismological Society of America*, 93, 2212–2221.

Hough, S. E., Bilham, R., Mueller, K., *et al.* (2005). Wagon loads of sand blows in White County, Illinois. *Seismological Research Letters*, 76, 373–386.

James, T. S., and Bent, A. L. (1994). A comparison of North American seismic strain rates to glacial rebound strain-rates. *Geophysical Research Letters*, 21, 2127–2130.

Johnston, A. C. (1996). Seismic moment assessment of earthquakes in stable continental regions III, New Madrid 1811–1812, Charleston 1886, and Lisbon 1755. *Geophysical Journal International*, 126, 314–344.

Johnston, A. C., and Schweig, E. S. (1996). The enigma of the New Madrid earthquakes of 1811–1812. *Annual Review of Earth and Planetary Science*, 24, 339–384, doi:10.1146/annurev.earth.24.1.339.

Kagan, Y. Y., and Knopoff, L. (1981). Stochastic synthesis of earthquake catalogs. *Journal of Geophysical Research*, 86, 2853–2862.

Keller, G. R., Lidiak, E. G., Hinze, W. J., and Braile, L. W. (1983). The role of rifting in the tectonic development of the midcontinent, U.S.A. *Tectonophysics*, 94, 391–412, doi:10.1016/0040-1951(83)90026-4.

Kelson, K. I., Simpson, G. D., VanArsdale, R. B., Haraden, C. C., and Lettis, W. R. (1996). Multiple late Holocene earthquakes along the Reelfoot fault, central New Madrid seismic zone. *Journal of Geophysical Research*, 101, 6151, doi: 10.1029/95JB01815.

Kenner, S. J., and Segall, P. (2000). A mechanical model for interplate earthquakes: application to the New Madrid Seismic Zone. *Science*, 289, 2329–2332.

King, G. (1983). The accommodation of large strains in the upper lithosphere of the earth and other solids by self-similar fault systems: the geometrical origin of *b*-value. *Pure and Applied Geophysics*, 121, 761–815.

Leon, E., Gassman, S. L., and Talwani, P. (2005). Effect of soil aging on assessing magnitudes and accelerations of prehistoric earthquakes. *Earthquake Spectra*, 21, 737–759.

Liu, L., Zoback, M.D., and Segall, P. (1992). Rapid intraplate strain accumulation in the New Madrid Seismic Zone. *Science*, 257, 1666–1669.

Liu, M., Stein, S., and Wang, H. (2011). 2000 years of migrating earthquakes in north China: how earthquakes in midcontinents differ from those at plate boundaries. *Lithosphere*, 3, 128–132, doi:10.1130/L129.1.

Magnani, M., McIntosh, K. D., and Guo, L. (2011). Paleotectonic control on distribution of long-term deformation in the central United States from high-resolution seismic data. Abstract S22A-03, American Geophysical Union Fall Meeting, San Francisco, CA.

Mazzotti, S., James, T. S., Henton, J., and Adams, J. (2005). GPS crustal strain, postglacial rebound, and seismic hazard in eastern North America: the Saint Lawrence valley example. *Journal of Geophysical Research*, 110, B11301, doi:10.1029/2004JB003590.

McMurtrie, H. (1819). *Sketches of Louisville and Its Environs; Including, Among a Great Miscellaneous Matter, a Florula Louisvillensis; or, a Catalogue of Nearly 400 Genera and 600 Species of Plants, That Grow in the Vicinity of the Town, Exhibiting Their Generic, Specific, and Vulgar English Names.* S. Penn, Jun. Main-street, Louisville.

Mitchill, S. L. (1815). A detailed narrative of the earthquakes which occurred on the 16th day of December, 1811. *Transactions of the Literary and Philosophical Society of New York*, 1, 281–307.

Mueller, K., Hough, S. E., and Bilham, R. (2004). Analysing the 1811–1812 New Madrid earthquakes with recent instrumentally recorded aftershocks. *Nature*, 429, 284–288.

Munson, P. J., Obermeier, S. F., Munson, C. A., and Hajic, M. R. (1997). Liquefaction evidence for Holocene and latest Pleistocene seismicity in the southern halves of Indiana and Illinois: a preliminary overview. *Seismological Research Letters*, 68, 521–536.

Newman, A., Stein, S., Weber, J., *et al.* (1999). Slow deformation and lower seismic hazard at the New Madrid seismic zone. *Science*, 284, 619–621.

Nuttli, O. W. (1973a). Seismic wave attenuation and magnitude relations for eastern North America. *Journal of Geophysical Research*, 78, 876, doi:10.1029/JB078i005p00876.

Nuttli, O. W. (1973b). The Mississippi Valley earthquakes of 1811 and 1812: intensities, ground motion, and magnitudes. *Bulletin of the Seismological Society of America*, 63, 227–248.

Nuttli, O. W. (1979). Seismicity of the central United States. *Geological Society of America, Reviews in Engineering Geology*, IV, 67–93.

Obermeier, S. F., Jacobson, R. B., Smoot, J. P., *et al.* (1990). Earthquake induced liquefaction features in the coastal setting of South Carolina and in the fluvial setting of the New Madrid Seismic Zone. U.S. Geological Survey Professional Paper 1504.

Odum, J. K., Stephenson, W. J., and Shedlock, K. M. (1998). Near-surface structural model for deformation associated with the February 7, 1812 New Madrid, Missouri, earthquake. *Geological Society of America Bulletin*, 110, 149–162.

Ogata, Y. (1988). Statistical models for earthquake occurrence and residual analysis for point processes. *Journal of the American Statistical Association*, 83, 9–27.

Omori, F. (1895). On the after-shocks of earthquakes. *Journal of the College of Science, Imperial University of Tokyo*, 7, 111–200.

Page, M. T., Custodio, S., Archuleta, R. J., and Carlson, J. M. (2009). Constraining earthquake source inversions with GPS data: 1. Resolution-based removal of artifacts. *Journal of Geophysical Research*, 114, B01314, doi:10.1029/2007JB005449.

Page, M., Felzer, K. R., Weldon, R. J., *et al.* (2010). The case for Gutenberg-Richter scaling on faults (abstract). *Seismological Research Letters*, 81, 330.

Page, M., Hough, S. E., and Felzer, K. (2012). Can current New Madrid seismicity be explained as a decaying aftershock sequence? Abstract S54D-07, American Geophysical Union Fall Meeting, San Francisco, CA, 3–7 December.

Petersen, M. D., Frankel, A. D., Harmsen, S.C., *et al.* (2008). *Documentation for the 2008 Update of the United States National Seismic Hazard Maps*. U.S. Geological Survey Open-File Report 2008–1128.

Pond, E. C., and Martin, J. R. (1997). Estimated magnitudes and accelerations associated with prehistoric earthquakes in the Wabash Valley region in the central U.S. *Seismological Research Letters*, 68, 611–623.

Pratt, T. L. (2012). Kinematics of the New Madrid seismic zone, central United States, based on stepover models. *Geology*, 40, 371–374, doi:10.1130/G32624.1.

Ramirez-Guzman, L., Graves, R. W., Olsen, K. B., *et al.* (2011a). Central United States earthquake ground motion simulation working group: the 1811–1812 New Madrid earthquake sequence (abstract). *Seismological Research Letters*, 82, 275.

Ramirez-Guzman, L., Graves, R. W., Olsen, K. B., *et al.* (2011b). Magnitude uncertainty and ground motion simulations of the 1811–1812 New Madrid earthquake sequence (abstract). American Geophysical Union Fall Meeting, abstract S22A-07.

Russ, D. P. (1982). Style and significance of surface deformation in the vicinity of New Madrid, Missouri. In *Investigations of the New Madrid Earthquake Region*, ed. F. A. McKeown and L. C. Pakiser, U.S. Geological Survey Professional Paper 1236, pp. 95–114.

Saucier, R. T. (1991). Geoarcheological evidence of strong prehistoric earthquakes in the New Madrid (Missouri) seismic zone. *Geology*, 19, 296–298, doi:10.1130/0091–7613(1991).

Schwartz, D. P., Hecker, S., Haproff, P., Beukelman, G., and Erickson, B. (2012). The Bear River fault zone, Wyoming and Utah: complex ruptures on a young normal fault (abstract). T31E-08, American Geophysical Union Fall Meeting.

Schweig, E. S., and Ellis, M. A. (1994). Reconciling short recurrence intervals with minor deformation in the New Madrid seismic zone. *Science*, 264, 1308–1311.

Seeber, L., and Armbruster, J. G. (1987). The 1886–1889 aftershocks of the Charleston, South Carolina, earthquake: a widespread burst of seismicity. *Journal of Geophysical Research*, 92, 2663–2696.

Stein, S., and Liu, M. (2009). Long aftershock sequences within continents and implications for earthquake hazard assessment. *Nature*, 462, doi: 10.1038/nature08502.

Stein, S., and Newman, A. (2004). Characteristic and uncharacteristic earthquakes as possible artifacts: application to the New Madrid and Wabash Valley seismic zones. *Seismological Research Letters*, 75, 173–198.

Street, R. (1982). A contribution to the documentation of the 1811–1812 Mississippi Valley earthquake sequence. *Earthquake Notes*, 53, 39–52.

Street, R. (1984). *The Historical Seismicity of the central United States: 1811–1928, Final Report*. Contract 14–08–0001–21251, Appendix A, Washington, D.C.: U.S. Geological Survey.

Stuart, W. D., Hildenbrand, T. G., and Simpson, R. W. (1997). Stressing of the New Madrid seismic zone by a lower crust detachment fault. *Journal of Geophysical Research*, 102, 27, 623–627, 633.

Sykes, L. R. (1978). Intra-plate seismicity, reactivation of preexisting zones of weakness, alkaline magmatism, and other tectonics post-dating continental separation. *Reviews of Geophysics and Space Physics*, 16, 621–688.

Talwani, P., and Cox, J. (1985). Paleoseismic evidence of recurrence of earthquakes near Charleston, South Carolina. *Science*, 229, 379–381, doi:10.1126/science. 229/4711.379.

Talwani, P., and Dura-Gomez, I. (2009). Finding faults in the Charleston Area, South Carolina 2. Complementary data. *Seismological Research Letters*, 80, 901–919, doi:10.1785/gssrl.80.5.901.

Talwani, P., and Schaeffer, W. T. (2001). Reccurence rates of large earthquakes in the South Carolina Coastal Plain based on paleoliquefaction data. *Journal of Geophysical Research*, 106, 6621–6642.

Talwani, P., and Sharma, N. (1999). Reevaluation of the magnitude of three destructive aftershocks of the 1886 Charleston earthquake. *Seismological Research Letters*, 70, 360–367, doi:10.1785/gssrl.70.3.360.

Tuttle, M. P., and Schweig, E. S. (1996). Archaeological and pedological evidence for large prehistoric earthquakes in the New Madrid seismic zone, central United States. *Geology*, 23, 253–256.

Tuttle, M. P., Schweig, E. S., Sims, J. D., *et al.* (2002). The earthquake potential of the New Madrid seismic zone. *Bulletin of the Seismological Society of America*, 92, 2080–2089.

Tuttle, M. P., Al-Shukri, H., and Mahdi, H. (2006). Very large earthquakes centered southwest of the New Madrid seismic zone 5000–7000 years ago. *Seismological Research Letters*, 77, 755–770, doi:10.1785/gssrl.77.6.755.

Van Arsdale, R. (2000). Displacement history and slip rate on the Reelfoot fault of the New Madrid seismic zone. *Engineering Geology*, 55, 219–226.

Van Arsdale, R., Bresnahan, R., McCallister, N., and Waldron, B. (2007). Upland complex of the central Mississippi River valley: its origin, denudation, and possible role in reactivation of the New Madrid Seismic Zone. In *Continental Intraplate Earthquakes: Science, Hazard, and Policy Issues*, ed. S. Stein, and S. Mazzotti. Geological Society of America Special Paper 425, 177.

Working Group on California Earthquake Probabilities (WGCEP) (2013). *Proposed Time-Independent Uniform California Earthquake Rupture-Forecast, Version 3.1 (UCERF3.1)*. Report delivered to California Earthquake Authority.

Wu, P., and Johnston, P. (2000). Can deglaciation trigger earthquakes in North America? *Geophysical Research Letters*, 27, 1323–1326.

13

Conclusions

PRADEEP TALWANI

This book presents the results of studies of intraplate earthquakes in eight diverse regions of the world (Chapters 2 to 9). Together they present an opportunity to confirm or modify our current ideas and to devise future studies aimed at improving our understanding of this phenomenon. These chapters primarily address the spatial and temporal pattern of intraplate earthquakes and, where possible, identify the seismic sources. In some cases the authors speculate on the mechanisms responsible for the seismicity. Chapter 10 confirms our understanding on how stresses originating at plate boundaries are responsible for mid-plate earthquakes. The unified model for intraplate earthquakes (Chapter 11) presents a synthesis of earlier models into a testable one, while Chapter 12 on seismic hazards cautions against the continued use of probabilistic seismic hazard analysis methodology for intraplate earthquakes in Central and Eastern United States.

Detailed seismic tomography studies along the eastern shore of the Sea of Japan and near the source zone of the 2001 Bhuj earthquake in India show that, with detailed data and improved analytical techniques, it is possible to delineate the structural features of the seismic zones and test our ideas and models of the genesis of these earthquakes and to formulate new approaches. For most seismic zones, such detailed information is lacking. The results of investigations presented here also help to raise questions about the validity of our current methods of investigation and illuminate the need to develop new ones. This chapter is divided into three parts that address the current ideas that have been confirmed by the results of these studies; those that have been contradicted and what new insights have been gained; and suggestions for future studies.

13.1 Observations that support earlier ideas

Sykes (1978) first suggested that there is a global pattern of intraplate earthquakes. Subsequently, several investigators recognized that intraplate earthquakes preferentially occur in old rift structures and along craton boundaries (Mooney *et al.*, 2012). The various examples presented in this book have confirmed these ideas. Ziegler (1987) and Zoback

(1992) suggested that mid-plate deformation resulted from stresses that originated at plate boundaries and were transmitted through the lithosphere. Numerical modeling (Chapter 10) has confirmed this premise.

Some of the earlier models in the 1970s and 1980s hypothesized that intraplate earthquakes occurred as a result of stress build-up on discrete structures in the upper and lower crust. These models were combined into a testable, unified model that postulates that these discrete structures act as local stress concentrators, a concept that was confirmed at two locations with the results of seismic tomography analysis. A comparison of Figures 6.3 and 9.5, which show the modeled structures at the Sea of Japan and Kutch Rift Basin, with Figure 11.3c, which shows the predictions of the unified model, illustrates this point. Such detailed information about the seismic structures is lacking in other regions. As newer and better data become available, this model will be tested and other models will evolve.

13.2 New observations

The temporal and spatial pattern of large historic and prehistoric earthquakes in China (Chapter 5) and the Kutch Rift Basin (Chapters 6 and 11) show an absence of repeat earthquakes on the same fault. In each case the earthquakes jumped from one location to another and were described as "roaming earthquakes" (Chapter 5). Although the spatial extent was different in these cases – widespread, continental scale in China and within a single Kutch Rift Basin in India – there was no evidence of repeat earthquakes. A similar pattern was observed for the morphogenic earthquakes (prehistoric events detected from paleo-seismological studies) in Australia (Chapter 2). These observations are consistent with a model that hypothesizes sequential reactivation on different structures. This lack of stationarity of the seismic source on a particular fault in an intraplate setting is at odds with the basic assumption of stationarity in probabilistic seismic hazard estimation methodology, which was developed for plate boundary earthquakes (Chapter 12). This dichotomy requires a reassessment of methods of seismic hazard estimation in intraplate settings and the development of new ones.

In mid-plate regions there should perhaps be a lower magnitude threshold used in seismic hazard estimation, as advocated by the authors of Chapters 4 and 8. They point out that because of low seismic-wave attenuation in mid-plate regions, the presence of poorly constructed buildings, and vulnerable local soil conditions, there may be significant damage potential due to earthquakes with magnitudes as low as 4.5.

The seismicity and geodetic observations in the decade following the 2001 M 7.7 Bhuj earthquake provide a unique dataset to study the seismogenesis of an intraplate earthquake. There was evidence of a migration of a seismic pulse away from the source of the 2001 event, reaching 250 km in about 10 years with a sequential activation of faults along the way. Monitoring of ground motion led to the detection of local, isolated pockets (hundreds of kilometers) of elevated strain rates 6 to 8 years after the 2001 earthquake. These anomalous pockets of elevated strain-rate build-up provide evidence that the spatial and temporal pattern of strain accumulation is different from that for plate boundaries.

To explain some of our current observations, different elements of earlier models were integrated into a unified model for intraplate earthquakes (Chapter 11). In this simple, testable model stresses can accumulate at local stress concentrators and react with, as well as locally rotate, the maximum horizontal stress direction. Seismic tomography has made it possible to identify these stress concentrators (Chapters 6 and 9). These local stress concentrators are of different kinds and of diverse sizes, varying from tens to hundreds of kilometers across (Figure 11.3c). Compressional jogs between en echelon faults also act as stress concentrators and have been identified in regions of large earthquakes, e.g., in the New Madrid seismic zone (Figure 7.10), the Kutch Rift Basin (between KMF and NWF in Figure 6.13) and in the Middleton Place Summerville Seismic Zone of the 1886 Charleston earthquake (Talwani and Dura-Gomez, 2009). The local stress build-up can, in some cases, be associated with a detectable rotation of the local maximum horizontal stress direction (Chapter 11). Future studies need to be designed to test this model and seek additional evidence for the local stress concentrator and the associated rotation of S_{Hmax}.

13.3 Future studies

The role of fluids as weakening agents was not addressed in this book. However, the presence of fluids has been detected by seismic tomography near two source zones (Chapters 6 and 9), and its role in the seismogenesis of intraplate earthquakes merits further study.

The unified model (Chapter 11) is based on a synthesis of current ideas and observations. It provides a framework for studying intraplate earthquakes. Both field studies and theoretical modeling are needed to test its viability and predictability. As additional and more focused data become available, new and better models will evolve to explain the phenomenon of intraplate earthquakes. How new data can help this evolution is shown by the following example.

A few years after the 2001 Bhuj earthquake, the large coseismic strain rates had returned to their background levels. More than 6 years after the mainshock, two earthquakes with magnitudes 4.1 and 4.7 occurred just south of the Kutch Mainland fault (KMF in Figure 6.13), about 25 to 30 km southwest of the 2001 earthquake. InSAR data collected between 2004 and 2007 and between 2007 and 2010 detected a local pocket of elevated vertical strain rate (16–27 mm/yr) in about a 200 sq. km area surrounding the 2007 epicenters (Figure 6.17). Does this pocket of increased strain rate surrounding the M 4+ epicenters signify that stresses are building up on a new, hitherto unidentified local stress concentrator and the potential location of a larger future earthquake? To test this idea, an array of additional GPS receivers and seismometers are being deployed in the vicinity of the region of increased uplift rate. This example shows that with further improvements in instrumentation and analysis techniques and with detailed, focused, complementary seismological and geodetic measurements, it may become possible to routinely detect potential locations of moderate to large intraplate earthquakes.

As intraplate earthquakes jump from one fault to another and there is an absence of a stationary source, we need to revise our strategies for studying them and for estimating the attendant seismic hazard. A new set of questions need to be addressed. Do we abandon the probabilistic seismic hazard analysis approach and are deterministic site-specific investigations the correct approach in estimating seismic hazards in intraplate regions? Do we examine the entire rift structure and look for local stress concentrators, or do we need to focus our efforts on areas of past large historical earthquakes?

Mooney *et al.* (2012) showed that in addition to old rift structures, many intraplate earthquakes are located near craton boundaries. Dedicated field studies are needed to identify the responsible seismic sources. A possible starting point may be based on Ziegler's (1987) observation that, in mid-plate regions, deep faults that extended to lower crustal depths are likely to be reactivated by the stresses originating at plate boundaries. Identification of such faults by geological and/or geophysical studies would be a good starting point for further investigations of the potential seismic source zones associated with craton boundaries.

Future theoretical studies could address the reasons why large earthquakes do not recur on the same fault but instead jump around, suggesting that after an earthquake most of the built-up seismic energy is expended, while at other locations it continues to build until the next earthquake. The observation of "roaming" or "wandering" earthquakes also suggests the presence of several potential stress concentrators that are sequentially reactivated. Theoretical studies are needed to estimate the magnitude of the stresses that accumulate at different stress concentrators and mechanisms to explain their temporal and spatial behavior.

The maximum magnitudes (M_{max}) of intraplate earthquakes vary from region to region. For example, in Western Europe it is rare to observe an earthquake with $M > 5$, or in Brazil with $M > 6$. Several earthquakes with $M_{max} > 6$, however, have occurred in Australia and Eastern Canada. Considerably larger earthquakes have been observed in northern China, the New Madrid seismic zone and in the Kutch Rift Basin. What controls M_{max} in intraplate regions? Along plate boundaries, M_{max} is related to the fault length, a correlation not seen in intraplate regions. Modeling suggests that M_{max} is related to the amount of stress that can build up on a stress concentrator, which in turn depends on the kind of stress concentrator present. Historically, shallow plutons have been associated with $M_{max} < 5$ and rift pillows with $M_{max} < \sim 5.5$. As mentioned earlier, for the larger earthquakes in the New Madrid seismic zone, Kutch Rift Basin, and in the source zone for the 1886 Charleston earthquake, M_{max} appears to be related to the width of the compressional step between the en echelon faults associated with the source zone. This speculative association needs to be confirmed with data from other source zones and by numerical modeling.

The diverse studies described in this book represent a step forward in our understanding of intraplate earthquakes. While they answer some questions about our understanding, they also raise many others that should be the focus of future research.

References

Mooney, W. D., Ritsema, J., and Hwang, Y. K. (2012). Crustal seismicity and earthquakes catalog maximum moment magnitude (Mcmax) in stable continental regions (SCRs): correlation with the seismic velocity of the lithosphere. *Earth and Planetary Science Letters*, 357–358, 78–83.

Sykes, L. R. (1978). Intra-plate seismicity, reactivation of pre-existing zones of weakness, alkaline magmatism, and other tectonics post-dating continental separation. *Reviews of Geophysics and Space Physics*, 16, 621–688.

Talwani, P., and Durá-Gómez, I. (2009). Finding faults in the Charleston area, South Carolina: 2. Complementary data. *Seismological Research Letters*, 80, 901–919.

Zeigler, P. A. (1987). Late Cretaceous and Cenozoic intra-plate compressional deformations in the Alpine foreland: a geodynamic model. *Tectonophysics*, 137, 389–420.

Zoback, M. L. (1992a). First- and second-order patterns of stress in the lithosphere: The World Stress Map Project. *Journal of Geophysical Research*, 97, 11,703–11,728.

Index